Encyclopedia of Technical Aviation

Gary V. Bristow

D1534547

McGraw-Hill

New York   Chicago   San Francisco   Lisbon   London   Madrid
Mexico City   Milan   New Delhi   San Juan   Seoul
Singapore   Sydney   Toronto

The McGraw·Hill Companies

Cataloging-in-Publication Data is on file with the Library of Congress

Copyright © 2003 by The McGraw-Hill Companies, Inc. All rights reserved. Printed in the United States of America. Except as permitted under the United States Copyright Act of 1976, no part of this publication may be reproduced or distributed in any form or by any means, or stored in a data base or retrieval system, without the prior written permission of the publisher.

1 2 3 4 5 6 7 8 9 0   AGM/AGM   0 8 7 6 5 4 3 2

ISBN 0-07-140213-6

The sponsoring editor for this book was Shelley Ingram Carr, the editing supervisor was Stephen M. Smith, and the production supervisor was Sherri Souffrance. It was set in Century Schoolbook by Paul Scozzari of McGraw-Hill Professional's Hightstown, N.J., composition unit.

Printed and bound by Quebecor/Martinsburg.

McGraw-Hill books are available at special quantity discounts to use as premiums and sales promotions, or for use in corporate training programs. For more information, please write to the Director of Special Sales, McGraw-Hill Professional, Two Penn Plaza, New York, NY 10121-2298. Or contact your local bookstore.

Preface

Throughout my flying career, from the early training days to light aircraft work and finally to the big shiny jets for the airlines, I have often heard colleagues and friends express a desire for a single reference book that covers all the main aviation-related subjects. Thus I conceived the idea for an A-to-Z–format book that provides quick and easy access to most of the technical aviation topics.

I hope readers will find this book to be an ideal companion throughout their careers and that it will provide either a first insight or a reminder of a topic forgotten.

Gary V. Bristow

Introduction

Technical aviation subjects are both varied and diverse, from aerodynamics to human factors, engines to meteorology, satellite navigation to rules of the air, instrument flight rules (IFR) to aircraft performance. Pilots, and indeed many professional aviation personnel, expand their knowledge on these subjects throughout their careers. This book covers all the main topics from these various subjects, and can be used as either initial learning material or a reminder for the more-experienced pilot. The A-to-Z format of this book will make it quick and easy for the reader to locate a particular topic.

For the training pilot, this book provides a useful reference source of information on topics that are being learned for the first time. It is often difficult for new pilots to locate a particular subject from the several training manuals they are given during their training. The format of this book, though, is designed so that a topic is found easily and quickly, even by the first-time reader.

For the early-career pilot (instructors, air taxi pilots, those involved in regional operations), this book will help you to refresh your knowledge of training topics, as well as provide information on new topics associated with a move up into more-complex aircraft, especially jet-powered machines, and commercial operations. Many of the topics that relate to this stage in your career are detailed in practical terms, and not just in theory.

For the experienced pilot, this book is especially ideal as a "revision" tool. For instance, you might hear on the grapevine that your next simulator exercise will include an engine failure at takeoff, at VMCA (minimum control speed in the air), and you cannot remember much about VMCA and its further implications given other factors in the scenario. Using this book to research the meaning of VMCA or to expand your knowledge of this topic can be a great help in achieving a successful simulator assessment. This book may also be used for any general revision during a pilot's career. I think we would all admit to having forgotten something we learned early in our flying careers that we still need to know today and every day.

As pilots we all live in a changing and technology-advancing world. Many of these developments are also addressed in this book, especially topics that relate to new aircraft types as well as new auxiliary technology and procedures, e.g., FANS (Future Air Navigation Systems).

It is hoped that this book will be invaluable to all pilots at some stage in their careers. In fact, it should serve as an ideal companion from your first flight to your last.

Abbreviations

AAL	Above aerodrome level
ABC	Auto boost control
AC	Alternating current
ACARS	Aircraft communications and reporting system
ACAS	Airborne collision avoidance system
ADC	Air data computer
ADF	Automatic direction finder
ADI	Attitude director indicator/instrument
ADS-B	Automated dependent surveillance—broadcast
AGC	Automatic gain control
AGL	Above ground level
AH	Artificial horizon; alert height
AIC	Aeronautical information circular
ALT	Altimeter
A(M)SL	Above (mean) sea level
A of A	Angle of attack
AP	Autopilot
APFDS	Autopilot flight director system
APS	Aircraft prepared for service
APU	Auxiliary power unit
ASDA(R)	Accelerate stop distance available (required)
ASI(R)	Airspeed indicator (reading)
AT	Auto thrust/throttle
ATC	Air traffic control
ATM	Aerodynamic turning moment; air traffic management
BDC	Bottom dead center
BEA	British European Airways
BFO	Beat frequency oscillator
BHP	Brake horsepower
C	Celsius
CAA	Civil Aviation Authority
CAS	Calibrated airspeed

CAT	Clear air turbulence
CB	Cumulonimbus cloud; circuit breaker
CDI	Course deviation indicator
CDU	Control display unit
CFIT	Controlled flight into terrain
CI	Cost index
CL	Coefficient of lift
CMD	Command
CN	Compass north
C of A	Certificate of airworthiness
c of g/cg	Center of gravity
c of p/cp	Center of pressure
CP	Critical point
CRT	Cathode-ray tube
CSD(U)	Constant-speed drive (unit)
CSU	Constant-speed unit
CTM	Centrifugal turning moment
CU	Cumulus cloud
CWS	Control wheel steering
DALR	Dry adiabatic lapse rate
DC	Direct current
DH(A)	Decision height (altitude)
DI	Directional indicator
DME	Distance-measuring equipment
EADI	Electronic attitude director indicator/instrument
EAS	Equivalent airspeed
EAT	Expected approach time
ED(R)	Emergency distance (required)
EFIS	Electronic flight instrument system
EGPWS	Electronic ground proximity warning system
EGT	Exhausted gas temperature
EHSI	Electronic horizontal situation indicator/instrument
ELR	Environmental lapse rate
EMDA(R)	Emergency distance available (required)
emf	Electromotive force
EPR	Engine pressure ratio
ETOPS	Extended twin operations
FAA	Federal Aviation Administration
FADEC	Full-authority digital engine control
FAF	Final approach fix
FANS	Future air navigation system
FCC	Flight control computer
FCU	Fuel control unit; flight control unit
FD(S)	Flight director (system)
FIR	Flight information region

FL	Flight level
FMC(S)	Flight management computer (system)
FMS	Flight management system
fpm	Feet per minute
g	Gram
GLONASS	Global orbiting navigation satellite system
GLS	Global landing system
GPS	Global positioning system
GPWS	Ground proximity warning system
GS	Glide slope
HF	High frequency
hPa	Hectopascal
HSI	Horizontal situation indicator/instrument
HUD	Head-up display
HUGS	Head-up guidance system
HWC	Headwind component
IAF	Initial approach fix
IAS	Indicated airspeed
ICAO	International Civil Aviation Organization
IFR	Instrument flight rules
ILS	Instrument landing system
IMC	Instrument meteorological conditions
INS	Inertia navigation system
IRS	Inertia reference system
ISA	International standard atmosphere
ITCZ	Intertropical convergence zone
IVSI	Inertia/instantaneous vertical speed indicator/instrument
L	Light
LDA(R)	Landing distance available (required)
(M)LW	(Maximum) landing weight
LF	Low frequency
LNAV	Lateral navigation
LOC	Localizer
LRC	Long-range cruise
LSS	Local speed of sound
μm	Micron
MABH	Minimum approach break-off height
MAC	Mean aerodynamic chord
MAP	Manifold absolute pressure; missed approach point
mbar	Millibar
MCP	Mode control panel
M_{Crit}	Critical Mach number
MDH(A)	Minimum decision height (altitude)
MEA	Minimum en route altitude
MEL	Minimum equipment list

MLS	Microwave landing system
MM	Mach meter
MN	Mach number; magnetic north
MRC	Maximum range cruise
MSA	Minimum sector altitude
MSL	Mean sea level
MSU	Mode selector unit
NAV	Navigation
NDB	Nondirectional beacon
Notams	Notices to airmen
NTOFP	Net takeoff flight path
OAT	Outside air temperature
OBI	Omni bearing instrument
OCH(T)	Obstacle clearance height (time)
PAPI	Precision-approach path light
PAR	Precision-approach radar
PNR	Point of no return
psi	Pounds per square inch
ρ	Density
RAS	Rectified airspeed
RAT	Ram air turbine
RBI	Relative bearing indicator/instrument
RMI	Radio magnetic indicator/instrument
RNAV	Area navigation
ROC	Rate of climb
ROD	Rate of descent
rpm	Revolutions per minute
RTO	Rejected takeoff
RVR	Runway visual range
RW	Ramp weight
s	Second
S	Span
s/kn	Seconds per knot
SALR	Saturated adiabatic lapse rate
SAT	Static air temperature
SFC	Specific fuel consumption
SG	Symbol generator; specific gravity
SIDs	Standard instrument departures
SL	Sea level
SRA	Secondary radar approach
SSA	Sector safe altitude
SSR	Secondary surveillance radar
STARs	Standard arrivals
SVFR	Special visual flight rules
SWD	Supercooled water droplet

TAF	Terminal aerodrome forecast
TAS	True airspeed
TAT	Total air temperature
TCAS	Traffic collision and avoidance system
TDC	Top dead center
TGT	Total gas temperature
TN	True north
TOD(A/R)	Takeoff distance (available/required)
TOGA	Takeoff go-around
TOR(A/R)	Takeoff run (available/required)
(M)TOW	(Maximum) takeoff weight
TRU	Transformer rectifier unit
TWC	Tailwind component
UHF	Ultrahigh frequency
V	Velocity
V_1	Decision speed
V_2	Takeoff safety speed
V_a	Maneuver speed
V_{app}	Approach speed
VASI/L	Visual approach slope indicator/lights
VAT	Velocity at the threshold
VDF	VHF directional finding
VFR	Visual flight rules
VHF	Very high frequency
VIMD	Minimum drag speed
VIMP	Minimum power speed
VLF	Very low frequency
VMBE	Maximum brake energy speed
VMC	Visual meteorological conditions
VMC(G/A)	Minimum control speed (on the ground/in the air)
V/MDF	Velocity/Mach demonstrated flight diving speed
V/MMO	Velocity/Mach maximum operating speed
V(M)RA	Rough airspeed
VMU	Minimum demonstrated unstick speed
VNAV	Vertical navigation
VNE	Never-exceed speed
VNO	Normal operating speed
VOR	VHF omni range
V_R	Rotation speed
V_{REF}	Reference speed (for approach)
V_S	Stall speed; vertical speed
V_{S0}	Full-flap stall speed
V_{S1}	Clean stall speed
WAT	Weight, altitude, and temperature
WED	Water equivalent depth
(M)ZFW	(Maximum) zero fuel weight

Absolute Ceiling The absolute ceiling is an aircraft's maximum attainable altitude or flight level at which the Mach number buffet and prestall buffet occur coincidentally. This scenario is known as *coffin corner*. Therefore an aircraft is unable to climb above its absolute ceiling. The absolute ceiling is determined during flight certification trials.

AC (Alternating Current) Alternating current continuously reverses its direction of flow in an electric circuit. The complete reversal of current flow is known as a *cycle* and occurs very rapidly. For example, an aircraft AC electrical system reverses the current flow 400 cycles per second.

   NB: 1 cycle/second is 1 hertz (Hz).

   The advantages of AC power are as follows:

1. AC generators (alternators) are simpler and more robust in construction than the direct-current (DC) machines.
2. The power-to-weight ratio of AC machines is much better than that of comparable DC machines.
3. The supply voltage can be converted to a higher or lower value with great efficiency by using transformers.
4. Any required DC voltage could be obtained simply and efficiently by using transformer rectifier units (TRUs).
5. Three-phase AC motors can be operated from a constant-frequency source (alternator).
6. AC generators do not suffer from commutation problems associated with DC machines and consequently are more reliable, especially at high altitudes.
7. High-voltage AC systems require less cable weight than comparable-power low-voltage DC systems.

   The typical sources of AC electric power for an aircraft are

1. AC generators (alternators)—engine, auxiliary power unit (APU), or ram air turbine (RAT) driven.
2. Inverters (convert DC to AC). Static inverters produce constant-frequency AC power. Rotary inverters produce power for frequency-wild AC system.
3. Ground AC supply.
4. Transformers (change the AC voltage level).

ACAS ACAS is a European Airborne Collision Avoidance System (ACAS II).

AC Generator An AC generator (alternator) produces electric current by electromagnetic induction (magnetic field excitement). There are two basic types of alternator:

1

1. The *rotating armature alternator* has an armature made from a conductive metal that is driven by a motor to rotate it in a stationary magnetic field. As the armature coil rotates, its sides cut the magnetic flux, which produces free electrons and thereby induces an electromotive force (emf). The emf then induces a current, which flows through the armature coil and is taken out as AC electric power through two slip rings.

   NB: Rotating armatures are only used in very small-output alternators.

2. *Rotating field alternators* are designed so that the magnetic field windings are the rotating part, with a stationary armature. Either a permanent magnet or a DC source, via the armature, is used to energize the field. They produce emf/current in the same manner as the rotating armature alternator, with the AC output being taken from the stator windings.

Aclinic Line   The aclinic line is the line joining points of zero dip, i.e., the magnetic equator.

Active Control   An active control is a surface that moves automatically/actively in response to nondirect inputs. For example, balance tabs, actively/automatically move in response to the associated control surface being moved. Or, auto slats actively/automatically move to their full extended position to provide a better coefficient of lift (CL) on a B737-300 if the aircraft senses a particular flight condition:

- Trailing edge flaps set 1 to 5 positions.
- Auto slats in normal position.
- Aircraft close to the stall.

ADF   *See* NDB/ADF.

Adiabatic Process   The adiabatic process is one in which heat is neither added nor removed from a system; but any expansion or compression of its gases changes the temperature of the system, with no overall loss or gain of energy. That is, compressing air increases its temperature, and decompressing (expansion) reduces its temperature. A common adiabatic process for pilots is the expansion of a cooling parcel of air when it rises in the atmosphere.

ADS-B   The Automated Dependent Surveillance—Broadcast—system is a traffic collision avoidance system, similar to TCAS. The ADS-B incorporates cockpit display of traffic information and is overall a more capable system than TCAS II.

Adverse Rolling Motion—Yaw-Induced   When the rudder induces the aircraft to yaw one way, it can inversely cause another form of *adverse rolling motion* in the opposite direction. This happens at high speeds (above VMO/MMO) because the deflected rudder experiences a sideways force, which causes the aircraft to roll in the opposite direction. That is, the right rudder experiences a sideways force from right to left, causing a rolling moment to the left.

   See Fig. 1 at the top of the next page.

Adverse Yaw   This is a yawing motion opposite to the turning/rolling motion of the aircraft. Adverse yaw is caused because the drag on the downgoing aileron is

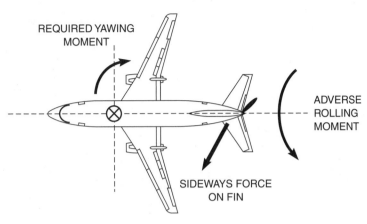

Figure 1 Adverse rolling motion with yaw.

greater than that on the upgoing aileron. This imbalance in drag causes the yawing motion around the normal/vertical axis. As this yaw is *adverse* (i.e., in a banked turn to the right the yaw is to the left), it is opposing the turn, which is detrimental to the aircraft's performance.

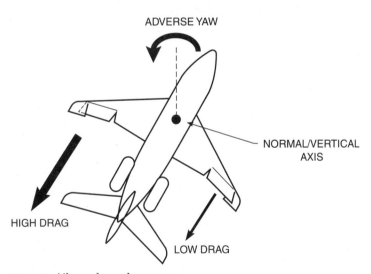

Figure 2 Aileron drag adverse yaw.

Adverse (aileron) yaw is corrected in the design by the use of one of these:

1. *Differential ailerons.*   The upgoing differential aileron moves through a greater angle than the downgoing aileron, which balances the aileron drag between both sides. This method is usually found on high-speed aircraft.

2. *Frise ailerons.*   The upgoing frise aileron has its "nose" protruding below the underside of the wing, creating a higher value of drag. The downgoing aileron's

drag value remains unchanged; thereby the drag between the two sides is balanced. This method is usually found on older, slower aircraft types. It is not suitable for high-speed aircraft because the protruding nose would cause unacceptable levels of high-speed buffet.

Figure 3 Frise aileron.

Aerodynamic Dampening Natural aerodynamic damping is reduced at high altitudes. The restoring force applied to a body which has been displaced from its neutral position can set an oscillation, e.g., an oscillating yawing aircraft. Here the fin (vertical tailplane), which acts toward the neutral position and is proportional to the displacement from the neutral position, provides the restoring force. The air loads opposing the oscillation motion, which the restoring force caused, provide the aerodynamic damping. Thereby the restoring force, which causes the oscillation, and the damping centering force, which opposes the oscillation once it exists, both come from the aerodynamic forces on the fin. The fin area and how it generates these aerodynamic forces, especially their magnitude, are very important in this equation.

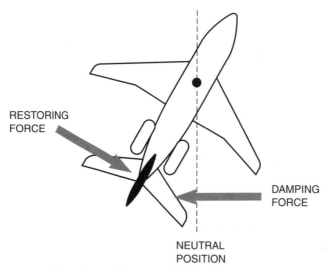

Figure 4 Restoring and damping effect on the vertical tailplane. When the aircraft has a rate of yaw to starboard, because of an oscillation, the fin has a sideways velocity to port. So the relative velocity of the restoring force is from port to starboard.

Now let us examine the effects of altitude. The frequency of the oscillation for a given indicated airspeed (IAS) (and therefore a constant dynamic pressure) is independent of altitude. So for a given amplitude of disturbance the fin will always have the same sideways velocity when passing through the neutral point at both low and high altitudes. However, the sideways velocity is compounded with the aircraft's forward velocity to give the fin its resultant angle of incidence and hence the magnitude of the sideways damping force opposing the direction of the oscillation motion. The higher the altitude, the greater the forward true airspeed (because dynamic pressure is constant); therefore, the smaller the resultant fin angle of incidence, and so the smaller the natural aerodynamic damping effect.

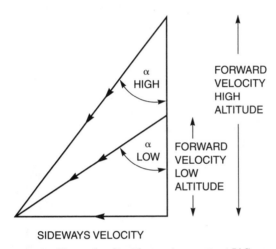

Figure 5 Fin angle of incidence at a constant IAS with altitude.

Aerodynamic Turning Moment (ATM) This is a force that tends to turn the blades to a coarse pitch.

Aerofoil An aerofoil is a body that gives a large lift force compared to its drag when set at a small angle to a moving airstream. Examples include aircraft wings, tailplanes, rudders, and propellers.

Afterburner An afterburner was introduced because it was realized that the compressor was working so efficiently that some compressed air was passing through the engine unburned. Therefore an afterburner was designed to utilize this compressed air as a pure resultant jet thrust.

The afterburner is situated in the exhaust nozzle, after any turbine section, and consists of

1. Igniter(s)
2. Fuel line/jets

Fuel is delivered via the line/jet to mix with the compressed unburned air, and the igniter provides an electric spark to ignite the mixture, thereby producing a jet *afterburn.* This is a pure jet afterburn that gives an instant increase in thrust, albeit rather uncontrolled, except for the on and off choices. An afterburner is usually found on military jets and in only a few airliners, such as the Concorde.

Agonic Line The agonic line is a line joining points of zero variation, i.e., where the true and magnetic headings are the same.

AICs (Aeronautical Information Circulars) Aeronautical information circulars are published monthly and concern administrative matters and advance warnings of operational changes, and they draw attention to and advise on matters of operational importance, such as the availability of aeronautical charts, correction of these charts, and amendments of the chart of airspace restrictions.

- AICs directly associated with air safety are printed on pink paper.
- Administrative AICs are on white paper.
- Operational and air traffic services AICs are on yellow paper.
- Restriction chart AICs are on mauve paper.
- Map/chart AICs are on green paper.

Aileron(s) These are control surfaces located at the trailing edge of the wing that control the aircraft's motion around its longitudinal axis, known as roll. They are controlled by the left and right movement of the control column, which commands the ailerons in the following manner: Moving the control column to the left commands the left aileron to be raised, which reduces the lift on the wing, and the right aileron is lowered, which increases the lift generated by this wing. Thus the aircraft rolls into a banked condition, which causes a horizontal lift force (centripetal force), which turns the aircraft. The ailerons are normally (hydraulically) powered on heavy/fast aircraft because of the heavy operating forces experienced at high speeds.

Aileron Reversal This occurs at high speeds when the air loads/forces are large enough that they cause an increase in lift. But because most of this lift is centered on the downgoing aileron at the rear of the wing, a nose-down twisting moment will be caused. This will cause a decrease in the incidence of the wing to the extent that the loss of lift due to the twisting cancels the lift gained from the aileron. At this point the aileron causes no rolling moment; and if the wing twisting is exaggerated (which a downgoing aileron can do), the rolling motion around the longitudinal axis can be reversed. Hence it is an adverse rolling motion.

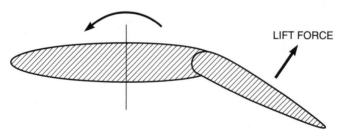

Figure 6 Aileron reversal—wing twisting moment.

Aircraft Weight Categories *See* Weight—Aircraft Categories.

Air Data Computer (ADC) Modern aircraft feed their static and pitot lines into an air data computer which calculates the calibrated airspeed/rectified airspeed (CAS/RAS), true airspeed (TAS), Mach number (MN), total air temperature (TAT), rate of climb (ROC), and rate of descent (ROD) and then passes the relevant information electronically to the servo-driven flight instruments, but not to the standby instruments, which retain their own direct static/pitot feeds.

The advantage of the ADC system is that the data calculated could be fed to the following:

1. Autopilot (AP)
2. Flight director system (FDS)
3. Flight management system (FMS)
4. Ground proximity warning system (GPWS)
5. Navigation aids
6. Instrument comparison systems

Airfields—Beacon Aeronautical light beacons are installed at various civil and military airfields in some countries. Their hours of operation vary, but generally they are on at night and by day in bad visibility whenever the airfield is in operation.

The *identification beacon* flashes a two-letter Morse identifier for the aerodrome.

NB: Usually it is green at civil aerodromes and red at military aerodromes.

They allow a visual identification, as well as a bearing to the aerodrome.

The *aerodrome beacon* flashes an alternating signal as a homing signal to the aerodrome.

NB: Usually it is white/white, or less commonly white/green.

They are not normally provided in addition to an identification beacon.

Air Law When flying over or even landing at a foreign state, you must obey the law of that country. Therefore your aircraft is not to be used for a purpose that is prejudicial to the security, public order, health, or safety of air navigation in relation to that country. However, whenever your own country's (i.e., the country of the aircraft's registration) legislation (on any particular issue) is more limiting, then this should take precedence and should be adhered to.

Air Law—Aerodrome Right-of-Way The right-of-way on the ground is as follows:

1. Regardless of air traffic control (ATC) clearance, it is the duty of the aircraft commander to do all that is possible to avoid a collision with other aircraft or vehicles on the ground.
2. Aircraft on the ground must give way to those taking off or landing, and to any vehicle towing an aircraft.
3. When two aircraft are approaching head-on, each must turn right to avoid the other.
4. When two aircraft are converging, the one that has the other on its right must give way—either by stopping or by turning to pass behind the other. Avoid crossing in front of the other unless passing is clear.
5. An aircraft that is being overtaken by another has the right-of-way, and the overtaking aircraft must keep out of the way by turning left until past and well clear.

NB: This enables the commander of the overtaken aircraft (who has the right-of-way) to have an unrestricted view of the overtaking aircraft.

The priority for all vehicles on an aerodrome is as follows:

1. Aircraft landing
2. Aircraft taking off
3. Vehicles towing aircraft
4. Aircraft taxiing
5. Other vehicles

Air Law—Rules of the Air   An aircraft landing or on final approach and cleared to land has the right-of-way over other aircraft in flight or on the ground.

NB: In the case of two or more aircraft approaching to land, the lower aircraft has the right-of-way, except when the commander of the lower aircraft is aware that another aircraft is making an emergency landing. Then he or she must give way; and at night, even if she or he has already been cleared to land, the commander must not attempt to land until given further permission from ATC.

An aircraft in the air must give way to other converging aircraft regardless of position as follows:

- Powered aircraft must give way to those towing objects such as banners.
- Flying machines must give way to airships, gliders, and balloons.
- Airships must give way to gliders and balloons.
- Gliders must give way to balloons.

With two aircraft converging in the air, i.e., at a constant relative bearing at the same altitude, the aircraft which has the other on its right must give way by turning right and passing behind the opposing aircraft.

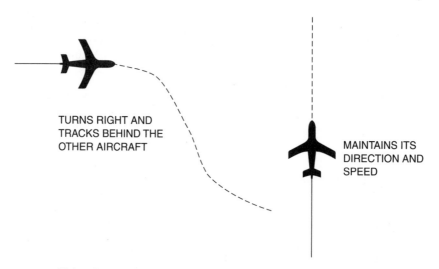

TURNS RIGHT AND
TRACKS BEHIND THE
OTHER AIRCRAFT

MAINTAINS ITS
DIRECTION AND
SPEED

Figure 7   Right-of-way.

When two aircraft are approximately approaching head-on at the same altitude, there is a definite danger of collision. Therefore both aircraft must turn right to avoid collision.

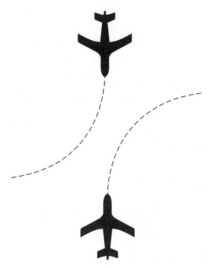

Figure 8 Turn right to avoid a head-on collision.

Whether climbing, descending, or in level flight, an aircraft overtaking another aircraft must turn right and keep right of the other aircraft during the overtaking maneuver.

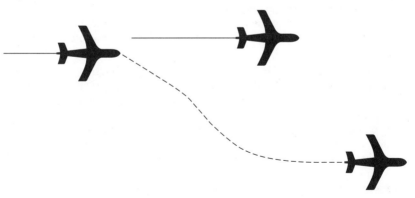

Figure 9 Keep right while overtaking.

The aircraft being overtaken in the air has the right-of-way.

Air Mass   An air mass is a large parcel of air with fairly similar temperature and humidity (moisture content) properties throughout. For this definition it follows that the environmental lapse rate (ELR) will be fairly constant throughout the air mass, and all the air will behave in the same way when lifted, heated, or cooled.

The origin/sources of all air masses are high-pressure systems, e.g., polar high, Azores high, and Siberian high. This is so because high-pressure systems are very stable. Therefore they allow the air mass to remain stationary over the same place for a period of time, which allows its properties of temperature, humidity, etc., to become uniform throughout. Eventually, though, the air mass equilibrium will break down, and the air mass will start to move. As it moves, it will be modified by the earth's surface over which it is traveling, so that it is heated or cooled, and gains or loses moisture. The amount of change will depend on the speed of its travels as well as the intensity of the factors influencing it. It is possible to classify an air mass and to generalize about its characteristics given its source region and path traveled.

An air mass is further influenced by one of two factors:

1. The divergence of air flowing out of a high-pressure system at the earth's surface causes the air mass to slowly sink or subside, and as a consequence to become warmer, drier, and more stable.

2. The convergence of air flowing into a low-pressure system at the earth's surface forces the air mass to rise slowly and as a consequence to become cooler, its relative humidity to rise, and it to become more unstable.

An air mass is usually classified according to the following:

1. Its origin/source, which can be categorized as one of the following:
   a. Maritime air, i.e., over an ocean that absorbs moisture and becomes rather saturated.
   b. Continental air, i.e., over land that remains rather dry since no or little water is available for evaporation.
2. Its path over the earth's surface, which can be categorized as
   a. Polar air, i.e., cold air flowing toward the warmer lower latitudes that is warmed from below and so becomes unstable.
   b. Arctic air, which is an extreme version of polar air.
   c. Tropical air, i.e., warm air flowing toward the cooler, higher latitudes that is cooled from below and so becomes stable.

Hence, examples of air mass classifications are polar maritime, polar continental, arctic maritime, tropical maritime, tropical continental, returning polar maritime, etc. Their names are self-descriptive; e.g., polar maritime describes cold air gaining moisture, which is warmed and becomes unstable. From this description we can conclude that this air mass will usually produce cumulus or even cumulonimbus clouds, especially over land that provides the additional lifting action and greater convective currents.

Airmet   An airmet is a recorded telephone message, which gives the met (meteorological) forecast for a particular area. It can be accessed by area, where telephone numbers for different areas are listed on an *Airmet areas chart.*

Airspeed Indicator (ASI)   This device measures dynamic pressure as the difference between the total pitot pressure measured in the instrument's capsule/diaphragm and the static pressure measured in the case. The dynamic pressure represents the indicated airspeed (IAS), in knots per hour (kn/h).

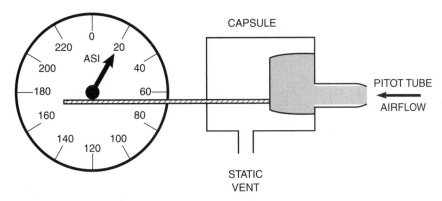

Figure 10 Airspeed indicator instrument.

The ASI instrument is calibrated to ISA mean sea-level (MSL) density of 1225 g/m^3. A red/black-striped pointer on the display indicates the VMO speed.

Airspeed Indicator—Errors   The ASI instrument suffers from the following errors:

1. *Instrument error,* which is the inaccuracy of the ASI instrument itself mainly due to the friction of its moving parts.
2. *Pressure error,* also known as *position* or *configuration error,* which is caused by variations of the static and pitot probe relative position with respect to the airflow over the aircraft's surface. This may lead to sensed pressure readings that are not representative of the free atmosphere when the airflow pattern is disturbed at different airspeeds, angle of attack, configuration, etc.
3. *Density error.*   *See* Density Errors.
4. *Compressibility error,* which affects the ASI at high speeds because the air comes to rest on the pitot probe surface and becomes "compressed," resulting in a rise in air density and therefore dynamic pressure. This results in a high indicated airspeed error.
5. *Maneuver error,* which occurs when a maneuver that disturbs the airflow pattern leads to incorrect sensed pressure readings, and is unpredictable.
6. *Blocked pitot static system.*   A static line blockage means that the static pressure in the ASI instrument case remains a constant value. Therefore at a constant altitude the ASI will read correctly, but in a descent the ASI will *overread,* due to an increase in the pitot pressure in the capsule, against the trapped low static pressure of the higher altitude. In a climb the ASI will *underread,* due to a decrease in the pitot pressure in the capsule, against the trapped high static pressure of the low altitude.

    A pitot line blockage means that the total pressure in the ASI instrument capsule remains a constant value. Therefore, at a constant altitude the ASI reading will not change, even if the airspeed does, due to the trapped pitot pressure in the capsule, against a constant altitude static pressure. In a descent the ASI will *underread,* due to an increase in the static pressure in the case, against a constant pitot pressure. In a climb the ASI will *overread,* due to a decrease in the static pressure in the case, against a constant pitot pressure.

The blockage is normally caused in flight by ice formation or small insects covering the pitot or static port. The actions taken for a blocked pitot/static system causing an unreliable ASI reading would be to

a. Ensure that the pitot static probe anti-ice heating (pitot heat) is *on*, if applicable.

b. Use an alternative—static source or an air data computer (if applicable).

c. Use a limited flight panel, i.e., standby ASI.

d. Fly at a correct attitude and power setting.

NB: If the blockage has been caused by ice formation and you have an unserviceable or *no* pitot heat system, then you should avoid icing conditions.

Air Temperature—Measured Air temperature is commonly measured by either a total head thermometer or a rosemount probe that is extended into the free airstream, and it is usually displayed to the pilot on a TAT (total air temperature) gauge.

The temperature probe is normally heated to prevent icing (pitot heat), and because of this you can have two types of air temperature probes:

■ *Aspirated.* Cooling air is circulated in the probe to compensate for the rise in the TAT due to the anti-ice heating of the probe. Thus on the ground with pitot heat on, TAT is equal to outside air temperature (OAT).

■ *Nonaspirated.* No cooling air is circulated in the probe. Therefore TAT is increased on the ground with the pitot heat on, and is only equal to OAT with the pitot heat off.

In flight TAT is only a function of the ram effect of the air entering the probe, and pitot heat is not considered when calculating OAT.

Alcohol A pilot cannot operate for a minimum of 8 h after consuming a small quantity of alcohol, to ensure that all the alcohol has left the individual's system.

NB: 1 unit of alcohol takes 30 min to leave the human system.

NB: However, some authorities and companies may require longer periods because after heavy drinking, alcohol can stay in the body for up to 30 h.

Alert Height Alert height is a specified radio height (typically approximately 100 ft) based on the characteristics of the aircraft and its fail operational landing system. Above this height if a failure occurred in one of the required redundant operational systems (fail operational systems) during a CAT II or III approach, the approach would be discontinued and a go-around would be executed.

The exception would be a reversion to a higher decision height, i.e., CAT I or II, if the aircraft were still above the approach-to-landing height, normally 1000 ft, whereby it was considered that enough time existed to safely transfer to a new approach limit.

Also, if a failure occurred in one of the required redundant operational systems below the alert height, the approach would be continued. The reason is that the probability and effects of a failure are such that if it should occur, the operation can be safely continued with the remaining system, "A second failure below the alert height is considered extremely improbable." Improbable but not impossible!

Altimeter—Pressure A simple pressure altimeter is designed to measure static air pressure, which it relates to an indicated altitude.

As the aircraft ascends, the static pressure in the instrument case decreases, which allows the enclosed capsule to expand; in turn, this moves the needle on the instrument face to indicate a corresponding altitude. For a descent the opposite function applies.

A subscale setting device is included so that the instrument can be zeroed to various datum elevations. The altimeter capsule is calibrated to full ISA MSL conditions, that is, +15°C, 29.92 in/1013 mbar and 1225 g/m^3.

A sensitive altimeter uses two capsules to increase the altimeter sensitivity to a change in static pressure conditions. Also it has vibrators fitted to overcome static friction, which further improves the rate of response to altitude changes.

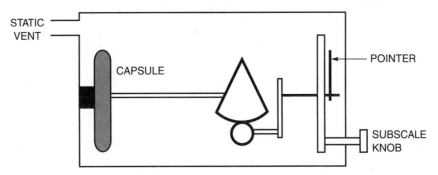

Figure 11   Simple pressure altimeter.

For a temperature deviation of ISA −36°C, the pressure altimeter overreads, because the temperature deviation is colder than ISA; i.e., the altimeter reads an altitude higher than the actual altitude of the aircraft.

Altimeter (Pressure Instruments) Error   The altimeter suffers from the following errors:

1. *Instrument error* has two sources. Friction of the instrument's moving parts causes an error in the indicated reading, and the instrument design is poor in reading and indicating any small capsule movements when a very small change in static pressure exists. This is particularly evident at very high altitudes where the change in static pressure is very small. Therefore the simple, single-capsule altimeter becomes increasingly unreliable with an increase in altitude.
2. *Pressure error* is caused by variations of the static and pitot probe relative position with respect to airflow over the aircraft's surface. This can lead to sensed pressure readings that are not representative of the free atmosphere when the airflow pattern is disturbed at different airspeeds, angles of attack, and configurations. (This is also known as *position* or *configuration error*.)
3. *Time lag error* occurs because the mechanical linkage, between the capsule and the dial face, fails to transmit static pressure changes instantly. A time lag is therefore present between the occurrence of an altitude change and its display on the altimeter dial face.
4. *Barometric error. See* Pressure Altimeter Errors.
5. *Density (air)/temperature error. See* Density (Air) Errors.

6. *Blocked static port* error occurs when the static pressure in the altimeter instrument case remains a constant value. Therefore the altimeter will display the altitude where the blockage occurred regardless of any actual change in the aircraft's altitude. The blockage is normally caused in flight by ice formation or small insects covering the static port.

The actions taken for a blocked static line causing an unreliable altimeter reading would be to

   a. Ensure that the pitot static probe anti-ice heat is on (pitot heat), if applicable.
   b. Use an alternative—static source or air data computer (if applicable).
   c. Use a limited flight panel, i.e., standby altimeter, VSI if available.
   d. Fly at correct attitude and power settings, especially for level flight. For climbs and descents calculate the rate of climb (ROC)/rate of descent (ROD), from either VSI or approximate values given known aircraft attitude and power performance, against time from known altitudes to determine the present altitude.

NB: If the blockage has been caused by ice formation and you have an unserviceable or no pitot heat system, then you should avoid icing conditions.

Altimeter—Servo-assisted   A servo-assisted altimeter increases the accuracy of a simple pressure altimeter, because its design no longer relies on a direct mechanical link between its capsule and the altitude pointer on the instrument's display dial. Instead the servo-assisted altimeter utilizes an electrically conducted E and I bar arrangement.

The servo-assisted altimeter has a capsule assembly, the movement of which is transmitted to a pivoted I bar which is adjacent to the E-shaped bar. The E bar has coils wrapped around its protrusions that are each fed with an AC electrical voltage. When the I bar is moved by the expansion or contraction of the capsule assembly, one limb of the E bar becomes closer to the I bar than the others, and this produces a greater voltage than in the other E bar protrusions. The resultant voltage difference is fed to a servomotor which through a feedback system repositions the I bar so that the air gaps are all equal again. Therefore the servomotor movement is a measure of the capsule's expansion and contraction, and this is used to drive the dial pointer and digital readout on the altimeter instrument face.

The advantage of the servo-assisted altimeter is its improved design feature between the capsule and the display pointer that removes both instrument error and time lag error, which are associated with a simple altimeter. As a result, the servo-assisted altimeter is more accurate, especially at high altitudes, due to the reduced friction experienced by its moving parts and its improved ability to read small movements of the capsules more accurately. Servo-assisted altimeters are accurate to ±30 ft at sea level and ±100 ft at 40,000 ft.

The servo-assisted altimeter's improved linkage design also greatly improves the response time of a measured change in the aircraft's altitude to its being displayed on the altimeter's face. In addition, the improved system allows for the altitude information to be digitally sent to the ATC transponder. This transponder altitude information is always relative to the 29.92-in, 1013-mbar datum.

Altitude   In aviation terms, altitude is the measured distance above the local pressure setting, that is, the QNH, or altitude above mean sea level.

Altitude—Pressure   *See* Pressure Altitude.

Ammeter (Gauge)   An ammeter indicates the amount of amperage (quantity of electrons flowing in a current) in an electric circuit.
   NB: A quantity of amperes is often referred to as the *current*.

Amp (Ampere)   One amp (ampere), denoted by 1 A, is a unit of electric current (expressed as *I*), i.e., the quantity of electrons flowing in a circuit when an electric pressure/force (volts) is applied.
   NB: 1 milliampere (1 mA) = 1/1000 A (0.001 A)

Anabatic Wind   An anabatic wind is a local valley wind that flows up the side of a hill. It is formed when the side of a hill is warmed by the sun's heat, which in turn heats up the layer of air next to the ground, making it lighter, resulting in an airflow up the side of the hill. However, because the air tends to rise vertically (away from the sloping hillside), anabatic winds are very slight.

Angle of Attack   Angle of attack is the angle between the chord line of the aerofoil and the relative airflow.

Angle of Incidence   The angle of incidence is the angle between the aerofoil's chord line and the aircraft's longitudinal datum. It is a fixed angle for a wing, but may be variable for a tailplane. (Sometimes it is called *rigging incidence.*)

Anhedral   This is the downward inclination of the wing from the root to the tip.

Anticollision Lights—on an Aircraft   *See* Navigation and Anticollision Lights—on an Aircraft.

Anticyclone   *See* High-Pressure System.

Anti-icing System   An anti-icing system is one that prevents ice buildup on a surface, e.g., thermal or electrical anti-icing systems on engine cowls.

Antiskid System   The purpose of an antiskid system is to sense when the wheels are "locked," i.e., not spinning, and to release brake pressure through the brake modulating/metering valve system to generate wheel spin. Wheel spin is required to maintain directional control and braking efficiency, which are necessary to ensure landing performance, which is especially important in the operation of modern aircraft with high speeds, low drag, and high weights, particularly when coupled with operation from short runways in bad weather.
   The antiskid system has a detection system that senses the moment a wheel stops rotating, which it interprets as meaning that the wheel is off the ground (hydroplaning). Once it senses the wheel is not rotating, the antiskid system releases brake pressure in the brake metering/modulating valve system, to release the brake pad from the brake wheel plate, and therefore reinitiates wheel spin.
   An antiskid system protects against the following:
1. *Locked wheel.* This can sometimes occur very rapidly, especially on wet or icy runways with a lightly laden aircraft, due to excessive brake pressure for the

prevailing conditions. An antiskid system will release excessive brake pressure and therefore prevent locked wheels.

2. *Wheel skid / slip.*   Whenever braking torque is developed, there exists a degree of slip between the wheel and the ground.

3. *Hydroplaning / Aquaplaning.*

4. *Touchdown.*   An antiskid circuit maintains zero brake pressure prior to touchdown. Therefore the brakes cannot be applied until the wheels have spun up initially, which thus protects against locked wheel at touchdown.

Apparent Wander   Apparent wander of a gyro is a natural phenomenon. A directional indicator (DI) gyro *appears* to wander, not because of any real changes in the direction of the gyro's spin axis alignment in space, but because its orientation has changed due to the earth's rotation. If this occurs, it gives rise to an incorrect heading display.

For example, a gyro not at the equator is aligned (pointed) to true north. As the earth rotates, the gyro maintains its alignment in space but is progressively misaligned to true north. This change in orientation is due to the earth's rotation and is known as *apparent wander (drift)*. The rate at which a gyro becomes disoriented is related to latitude. At the equator no apparent wander (drift) occurs; but as one moves toward the poles, the apparent wander (drift) increases.

The amount of apparent wander (drift) can be calculated from

$$\text{Apparent wander (degrees per hour)} = 15 \times \sin(\text{latitude})$$

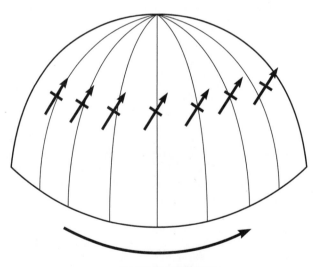

EARTH'S ROTATION

Figure 12   Apparent wander.

Depending on the sophistication of the directional indicator, apparent wander can be corrected in one of two main ways:

1. Periodically (in 10- to 15-min intervals) realign the DI to the magnetic compass heading, using the slaving knob on the DI. As the slaving knob is pulled out, the gyro becomes caged and is realigned to magnetic north by turning the slaving knob until the DI scale and the magnetic compass readings are coincident.

NB: This procedure covers any type of wander realignment.

2. If the DI instrument is fitted with a latitude nut, it produces an opposite error to the earth's rotation [that is, 15 × sin (latitude) in degrees per hour] to give an adjusted heading. In early instruments, this was a real nut that was screwed in and out to produce the necessary imbalance in drift on the gyro's spin axis for the aircraft's actual latitude. Only maintenance crews on the ground carried out the setting for the latitude. If, however, the aircraft was moved away from the latitude for which the latitude nut was set, then an error of either +*ve* or −*ve* would occur because the degree of apparent wander varies with latitude, namely, it increases toward the poles. On modern aircraft this effect is accomplished by onboard computer software adjustments.

Approach—Clearance   If you are cleared by ATC for an approach, this means you have been cleared for the complete procedure—in terms of lateral navigation and vertically to any altitude/height constraints in the procedure from your present altitude/height, and then onto the approach descent to the runway. It does *not* mean you have been cleared to land. Landing permission is a distinct and separate clearance.

Approach Lights for an ILS Runway   A typical approach lighting arrangement for an instrument landing system (ILS) equipped runway would consist of the following:

1. Extended centerline white lights.

2. Up to three crossbars sited at 300-m intervals back along the approach path from the threshold.

NB: For increased approach lighting more crossbars can be used, which may be sited at 150-m intervals.

NB: Approach lights simply act as a lead into a runway, and they do not mark the boundaries of a suitable landing area.

In addition, an advanced precision approach (CAT II and III) lighting arrangement could be never-ending, but as a minimum you could expect the following:

3. An all-round increase in white approach centerline and crossbar lights, at reduced spacing intervals and higher intensity.

4. High-intensity red approach lights, both sides of the centerline white approach light and in between the white crossbars. In addition, these lights can be found on the runway prior to the touchdown zone.

NB: These would be listed on the runway approach plate.

5. Extra green threshold wing bar lights, which are extended outboard of the runway edge threshold lights or at the displaced threshold.

Approach Lights for a Nonprecision Runway   The approach to a nonprecision approach runway may not be continued below the minimum decision height (altitude) [MDH (A)] unless at least one of the following visual references for the intended runway is distinctly visible to and identifiable by the pilot.

- Element of the approach light system
- The threshold
- The threshold markings
- The threshold lights
- The threshold identification lights
- The visual glide slope indicator
- The touchdown zone lights
- The touchdown zone or touchdown zone markings
- The runway edge lights

Approach—Separation   You are required to maintain the minimum approach separation, so as to avoid the wake vortex/turbulence from the preceding aircraft.

NB: Wake turbulence can be a serious hazard to lighter aircraft following heavier aircraft.

The ICAO final approach spacing minima are as follows:

Leading aircraft	Following aircraft	Separation	
		nm	min
Heavy	Heavy	4	—
Heavy	Medium	5	2
Heavy	Light	6	3
Medium	Heavy	3	—
Medium	Medium	3	—
Medium	Light	5	3
Light	Heavy	3	—
Light	Medium	3	—
Light	Light	3	—

The final approach spacing minima in some other states, including the United Kingdom, vary from the ICAO minima as follows:

Leading aircraft	Following aircraft	Separation	
		nm	min
Heavy	Heavy	4	2
Heavy	Medium	5	3
Heavy	Small	6	3
Heavy	Light	8	4
Medium	Heavy	*	*
Medium	Medium	3	2
Medium	Small	4	2

Medium	Light	6	3
Small	Heavy	*	*
Small	Medium	3	2
Small	Small	3	2
Small	Light	4	2
Light	Heavy	*	*
Light	Medium	*	*
Light	Small	*	*
Light	Light	*	*

*Separation for wake vortex turbulence reasons alone is not necessary.

NB: These minima may be applied when an aircraft is operating directly behind another aircraft and when crossing behind at the same altitude or less than 1000 ft below, whenever a controller believes there is a potential hazard due to wake vortex turbulence. The minima may even be increased at the discretion of the controller, or at the request of the pilot. Such requests must be made before entering a runway or commencing a final approach.

Many of the very busy commercial airfields may space aircraft closer than these separation limits; therefore, you need to be aware of the increased possibility of encountering wake vortex turbulence.

The separation minima stated cannot entirely remove the possibility of a wake turbulence encounter. The objective of the minima is to *reduce* the probability of a vortex wake encounter to an acceptably low level, and to minimize the magnitude of the upset if an encounter does occur. Care should be taken when following any substantially heavier aircraft, especially in conditions of light winds. *The majority of serious incidents occur close to the ground, especially when the winds are light.*

Approach—Sink Rate   If you need extra lift to correct for the unique problem of a high sink rate on the approach, you can create the extra lift required in one of two ways:

1. *By increasing incidence.*   In this case you must increase thrust to counter the extra drag from the higher incidence, or else the resulting sink rate will be higher. The exception occurs when you are trading an excess of airspeed for height, but then be careful to monitor your speed in case it decreases below your approach speed.

2. *By increasing airspeed.*   This is obtained by increasing thrust while keeping all other parameters substantially constant. A heavy aircraft takes a lot of accelerating; so if this option is exercised, a lot of thrust will be needed.

NB: If you see the need for more thrust for any reason, then apply enough of it early enough, especially for jet engines. This is necessary because of the poor response from a low rpm setting and the comparatively small amounts of thrust produced for a given power lever movement in the lower ranges of power settings.

Approach—Speed Control   It is important not to lose speed on the approach, especially in jet swept-wing aircraft. This is true because the approach speed is only just above VIMD (the minimum drag speed); and because of the relative flatness of the drag curve, especially around VIMD, the jet aircraft does not produce any noticeable changes in flying qualities. Speed is unstable below VIMD, where an increase in thrust has a greater drag penalty for speed gained, with a net result of losing speed for a given increase in thrust. Therefore if you lose speed below your approach speed, you could find yourself in this situation, which results in a decreasing speed and altitude.

NB: VIMD is a higher speed on a jet aircraft because the swept wing is more efficient against profile drag, and therefore the minimum drag speed is typically a higher value.

Recovery from this situation is further compounded by the jet engine's slow response from idle, and if you have slipped up the back of the drag curve, you need a lot of thrust to counter its effect and quickly. This all makes a recovery rather marginal when close to the ground, and therefore it is doubly important not to lose speed on the approach.

For a propeller-driven aircraft, with a well-defined VIMD (minimum drag speed) point on its drag curve (which is due to the marked increase of profile and induced drag on either side of VIMD), any deviation of airspeed below VIMD on the approach is very well defined by the steep increase of the drag curve.

In addition, the slipstream effect from the propeller makes a recovery from a falling airspeed on the approach much better. Therefore although it is always important to maintain a stable speed on the approach, it is not as marginal on a propeller-driven aircraft as it is on a jet.

Approach Speed—Performance   The landing performance of an aircraft is based on specified approach speeds VAT/VREF. If you approach to land at a speed higher than that specified, the landing distance will probably exceed that calculated. This can be very important if the LDA is limiting.

NB: Actual approach speed below the specified approach speed should not be flown for safety reasons.

APS Weight   Aircraft prepared for service (APS) is

$$APS = \text{basic weight} + \text{variable load}$$

Aquaplaning/Hydroplaning   Aquaplaning is the phenomenon of a tire skating (not rotating) over the runway surface, on a thin film water. It is caused by the buildup of a layer of water beneath the tire, which becomes a wedge between the tire and the runway. This wedge of water effectively lifts the tire off the ground, reducing the friction forces to practically zero. Therefore wheel braking has no effect, and directional control may also be lost. Aquaplaning is extremely significant on jet transport aircraft, and it can significantly increase the stopping distance required.

There are three types of aquaplaning:

1. *Dynamic.*   This is due to standing water on the runway when the tire is lifted off and completely supported. That is, it is a function of fluid density.

2. *Viscous.*   This occurs when the surface is damp and provides a very thin film of fluid which cannot be penetrated by the tire. Viscous aquaplaning can occur at, or persists down to, much lower speeds than simple dynamic aquaplaning. It is particularly associated with smooth surfaces and is quite likely to occur in the

touchdown area, which is often liberally smeared with rubber deposits. It is a function of fluid viscosity.

3. *Reverted rubber, or rubber reversion.*   This refers to the tire becoming tacky. It requires a long skid, rubber reversion, and a wet surface. The heat from the friction between the tire and the wet runway surface boils the water and reverts the rubber, which forms a seal and delays water dispersal. The steam then prevents the tire from contacting the runway. That is, melting rubber traps the steam, which causes aquaplaning.

NB: All three types of aquaplaning can occur during one landing run if the conditions are correct.

The approximate minimum true ground speed, in knots, at which aquaplaning can be initiated is calculated from the following formula:

$$9 \times \sqrt{\text{tire pressure, psi}} \qquad \text{for takeoff}$$

$$7.6 \times \sqrt{\text{tire pressure, psi}} \qquad \text{for landing}$$

However, it is generally regarded that $9 \times$ square root should be used also for the landing, although in theory only $7.6 \times$ square root is required at landing followed by $9 \times$ square root during the rollout.

NB: Aquaplaning occurs at any speed above this minimum speed.

On takeoff, there is normally a considerable gap, that is, 20 to 50 kn, between the minimum aquaplane speed and V1, for example, minimum aquaplane speed of 115 and a V1 speed of 145. On landing, or a rejected takeoff, if aquaplaning occurs, it may continue at speeds lower than the calculated aquaplane speed. This is so because the calculation gives us a speed below which aquaplaning will not *start*.

Tire wear is obviously an important factor; the more worn the tire, the more likely it will be to aquaplane. A general awareness of this fact should be retained.

You can control an aquaplane condition by using the antiskid braking system which releases the brakes if it senses a skid. This allows the wheel to rotate and to push through the water layer, making contact with the runway again.

If you have no antiskid systems, or they are inoperative, then manually brake, using the same principle. Namely, if you feel a slip, then release the brakes, allow the skid to stop, and then reapply the brakes.

Arm   *See* Component Arm.

Artificial Feel System   An artificial feel system is required because power-operated flying controls are irreversible; i.e., they do not feed back to the pilot any sensory information about how hard the control surface is and thus what aerodynamic air forces it is coping with. Therefore there is a need to give this information to the pilot so that he or she is aware of the control angles being applied and their effect on the aircraft. In short, keep the pilot in the sensory loop, which allows him or her to guard against overstressing the control surface.

The simplest form of artificial feel consists of a spring box fitted into the control run. This provides a feel and self-centering action, but the stick forces are constant and are therefore only suitable for aircraft types with a limited altitude and speed range.

A progressive feel system, *Q feel,* based on $\frac{1}{2}R(V^2)$ is required for aircraft that use powered flying controls and have a large speed range. It works by adjusting the feel to match the variable control surface deflection at a constant speed, or for a variable speed with a constant control surface angle.

There are many variations of this system, but basically it senses static and dynamic pressures and feeds this information into a computer. The computer then feeds into the flight control system a force which duplicates a proportion of the normal resistant feel which would be obtained from a conventional aerodynamic control working under an ideal condition.

See Elevator Artificial Feel System.

Artificial Horizon (AH) Instrument—Air-Driven   The artificial horizon is the primary attitude instrument that measures and displays the pitch and roll of the aircraft, about the horizon level. The artificial horizon consists of the following:

1. An earth gyro
2. The gyro that rotates about a vertical axis
3. Two gimbals
4. Three planes of freedom
   a. The gyro's spin axis
   b. Pitch and roll axis of the gimbals
5. The gyro's axis that is aligned to the earth's vertical

Simply the aircraft moves around the artificial horizon gyro, and the gimbals measure the aircraft's pitch and roll maneuvers.

Simple air- (vacuum-) driven artificial horizons are common on older aircraft types. The gyroscope assembly is directly behind the instrument face, and air drives the rotor anticlockwise, when viewed from above, around its vertical axis by impinging on buckets on its outer edge. The rotor case, which is the inner gimbal, is connected to a guide pin that moves in a slot to move the horizon bar up and down on the instrument face, which displays the aircraft's pitch when a precessed force in this plane of freedom is sensed. The outer gimbal is free to swivel around the aircraft's fore and aft axis and is connected to the bank angle pointer, which displays the aircraft's roll information on the instrument face whenever a precessed force in this plane of freedom is sensed.

The air-driven artificial horizon is typically free in pitch through ±60° and in roll through ±110°. Mechanical stops prevent movement outside these limits, past which the gyro would topple. Artificial horizons are usually caged when not being used, and they should be uncaged only when the aircraft is straight and level and the gyro is up to speed. This should be done shortly before takeoff or in straight and level, constant-speed flight.

An air-driven artificial horizon erect position is achieved when its spin axis is maintained in the earth vertical. Air-driven artificial horizons are made pendulous, with their center of gravity below their suspension point, so that they settle in their gimbals in a nearly erect position when not working and reduce start-up time. Once the gyro is rotating, its erect position, which is manifested as its vertical axis being tied to the earth vertical, is refined by a system of "pendulous vanes and air jets" which makes the gyro precess back to the vertical if it is displaced. This is accomplished at the bottom of the rotor case with four air exhaust vents, which are each normally half covered by an outer flap which is hinged at the top, a "pendulous vane." When the gyro is erect (vertical), the air that has driven the rotor escapes from all four vents equally. However, when the gyro is even slightly displaced from the vertical, one of the pendulous vanes will cover its vent and the opposite vent will open by an equal amount. This sets up an imbalance reaction from the four air vents with the greatest of the four vent forces being precessed by 90° in the direction of the rotation (normally anticlockwise when viewed from above). This rotates the gyro's spin axis back to the earth vertical, which is its erect position. Once a gyro has toppled, it will become erect again at a rate of 2° to 4° per minute.

The indications of a failed air-driven artificial horizon instrument are (1) a low reading on the suction gauge and (2) a possible warning flag on some instruments. Because the air-driven artificial horizon instrument is, as the name suggests, driven by air, a blocked air supply line will be shown on the suction gauge as a low reading. This means that the airflow to the instrument rotor is low, and therefore the rotor will not be up to speed; this results in the gyro being unstable and responding slowly or not at all to the aircraft's maneuvers. Therefore the instrument is unreliable and should not be relied upon for accurate attitude information. The action required is to reerect the gyro. This is accomplished in flight by caging and uncaging the instrument when the aircraft is straight and level and at a constant speed, to achieve an approximate reerection. However, if the suction reading is still low, the artificial horizon will still be unstable, and therefore secondary flight instruments, e.g., the turn coordinator, VSI, etc., should be monitored. The indication of a failed electrically driven artificial horizon is a *power failure* warning flag on the instrument face. This will be indicated whenever the instrument senses the supply of electric power to the instrument is either low or nonexistent. In these circumstances a backup power supply may be automatically or manually selected; but if this is not available, then the instrument will be unreliable and should not be depended upon for accurate attitude information.

Artificial Horizon (AH) Instrument—Electrically Driven Electrically driven artificial horizons use the same basic principles as the air-driven instruments with a gyro tied to the earth's vertical and two gimbals. There are, however, some fundamental differences:

1. Most gyros/rotors of electrically driven artificial horizons rotate clockwise when viewed from above.
2. Electrically driven artificial horizons use an electric squirrel cage motor to drive the rotor at approximately 22,000 rpm, which is about twice the speed of an air-driven instrument. Therefore the electrical artificial horizon is more rigid.
3. Electric erection systems are very fast, and because of this there is no need for the gyro to be pendulous, although some do remain to some extent pendulous. Electrical erection systems can also be cut out at will.
4. Acceleration and turn errors are minimized or completely eliminated; because the instrument has little or no pendulousity, its normal erection system can be cut out at certain values of longitudinal or lateral (balanced turn) acceleration.
5. The electrically driven artificial horizon has a freedom of pitch of ±85° and unlimited roll.

The electrically driven artificial horizon has mercury tilt switches mounted on the rotor case (inner gimbal), which have replaced the pendulous vanes on the air-driven instrument. Any displacement of the gyro axis from the earth vertical is sensed by the tilt switch, which makes and breaks electric circuits connected to torque motors on the gimbals that reerect the system, at approximately 5° per minute.

Most electrically driven artificial horizons have roll cutout switches fitted on the roll axis to prevent the gyro from reerecting to a false datum in an extended turn.

Artificial Horizon (AH) Instrument—Errors   The artificial horizon experiences the following errors:

1. *Turning error.*   In a turn the erection system will try to erect the gyro to the resultant acceleration, which in a turn is the aircraft's vertical, and not the earth's vertical, which is the artificial horizon gyro's correct spin axis. This will produce an initial roll error, which changes to a pitch error as the aircraft turns. Furthermore, the pendulous lower part of the gyro will attempt to line up the resultant acceleration, but this effect is precessed through 90° and will give an initial pitch error that changes to a roll error as the turn continues. The overall effect is that the gyro's axis is offset a few degrees from its correct position during a 360° turn, which is the core of any turning error.

NB: Some instruments have the vanes adjusted to keep the erect gyro 2° to 2.5° from the true vertical at all times, to minimize turning errors. This system is called *compensation tilt,* but it only works for one specified rate of turn and one set TAS, usually a rate of 1 turn at 250 kn.

2. *Acceleration error.*   The air-driven artificial horizon's pendulous vanes of the erection system are displaced by a marked fore and aft acceleration, e.g., on takeoff. This produces a false pitch-up indication. In addition, acceleration affects the pendulous lower element of the gyro, and this force is precessed through 90° to indicate a false bank to the right.

The acceleration error can be summarized as follows: Acceleration error indicates a false climbing turn to the *right,* and a deceleration error indicates a false descending turn to the *left.*

NB: Because of these acceleration errors, air-driven artificial horizons are restricted to aircraft that have a relatively slow acceleration performance.

Turning and acceleration errors in the artificial horizon gyro are caused by lateral acceleration in turns and by the aircraft's acceleration and deceleration forces that induce a false indication position of the gyro's vertical axis.

NB: Remember the artificial horizon is an "earth gyro" with its spin axis in the earth vertical.

4. *Real wander* of the gyro's spin axis away from its alignment to the earth's vertical. *See* Real Wander.

Aspect Ratio   This is the ratio of the wing's span to its geometric chord, for example, 4:1. High aspect ratio = high lift (gliders), and low aspect ratio = lower lift but capable of higher speeds.

Assumed (Flexible) Temperature   An assumed (flexible) temperature is a performance calculation technique used to find the takeoff engine pressure ratio (EPR) setting for an aircraft's actual takeoff weight. This is known as a variable or reduced thrust value.

First, it should be clearly understood that the full takeoff thrust is calculated against an aircraft's performance-limited [by field length; weight, altitude, and temperature (WAT); tire or net flight path; or obstacle clearance climb profile] maximum permissible takeoff weight (MTOW), which itself is calculated from the ambient conditions of aerodrome pressure altitude and temperature. However, in many cases the aircraft takes off with a weight lower than the MTOW. When this happens, an assumed temperature performance technique offers a method of calculating a decreased takeoff thrust that is adapted for the aircraft's actual takeoff weight. This is done by calculating the corresponding *assumed / flexible temperature* (higher than the actual air temperature) from the WAT performance graph. This is done by using the aircraft's actual takeoff weight, as if it were the performance-limiting MTOW, against the actual aerodrome altitude to find the limiting temperature.

NB: Temperature increases as the maximum permissible weight decreases. So it is possible to assume a temperature at which the actual takeoff weight would be the limiting one.

Then by using this calculated assumed/flexible temperature you can calculate the aircraft's correct takeoff EPR thrust from the takeoff EPR graph. The assumed temperature (actual aircraft weight) EPR is a reduced thrust value from the MTOW full-thrust setting.

NB: The assumed/flexible temperature setting for lower-weight aircraft meets the same takeoff run (TOR) performance (i.e., rotates at the same point) as the maximum thrust setting for MTOW aircraft.

Assumed temperatures can be used to (1) calculate on paper the required takeoff EPR or (2) enter into the aircraft's engine computer to calculate the required takeoff EPR.

Finally, remember that the performance condition that was limiting at full thrust (field length, WAT curve, net flight path) is not necessarily the limiting condition at a reduced thrust.

ATIS   ATIS or Automatic Terminal Information Service, is a prerecorded tape, broadcast on an appropriate VOR or VHF channel to reduce the workload on ATC communication frequencies, that gives current aerodrome information on its operations and weather.

NB: Some of the larger aerodromes have both an arrival and a departure ATIS. It is changed with any significant change in the reported conditions, and each new message has a new alphabetical designator prefix, e.g., alpha, beta, etc., to distinguish the current from the old. ATIS information commonly includes the following:

1. Designator prefix.

2. Runway in use.

3. Wind direction and speed. The wind is measured in relation to magnetic direction to allow the pilot to relate the direction of the wind to the runway direction (which is also measured in relation to magnetic north).

4. Cloud base, runway visual range (RVR), visibility.

5. QNH.

6. Temperature and dew point.

7. Severe weather warning.

8. Contact frequency and instructions.

Atmosphere   The atmosphere consists of the following gas components: 78 percent nitrogen, 21 percent oxygen, and 1 percent carbon dioxide, water vapor, and other gases. These proportions remain the same irrespective of altitude. However, the earth's atmosphere is divided into the following;

Troposphere is the lower region of the atmosphere, where the majority of the earth's weather occurs and within which most aircraft operate.

Tropopause marks the top of the troposphere region at approximately 20,000 ft over the polar regions and 60,000 ft over the tropics. An ISA average is assumed to occur at 36,000 ft.

Stratosphere is the region above the tropopause. Some military jets and the Concorde may cruise in the stratosphere.

Stratopause marks the top of the stratosphere region at approximately 160,000 ft.

Meosphere is the region above the stratopause.

Meopause marks the top of the meosphere region at approximately 320,000 ft.

Thermosphere is the region above the meopause.

Attitude Directional Indicator (ADI)—Servo-Driven A servo-driven attitude directional indicator, also called a *remote artificial horizon,* is used on modern aircraft to display its attitude information (pitch and roll), which has been calculated by the aircraft's inertia navigation system (INS)/inertia reference system (IRS) platforms.

The system has the advantage of being free of turn and acceleration errors. As a dual system (INS #1 feeds the captain's instruments and INS #2 feeds the copilot's instruments), the displayed information can be compared with itself and with the separate standby instrument to detect errors and deselected incorrect systems.

Information from other aircraft systems can also be displayed on servo-driven attitude directional indicators, such as ILS localizer and glide slope deviation, flight director, radio altimeter, airspeed, and decision height.

Automatic Boost Control *See* Superchargers.

Automatic Dependent Surveillance—Broadcast (System). *See* ADS-B.

Automatic Gain Control (AGC) This makes the weather radar system less sensitive to short-range returns so that it provides an even overall picture. Without AGC you would expect the closer targets to produce a strong pulse return, and therefore the closer targets would be brighter on the screen than the targets farther away. AGC removes this misleading inaccuracy.

Automatic Igniters. These are used in gas-turbine engines to protect against disturbed/turbulent airflow upsetting the engine. This condition is particularly common with rear-mounted engines during some abnormal and even some rather normal flight maneuvers, because rear-mounted engines are ideally placed to catch any disturbed airflow generated by the wing when the airflow pattern breaks down because of

- High incidence of attack, e.g., prestall buffet
- High-g maneuvers, e.g., steep turns
- High Mach number effect

The engine is sensitive to airflow characteristics, and disturbed air entering the engine intake can completely upset the engine running condition by putting the fire out and causing the engine or engines to run down, which is obviously unacceptable. Therefore auto igniters are used to protect against this occurrence.

Auto igniters work by sensing a particular value of incidence of the aircraft, via the incidence sensing (probe) system (which is also used to activate the stick shaker and pusher), and it automatically signals "on" in the ignition system before the disturbed airflow generated by the wing affects the engines. This ensures that the engines at least continue to run, although in some cases they might surge a little.

On most aircraft, the use of auto igniters is solely to protect against the engine being upset from disturbed air, and the normal ignition system is used for the following:

- Ground starting
- In-flight starting
- Maintaining combustion during takeoff and landings from contaminated runways
- Selection to override, e.g., flight through precipitation

However, some engine types might transfer some normal ignition functions to the auto ignition system; but in general terms, the protection of the engine from disturbed airflow is the main motivation commonly associated with an auto ignition system.

Automatic Landing System An automatic landing system is a function of the autopilot flight director system and automatic throttle system, which is engaged using the approach mode of operation. It is designed to carry out automatic landings under all visibility conditions by providing guidance and control better than that provided by a pilot, from the interception of the localizer until touchdown. It controls the aircraft about all three axes simultaneously. This includes yaw to control drift and, by using the autopilot's flight control computers, including sensors and ILS radio couplings, to accurately track the localizer and glide path approach down to minima which can include a full CAT III automatic landing.

The auto throttle is also used by the auto landing system to control and maintain the correct airspeed, through engine power changes during the approach, to retard the thrust during the flare, and finally to automatically disconnect at touchdown (type specific).

The approach and landing maneuver is the most difficult phase of flight to control. Therefore, as a prelude to a "blind landing," automatic landing systems have always been the ultimate aim of the designers and aircraft operators. It was found that many problems that had to be solved in its development could be grouped into the following three main categories:

1. Need to achieve the highest reliable auto landing system possible, bearing in mind the need for the system to be entrusted with a very considerable authority over the controls of the aircraft, including the auto throttle, and in the presence of the ground
2. Provision to adequately monitor information on the progress of the approach and landing maneuvers
3. Ability to substitute the pilot's direct vision with an automatic ground guidance system, which is as reliable as the control system's

Furthermore, the operation of an automatic flight control system must also be such that it

- Doesn't disturb the flight path as a result of an active malfunction

- Has adequate authority for sufficiently controlling the aircraft accurately along the required flight path
- Is capable of completing the intended flight maneuver following an active or a passive failure

Therefore to solve these problems and to attain these capabilities, it was considered necessary to adopt the concept of *system redundancy,* i.e., to use multiple autopilot systems/channels (dual, triple, or quadruple), so that they operate in such a manner that a failure within a single system/channel would have an insignificant effect on the aircraft's performance, especially during the automatic approach and landing.

During the cruise and initial stages of approach to land, the control system operates as a single-channel system, controlling the aircraft about its pitch, roll, and yaw axes and providing the appropriate flight director commands. However, the multichannel operation is required for an automatic landing, and at a certain stage of the approach, the remaining one, two, or three channels are armed by selecting the approach operating mode, which allows the autoland system to work in a fail passive or operational manner.

In addition, altitude information essential for vertical guidance to touchdown is always provided by signals from a radio altimeter which becomes effective as soon as the aircraft's altitude is within the altimeter's operating range (typically 2500 ft).

The requirement for automatic landing systems (by the United States and many other certification authorities) is a minimum reliability value of 1 in 107; or in other words, the system should only suffer a fatal failure no more than 1 in 10,000,000 automatic landings.

A typical auto landing sequence (based on the B737-300) would be as follows:

1. At 1500-ft radio altitude, the auto landing system commands
   - The capture of the localizer and the glide slope beams
   - The remaining armed flight control channels to be automatically engaged
   - The computerized landing flare control to be armed

The localizer and glide slope beam signals control the aircraft about the roll and pitch axes so that any deviations are automatically corrected to maintain alignment with the runway. At the same time, the autoland status annunciation displays "LAND 2" or "LAND 3," depending on the number of channels of the autoland system.

2. At 330-ft radio altitude, the auto landing system commands
   - The aircraft's horizontal stabilizer to be automatically repositioned to begin trimming the aircraft to a nose-up attitude
   - The elevators to be deflected to counter the trim and to provide subsequent pitch control in the trimmed attitude

3. At 45 ft (gear altitude) above the ground, the auto landing system commands
   - The flare mode to be automatically engaged

NB: The flare mode takes over the pitch control from the glide slope, at approximately a 2 ft/s descent path.

   - The throttle retard to be engaged through the auto throttle system to reduce engine thrust to match the landing flare path

4. At 5-ft gear altitude, the auto landing system commands
   - The flare mode to be disengaged

- A transition to the touchdown and rollout mode
- The pitch attitude to be reduced to 2° at about 1-ft gear altitude

5. At touchdown

- A command signal is supplied to the elevators to lower the aircraft's nose and so bring the nosewheel into contact with the runway
- The auto throttle system is commanded to be disengaged when reverse thrust is engaged

The automatic flight control system remains on until disengaged by the flight crew.

Automatic Pilot System   An automatic pilot system is an integrated flight control system that enables an aircraft to fly a prescribed route and to land at a designated airport without the aid of a human pilot. It is usually part of an *automatic pilot flight director system* (APFDS).

The purpose of an automatic pilot system is to relieve the pilot of the physical and mental fatigue of flying the aircraft, especially during long flights. This will result in the pilot's being more alert during the critical phase of landing the aircraft, thus improving safety. The response of an autopilot is also substantially faster than that of a human pilot. A human pilot takes approximately ⅕ s to detect a change in the aircraft's attitude, and then there is a further delay in deciding which control to apply to oppose the disturbance. An autopilot, however, will detect a disturbance and put on the required control to correct the disturbance in approximately 50 ms.

When engaged, the autopilot is responsible for the following:

1. *Stabilizing the aircraft.*   For practical stabilization reasons, the autopilot is broken down into three basic control systems:

Pitch   to control the elevators

Roll   to control the ailerons

Yaw   to control the rudder

2. *Maneuvering the aircraft.*   The number of axes that the autopilot controls classifies the autopilot's maneuver capabilities.

A single-axis autopilot commands the ailerons that control the aircraft about the roll axis only. In its most basic concept it will only give lateral stability or a wings-leveling function. Its modes of operation are

- Manual commands, which allow the pilot to manually turn the aircraft
- Heading, compass-derived information of heading hold, or preselect heading modes
- Radio (some single-axis systems have VOR or localizer signals supplied to allow the aircraft to track, but not to intercept, radio beams)

A two-axes autopilot commands the ailerons and elevator that controls the aircraft about the roll and pitch axes. Two-axes systems could range from fairly simple systems with only a few basic modes of operation, found in light aircraft, up to the most complex integrated system with full-flight profile modes of operation, including an autoland mode.

NB: In many large aircraft with a two-axes autopilot, a yaw damper is used to command the rudder to control the yaw/vertical axis, which would be the third axis. However, because the yaw damper is classified as a separate system, the autopilot is still classified as a two-axes system.

A three-axes autopilot commands the ailerons elevator and the rudder that controls the aircraft about the roll, pitch, and yaw/vertical axes. The roll and pitch channels are used as the primary control channels, and the rudder channel is basically a stability channel, i.e., a more active yaw damper. It is common to have an interaction between the roll channel and the rudder channel to assist in coordinated turns and for faster stability response. A three-axes autopilot system provides a full-flight profile mode of operation.

Note: The following example is based on the B737-300 autopilot system:

First, the autopilot has different modes of engagement, which depend upon the manufacturer's design, but in general there are typically three:

- *Command (CMD)* engages the autopilot to fully control the aircraft, either from the flight management computer (FMC) stored route ("managed" for Airbus aircraft) or from the mode control panel (MCP) selected modes of operation, i.e., heading, vertical speed, etc. ("selected" for Airbus aircraft).

- *Control wheel steering (CWS) or manual* is provided in some automatic flight control systems, its purpose being to enable the pilot to maneuver the aircraft in pitch or roll, through the automatic control system by exerting normal maneuvering forces on the control wheel. In adopting this mode, it is not necessary for pitch and roll autopilot modes of operation to be selected on the MCP. When the control wheel is released, the automatic control system holds the aircraft at the newly established attitude.

- *Off.*

Second, the autopilot has different modes of operation, usually located on the MCP. These can command an autopilot either to follow manually inputted flight profiles (i.e., headings, vertical speed, etc.) or to follow automatically an FMC flight profile that also includes the automatic selection of navigation aids, with the selection of LNAV and VNAV operation modes. The autopilot modes of operation are normally the following.

1. *Heading* is an autopilot flight director system (APFDS) roll control mode that turns and holds the aircraft onto the MCP selected magnetic heading, by using the roll control channel of the flight control computer (FCC) to command the ailerons to move or hold the aircraft about the longitudinal axes. The heading mode is normally used to intercept a radio course or to following ATC heading assignments.

2. *Lateral navigation* (LNAV) is an APFDS roll control mode that commands or guides (if the autopilot is not engaged) the aircraft's lateral movement to match a predetermined lateral route profile in the FMC. This is accomplished by using the FCC's roll control channel to command the ailerons to move or hold the aircraft about the longitudinal axes. It is available from takeoff to localizer capture and is used to steer the aircraft to intercept and track the active route.

NB: The FMC lateral navigation route can include standard instrument departures (SIDs), STARs, and instrument approaches.

NB: It is known as *managed* navigation for Airbus aircraft.

3. *VOR/LOC (VOR/localizer) mode* is an APFDS roll control mode that captures and tracks a selected VOR or localizer course, which is either part of an FMC LNAV route or an MCP selection. This is accomplished by using the FCC's roll control channel to command the ailerons to move or hold the aircraft about the longitudinal axes.

NB: The localizer mode is often used as a prelude to an autoland approach mode capture.

4. *Altitude hold* is an APFDS pitch control mode that acquires and holds the MCP selected altitude when the altitude hold mode is engaged, by using the FCC's pitch control channel to command the elevator to move or hold the aircraft about the lateral axes.

5. *Vertical speed* (VS) is an APFDS pitch control mode that acquires and holds the MCP selected foot-per-minute rate of climb/descent when the VS mode is engaged. This is accomplished by using the FCC's pitch control channel to command the elevator to move and/or hold the aircraft about the lateral axes.

NB: A vertical speed needs to be selected, say, 500 ft/min, before the VS mode can be engaged. Normally VS is used with the auto throttle (AT) to hold an MCP selected airspeed.

6. *Level change* is an APFDS pitch control mode that coordinates pitch and automatic thrust commands to make automatic climbs and descents to preselected MCP altitudes at the MCP selected airspeed to give the maximum ROC/D for the selected airspeed when the level change mode is engaged. The FCC's pitch control channel commands the elevator to move and/or hold the aircraft about the lateral axes to gain and/or hold the MCP selected airspeed. While the auto throttle is coupled to this mode, it holds the N1 limit in the climb and idle thrust in a descent.

NB: This is known as *open climb or descent* for Airbus aircraft.

7. *Vertical navigation* (VNAV) is an APFDS pitch control mode that commands or guides (if the autopilot is not engaged) the aircraft's vertical movement to match a predetermined vertical route profile in the FMC. This is accomplished by using the FCC's pitch control channel to command the elevator to move or hold the aircraft about the lateral axes. The auto throttle is coupled to this mode to provide either thrust or speed control, depending on the stage of flight, either to match the FMC speed or to provide climb or descent thrust when the VNAV speed is controlled by pitch control. To summarize:

VNAV climb	APFDS pitch holds FMC speed, and the AT holds FMC thrust climb limit.
VNAV cruise	APFDS pitch holds altitude, and AT the holds FMC speed.
VNAV descent	APFDS pitch holds FMC speed, and the AT retards to idle descent thrust.
VNAV path descent	APFDS pitch holds descent path, and the auto thrust holds FMC speed down to idle thrust settings.

VNAV is available from takeoff and may end with an ILS approach to a destination airport. Any profile changes, i.e., speeds or altitudes, can be changed by using the FMC control display unit (CDU).

NB: The FMC vertical navigation route can include SIDs, STARs, and instrument approaches.

NB: It is known as *managed* vertical navigation for Airbus aircraft.

8. *ILS/approach* is an autoland operating mode, using FMC and MCP selections. It arms the remaining FCC channels and the localizer and glide slope modes; then when it is within system limits, it engages pitch and roll commands to capture and track the ILS localizer and glide slope, and to auto flare the aircraft on landing. It uses the FCC's pitch control channel to command the elevator to move and/or hold the aircraft about the lateral axes, the ailerons to move and/or

hold the aircraft about the longitudinal axes, and the rudder to hold the aircraft about the vertical (yaw) axes. The auto throttle is coupled to this mode to hold the MCP selected airspeed and to retard during the landing flare.

An active autopilot has several integrated subsystems. It responds to various sensors that feed navigation (radio couplings, i.e., selected VOR/ILS signals), altitude, heading, IRS (aircraft movement), and speed information into a FCC. This generates steering, pitch, and throttle (if the auto throttle is engaged) commands to the aircraft's control surfaces and throttle levers, to automatically maneuver and stabilize the aircraft along the programmed route, selected on the MCP and/or the FMC.

A digital FCC and its associated sensors and command line are known as a *channel,* which is in effect a single individual autopilot system, which is normally engaged by an associated autopilot engagement switch. The number of channels/autopilot systems employed determines whether the system is in a fail operational, fail passive, or fail soft status. As a minimum, during the cruise and the interception of the approach mode of operation, a single autopilot/channel is used, and therefore it is fail soft; i.e., it does not necessarily maintain the aircraft's trimmed condition and reverts to a manual flying mode if the autopilot fails. During the approach/land and go-around modes of operation, the autoland system uses either a dual or triple autopilot/channel system that establishes the system in either a fail passive or fail operational status, which allows the autoland system to continue to operate safely during this crucial stage of flight, because of its redundant channel capabilities.

See Fig. 13 at the top of the next page.

The APFDS and auto throttle work together to maintain the aircraft's airspeed and vertical path in different variations for the phase of flight and/or the mode of operation selected.

Note: This example is based on the B737.

1. *Takeoff.*　The AT sets takeoff thrust, and the APFDS adjusts pitch to maintain the airspeed of V2 + 10 to 20 kn.
2. *Climb.*
   a. Either the APFDS controls airspeed with pitch commands and the AT controls engine thrust to a specific N1, climb thrust value, e.g., level change or VNAV climb.　*or*
   b. The APFDS controls the vertical path attitude, and the AT maintains airspeed through the engine thrust control, e.g., vertical speed.
3. *Cruise.*　The AT controls engine thrust to maintain airspeed, and the APFDS controls attitude to maintain altitude.
4. *Descent*
   a. Either the throttles retard to idle to gain the attitude that gives the maximum descent rate, and the APFDS controls the pitch to maintain airspeed, e.g., level change/VNAV descent.　*or*
   b. The AT controls airspeed, and the APFDS controls the vertical flight path, e.g., vertical speed/VNAV path descent.
5. *Approach.*　The AT controls airspeed, and the APFDS controls the attitude to maintain a vertical path profile, e.g., autoland approach mode.

Auto Throttle System　An auto throttle system is designed to control and maintain thrust and/or airspeed, especially during an automatic approach and landing, by

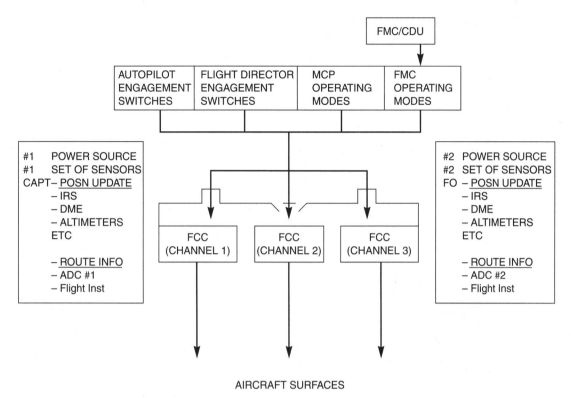

Figure 13 Autopilot schematic.

changing engine thrust. It may also provide a constant closure rate of the throttle levers during the auto flare phase, on some designs.

The auto throttle is part of the autopilot flight director (FD) system and is available with or without the AP or FD being engaged as long as a thrust or speed operating mode has been engaged. An AT computer receives inputs from various sources, especially (1) the FMC, which calculates EPR/N1, limits, and targets; and (2) APFDS pitch and speed targets and limits, which it uses to calculate the engaged operating mode's engine thrust targets and associated throttle lever positions, which are moved and held by servo-driven motors and a clutch system to attain the desired engine thrust level.

The AT is normally engaged (type-specific) by a master switch on the mode control panel being selected to ARM and then an operating mode being selected either automatically by the autopilot or manually through switches on the MCP. The AT is normally disconnected (type-specific) by the following:

- The pilot can override the AT by applying normal pressure to the throttle levers. This applies an opposing force to the servo drive, which will automatically disengage the clutch.

- The AT disengage switch, located on the end of each throttle lever, is thrown.

- The AT engagement switch located on the MCP is set to off.

■ The AT automatically disengages after touchdown and for various type-specific abnormal situations.

An auto throttle warning light flashes whenever the auto throttle is disengaged, particularly as a result of an invalid condition. The warning light is reset by pressing the AT warning light or by pressing an AT disengage switch.

The normal auto throttle's modes of operation are as follows (based on the B737-300):

1. *Takeoff thrust.*   This is a maximum thrust limit, usually restricted to a 5-min operating limit, that is either provided by the FMC or manually calculated, and is engaged by pressing a takeoff go-around (TOGA) switch. During the takeoff phase of flight, engine thrust provides the aircraft's rate of climb; therefore, the aircraft's best climb performance is acquired by using takeoff thrust. However, *derated* takeoff thrust limits are often used to protect the engine when climb performance is not limiting. During takeoff, the aircraft's speed is controlled by the APFDS's pitch control.

2. *Go-around thrust.*   This is a thrust limit that is either provided by the FMC or manually calculated, and it is engaged by pressing a TOGA switch. During the go-around phase of flight, engine thrust provides the aircraft's rate of climb; therefore the aircraft's best climb performance is acquired by using maximum go-around thrust. During the go-around, the aircraft's speed is controlled by the APFDS's pitch control.

3. *Maximum continuous thrust.*   This is the maximum thrust limit that the engine can be kept at indefinitely during flight. The auto throttle maintains either an N1 or EPR maximum continuous thrust (MCT) limit for a stage of flight and selected mode of operation, i.e., altitude hold, airspeed hold, level change, etc. The N1 or EPR thrust limit is an auto throttle mode, and it is normally calculated by the FMC or an engine computer. An N1 or EPR MCT limit can be selected by engaging a MCT/N1 switch, usually found on the MCP, and the auto throttle will maintain this thrust limit. With the auto throttle not engaged, the MCT N1 limit can be selected as a "bug" setting on the engine instrument for the throttle levers to be manually set to.

NB: Maximum continuous thrust cannot be engaged with an incompatible APFDS pitch mode of operation, e.g., vertical speed.

4. *Airspeed.*   Airspeed can be attained by the selection of an attainable indicated airspeed or Mach number and engagement of the speed mode of operation, usually on the MCP. The auto throttles will deliver a thrust, providing it is within the engine's thrust limits, to gain the selected airspeed either when coupled with a compatible APFDS pitch mode of operation, e.g., altitude hold; or, by the APFDS selecting a pitch attitude that will deliver the selected airspeed, providing it is within the aircraft's pitch limits, when it is coupled with an AT mode of operation, e.g., level change.

5. *Airspeed/Mach hold.*   The auto throttles maintain the FMC or MCP indicated airspeed/Mach number when within the engine's thrust limits and when coupled with a compatible APFDS pitch mode of operation, e.g., altitude hold, by pressing the airspeed hold engagement switch on the MCP.

Balanced and Unbalanced Field Performance A balanced field exists when takeoff distance available (TODA) = emergency distance available (EMDA) or accelerate stop distance available (ASDA), or, in other words when the end of the clearway is the end of the stopway, and the aircraft achieves the screen height over the end of the runway in all cases.

NB: A balanced field may be assumed to exist if that part of the clearway which extends beyond the stopway is ignored; therefore, the lower of TODA or EMDA is the balanced field length.

Ergo, an unbalanced field exists when TODA is greater than EMDA.

NB: TORA (runway length) does not feature in balanced field calculations. A balanced field length determines the MTOW.

The purpose of using a balanced field calculation is to optimize the V2 climb performance (second segment) with a correct V1/VR speed, from a single performance calculation/chart, without having to perform a second and separate increased V2 calculation and then to readjust the VR calculation.

Balance Tab A balance tab is a form of aerodynamic control balance on a control surface. A control balance tab balances the main aerodynamic lift force load on a control surface with an opposing force, which thereby reduces the overall hinge moment (air load force). Hinge moment (air loads) = lift force × arm. This is reflected by the stick *control force* that the pilot experiences being reduced to a manageable level.

A balance tab is connected to the trailing edge of the control surface via a linkage, and so as the control surface is moved, the balance tab moves in the opposite direction. The balance tab produces a lift force in the opposite direction to the control surface, which balances/opposes the main lift force and has the effect of reducing the overall hinge moment/air load force on the control surface to a manageable level. In this way the pilot provides a proportion of the control force, and the balance tab supplies the rest, which enables the control surface to be moved. Balance tabs are commonly found on ailerons, elevators, and rudders; and they actively move proportionally in response to the movement of the control surface.

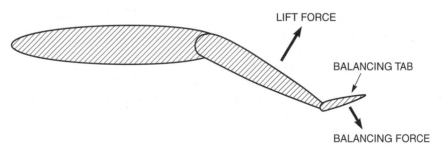

Figure 14 Balance tab.

Barometric Pressure Error   *See* Pressure Altitude—Error.

Basic Weight   This is the weight of the empty aircraft with all its basic equipment plus a declared quantity of unusable fuel and oil.

NB: For turbine-engine aircraft and aircraft not exceeding 5700 kg, the maximum authorized basic weight may include the weight of its usable oil.

Battery   A battery uses chemical action to separate electrons from their parent atoms and thus generate DC electricity. The material contained in battery cells determines the voltage output, while the amount of current is dependent upon the size of the plates that store the electric current.

A primary cell battery uses up the chemicals that produce the electric energy. Eventually the cell runs out of one or more of the chemicals, and the battery is said to be discharged. A secondary cell battery can reverse the chemical changes which take place in its discharge. The secondary cell can convert electric energy back to chemical energy, which is known as *charging*. This reconverted chemical energy can then be used to create electric energy. By careful handling a secondary cell can be charged and discharged many times.

BCF   *See* Fire—Extinguisher.

Beat Frequency Oscillator (BFO)   This is selected by the pilot on the automatic direction finder (ADF) navigation box and used when identifying any nondirectional beacons (NDBs) that use unmodulated carrier waves, i.e., non-A1A transmissions. This is needed because no audio message (Morse) can be heard by the pilot on an unmodulated carrier wave, so the BFO imposes a tone onto the carrier wave to make it audible to the pilot. Thus the NDB signal can be tuned and identified. However, it should not be left on when one is using the ADF to navigate.

NB: NDBs with A2A idents already modulate to an audible frequency. Therefore the BFO is not needed and should always be switched off.

BECMG   BECMG (becoming) followed by a four-figure time group, which is two different whole coordinated universal time (UTC) hours, indicates a permanent change in the forecasted conditions, occurring sometime during the specified period.

Bernoulli's Theorem   The total energy in a moving fluid or gas is made up of three forms of energy:

1. Potential energy (the energy due to the position)

2. Pressure/temperature energy (the energy due to the pressure)

3. Kinetic energy (the energy due to the movement)

When one is considering the flow of air, the potential energy can be ignored; therefore, for practical purposes it can be said that the kinetic energy plus the pressure/temperature energy of a smooth flow of air is always constant. Thus if the kinetic energy is increased, the pressure/temperature energy drops proportionally and vice versa, so as to keep the total energy constant. This is Bernoulli's theorem.

Black Hole Effect   The black hole effect occurs when a runway, in virtual pitch darkness (except maybe the odd aerodrome light around) and in the middle of nowhere, gives the visual illusion of the aircraft being at a greater height than its actual height.

Bleed Valves—Jet/Gas Turbine Engine. Bleed valves are fitted on gas turbine engines for two main reasons:

1. To provide bleed (tap) air for auxiliary systems. The following examples are systems that universally require a bleed air supply from most engine types.
   a. Air conditioning and cabin heating, pressurization, EFIS cooling, cargo heating.
   b. Engine cooling, especially the combustion chamber and the turbine section.
   c. Accessory cooling, e.g., generator, gearbox, and other engine-driven systems.
   d. Engine and wing anti-icing systems.
2. To regulate the correct airflow pressures between different engine sections. However, on modern gas turbine engines, their use is now usually limited to only the provision of an extra safety margin while the engine is being accelerated. That is, interstage bleed valves act as pressure relief valves, which drain away excessive volumes of air that protect the rear sections of the engine and prevent the engine from high-pressure compressor stalls and surges.

Brake Energy Limits   The brake energy capacity limits the aircraft's MTOW, given the ambient aerodrome pressure altitude, temperature, wind, and runway slope conditions. This is done so that the V1 speed does not exceed the maximum brake energy speed, denoted by VMBE, to ensure that the aircraft's brake system has sufficient energy to dissipate and stop the aircraft's inertia from V1 under most operating conditions.

NB: The greater the aircraft weight, the greater the V1 speed and the greater the brake energy required to stop the aircraft. If VMBE is exceeded (that is, V1 > VMBE) and a stop is initiated, then the brakes will experience a high level of friction beyond their design capabilities. This will generate heat in the brake system that can be high enough to melt the brake pads and cause them to seize onto the brake disk, and possibly catch fire, causing possible tire failure and resulting in the stopping distance being compromised.

Brake System   Most aircraft braking systems use hydraulic fluid pressure to move friction brake pads (that are connected to small hydraulic pistons) against rotating brake plates, to slow down the plates and therefore the wheel. The brake system is engaged by either manual application of the brake pedals or the automatic braking system which controls a brake metering valve for each wheel and adjusts the amount of piston movement and thereby the amount of pressure applied against each rotating wheel plate.

Brake System—Automatic   An automatic brake system regulates the amount of brake pressure by controlling the metering valve in the hydraulic brake line, so as to maintain a constant deceleration rate until the aircraft reaches a complete stop. Thus brake application is regulated with the reverse thrust applied, to maintain the selected deceleration rate.

NB: The auto brake system usually monitors the reverse thrust (where available and/or applied) and applies the auto brakes as a combined system to produce a constant deceleration rate to stop the aircraft. Thus the reverse thrust is used at high speed to decelerate the aircraft (where it is most efficient), and the brakes take over and are used at low speed to bring the aircraft to a stop. Many modern aircraft also use ground spoilers to assist in stopping the aircraft on landing or during an aborted takeoff.

Auto brake selection controls the magnitude of the deceleration rate. Typically a system would have landing settings of 1, 2, 3, and maximum and a takeoff setting of RTO (rejected takeoff).

Brake—Temperature   Prior to the takeoff run, the following factors affect the brake temperature: taxi time (distance) and the amount of brake applications used. During the takeoff run, the following factors affect the brake temperature:

- Aircraft takeoff weight
- Pressure altitude
- Outside air temperature (OAT)
- Runway slope
- Tailwind/headwind

All these factors generate a temperature rise in the brakes.

It is important to monitor the brake temperature, especially prior to commencing the takeoff run. This is to ensure that the aircraft's inertia, which has to be dissipated through the wheel brakes, will not cause the brakes to overheat and lose efficiency, bind, and even cause tire failure or fire, if the takeoff is abandoned at V1. Therefore, the brake temperature must be below a certain limit prior to the commencement of the takeoff run, to ensure the maximum brake energy limit is not exceeded during an aborted takeoff run.

During a landing, the downwind (with a crosswind) wheel brakes will generate the highest temperatures (i.e., are the hottest).

Braking Thrust—Propeller   *See* Propeller—Braking Thrust.

Buys Ballot's Law   Buys Ballot's Law is this: *If you stand with your back to the wind in the northern hemisphere, the low pressure (temperature) will be on your left.*

NB: Conversely, in the southern hemisphere the low pressure (temperature) will be on your right.

Bypass Engine   The principle of the bypass engine is an extension of the gas turbine engine that permits the use of higher turbine temperatures to increase thrust without a corresponding increase in jet velocity, by increasing the air mass volume intake and discharge to atmosphere, via the bypass ducts. Remember:

$$\text{Thrust} = \text{air mass} \times \text{velocity}$$

The bypass engine involves a division or separation of the airflow. Conventionally all the air entering the engine is given an initial low compression, and then a percentage is ducted to bypass the engine core. The remainder of the air is delivered to the combustion system in the usual manner. The bypass air is then mixed with the hot airflow from the engine core either in the jet pipe exhaust or immediately after it has been discharged to atmosphere to generate a resultant forward thrust force.

The term *bypass* is normally restricted to engines that mix the hot and cold airflows as a combined exhaust gas. This improves (1) propulsive efficiency, (2) specific fuel consumption, and (3) reduction in engine noise, which is due to the bypass air lessening the shear effect of the air exhausted through the engine core.

Bypass Ratio The bypass ratio, in an early single- or twin-spool bypass engine, is *the ratio of the cool air mass flow passed through the bypass duct to the air mass flow passed through the high-pressure system.* Typically this early evolution of the bypass engine has a low bypass ratio, that is, 1:1.

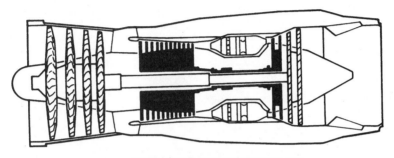

TWIN-SPOOL BYPASS TURBOJET
(low bypass ratio)

Figure 15 Bypass airflow for a single- or twin-spool gas turbine engine. (*Reproduced with kind permission of Rolls-Royce Plc.*)

Or the bypass ratio for a fan-ducted bypass engine is *the ratio of the total air-mass flow through the fan stage to the air mass flow which passes through the turbine section / high-pressure (engine core) system.* A high bypass ratio, say, 5:1, is usually common with ducted fan engines.

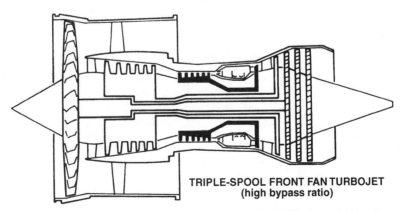

TRIPLE-SPOOL FRONT FAN TURBOJET
(high bypass ratio)

Figure 16 Triple-spool turbofan engine. (*Reproduced with kind permission of Rolls-Royce Plc.*)

C

Cabin (Pressurization) Pressure Cabin pressure is measured as a differential between the ambient atmospheric air pressure outside the aircraft and the air pressure inside the cabin. This pressure measurement is known as the *differential pressure,* say 8.21 psi. The cabin pressure is normally greater than the external atmospheric pressure, and therefore the cabin is often referred to as being blown up. The differential pressure value relates to the psi force on the inside of the fuselage pushing outward.

For a given aircraft altitude, a differential pressure value will represent a set cabin altitude. For example, an aircraft altitude of FL370 and a differential pressure of 7.80 psi will represent a cabin altitude of 8000 ft.

Typically an aircraft has standard differential pressure values for

- Negative maximum
- Positive operating maximum
- Positive structural maximum

The aircraft's (cabin) pressurization is controlled by the cabin's outflow air valve. Conditioned air is pumped into the cabin, as the air conditioning system dictates. However, the exhaustion of the air from the cabin is regulated by the main outflow valve to build up, reduce, or maintain an internal pressure, per the requirement of the pressure controller to a given differential pressure value. That is, on the ground with the outflow valve open, the internal cabin pressure is the same as the external atmospheric pressure. Thus the differential pressure is 0 psi.

During a climb as the external atmospheric pressure decreases, the cabin outflow valve is opened above the equilibrium setting so that more air is discharged than is delivered into the cabin. Thus the internal cabin pressure decreases, and a constant differential pressure is maintained. If during the climb the pressure controller system requires the differential pressure to increase, then the outflow valve is simply closed more, thereby decreasing the discharge so that the internal pressure becomes greater than the external atmospheric air pressure, resulting in the differential pressure's increasing.

To maintain a differential pressure, the outflow valve is set at an angle to maintain a discharge of air equal to the conditioned air delivered to the cabin.

During a descent as the external atmospheric pressure increases, the cabin outflow valve is closed above the equilibrium setting, so that less air is discharged than is delivered into the cabin, thereby increasing the internal cabin pressure to maintain a constant differential pressure. If during the descent the pressure controller system requires the differential pressure to decrease, i.e., back to zero for landing, then the outflow valve is simply closed more, thereby decreasing the discharge and increasing the internal pressure, which results in the differential pressure's decreasing.

Caging System A caging system locks the gyro, i.e., the artificial horizon, in a fixed position, especially when the gyros are not being used (i.e., when parked), and for some aircraft it is recommended during aerobatic maneuvers. Caging a gyro in this manner will prevent it from toppling (rigid in space), and thus when restarted, the instrument reaches its fully erect position very quickly.

Calibrated Airspeed (CAS) This term is commonly used in the United States, Australia, New Zealand, and many European countries, whereas *rectified airspeed* (RAS) is the term commonly used in the United Kingdom. CAS/RAS is the indicated airspeed (IAS) corrected for instrument and pressure errors. *Instrument error* is caused by the inaccuracies in the construction of the airspeed indicator (ASI) itself, especially the friction of its moving parts. *Pressure error,* also known as *position* or *configuration error,* is caused by variations of the static vent and pitot tube relative position with respect to the airflow over the aircraft's surface. This leads to sensed pressure readings that are unrepresentative of the free atmosphere whenever the airflow pattern is disturbed at different airspeeds, angles of attack, configuration, etc.

The calculated CAS/RAS figure is what the ASI would read if the particular ASI system were perfect. CAS/RAS is therefore more accurate than IAS and should be used in preference to IAS in navigation calculations, if you know it. However, the instrument and pressure errors of the ASI system are usually only no more than a few knots.

With a constant CAS (IAS), the true airspeed (TAS) will increase with altitude, but the Mach number (MN) will increase more.

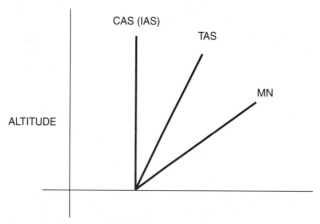

Figure 17   Constant CAS (IAS) versus true airspeed and Mach number.

If you descend at a constant TAS, your CAS (IAS) will increase. This is so because pressure increases with a decrease in altitude, due to the gravity effect. Density is proportional to pressure; therefore, density increases with a decrease in altitude, which means the CAS (IAS) increases in a descent at a constant TAS. Or this can be expressed as

$$\text{CAS (IAS)} = \tfrac{1}{2}R(V^2)$$

where R (density) *increases* with a decrease in altitude and V (TAS) is *constant.* Therefore CAS increases.

If you fly at a constant CAS (IAS) into a warmer area, your CAS (IAS) remains constant, because CAS (IAS) is unaffected by temperature while the TAS will increase because of the decrease in air density associated with warmer temperatures.

When you are climbing through an isothermal layer at a constant TAS, your CAS (IAS) will decrease. This is so because pressure decreases with altitude due to the gravity effect; therefore, density decreases, which means the RAS decreases. Or this can be expressed as

$$CAS\ (IAS) = \tfrac{1}{2}\,R\,(V^2)$$

where V (TAS) is *constant* and R (density) *decreases* with an increase in altitude. Therefore, CAS (IAS) decreases.

If an aircraft descends through an isothermal layer at a constant TAS, the CAS (IAS) will increase. This is so because pressure increases with a decrease in altitude due to the gravity effect; therefore, density increases, which means RAS (IAS) increases. Or this can be expressed as

$$CAS\ (IAS) = \tfrac{1}{2}R(V^2)$$

where V (TAS) is *constant* and R (density) *increases* with a decrease in altitude, due to increased pressure. Therefore CAS (IAS) increases.

If an aircraft climbs at a constant TAS through an inversion layer, then its CAS (IAS) will decrease more than normally. This is so because an inversion layer sees an increase in temperature with altitude; therefore the air density decreases greater than normally. CAS (IAS) $= \tfrac{1}{2}\,RV(\text{TAS})^2$. Therefore R decreases more than normally which results in CAS (IAS) decreasing more than normally when climbing at a constant TAS.

Carbon Monoxide (CO)   Carbon monoxide is a highly toxic, colorless, odorless, and tasteless gas for which hemoglobin in the blood has an enormous affinity. The main function of hemoglobin is to transport oxygen from the lungs throughout the body. Carbon monoxide molecules breathed into the lungs along with air will attach themselves to the hemoglobin, starving the brain and body of oxygen even though oxygen is present in the air.

Carbon monoxide is produced during the combustion of fuel in the engine. It may enter the cabin if there are any leaks in the exhaust system, when warm air from around the engine or the exhaust manifold is used to provide cabin heating. It is very difficult to detect, other than by two means:

- Cherry red skin pigmentation results from prolonged exposure.
- Specific carbon monoxide indicator systems detect it. (For example, cards that change color, or aural alarms that sound upon contact with carbon monoxide) detect CO.

The other symptoms of carbon monoxide poisoning are headaches, breathlessness, dizziness, nausea, deterioration of vision, impaired judgment, and eventually a loss of consciousness.

To minimize the effects of any carbon monoxide that enters the cockpit in this way, fresh air should be used in conjunction with cabin heat. However, if carbon monoxide is suspected, you should

- Shut off cabin heat.
- Stop all smoking.

- Increase the supply of fresh air.
- Don oxygen masks, if available.

Carburetor Icing   Ice formation can occur in the engine induction system and in the carburetor of piston engines, particularly in the venturi and around the throttle valve, where the acceleration of the air can produce a temperature fall by as much as 25°C. This, combined with the heat absorbed as the fuel evaporates, can cause serious icing, even when there is no visible moisture present. Such a buildup of ice in a carburetor can disturb or even prevent the flow of air and fuel into the engine manifold, causing it to lose power, run roughly, and even stop the engine, in extreme circumstances. This effect is called *carburetor icing.*

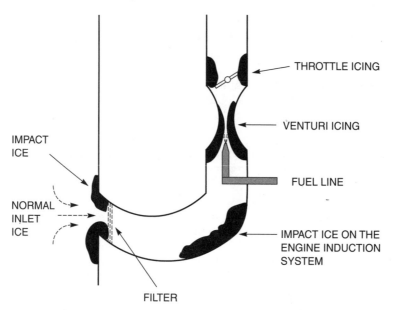

Figure 18   Carburetor icing.

Throttle icing (i.e., around the throttle valve) is more likely to occur at low-power settings, i.e., descents, when the partially closed butterfly creates its own venturi cooling effect.

Carburetor icing should be expected in a piston engine when the outside air temperature (OAT) is between −10 and +30°C with high humidity and/or visible moisture present in the air. However, carburetor/throttle icing is most likely to occur between +10 and +15°C with a relative humidity greater than 40 percent.

NB: Carburetor icing *can* be found on a warm day in moist air, especially with descent power settings.

You should use the carburetor heat system (hot air) at regular intervals to treat icing in the carburetor, when in carburetor icing conditions. The carburetor heat system delivers hot air from the engine compartment into the carburetor, which melts the buildup of ice.

NB: A slight loss of power is associated with the use of carburetor heat. Due to the slight risk of the melted ice being ingested into and saturating the engine combustion, causing an engine to flame out, carburetor heat should be applied to one engine at a time on multiengine aircraft.

NB: It is also advisable to apply carburetor heat at the start of a descent to prevent throttle icing and thereby ensure that full power is available in the event of a go-around engine application. However, carburetor heat should always be switched off before the throttle is advanced, to ensure maximum thrust delivery; therefore, it should be off in the final landing approach, in anticipation of an emergency go-around near to the ground. However, the use of carburetor heat should be avoided when the OAT is colder than $-10°C$, because by applying carburetor heat you raise the air temperature in the carburetor into the icing temperature range of -10 to $+30°C$.

CAT (Clear Air Turbulence)   CAT refers to turbulent air with no signs of visible moisture content. These are examples of clear air turbulence:

1. *Low-level CAT* is caused by the following:
   a. Temperature inversions near the surface, which are caused by radiation cooling overnight and/or subsidence in stable high-pressure systems, give rise to a marked vertical windshear (CAT) at the temperature inversion layer. This is especially true at night or in the early morning when there is little or no mixing of the lower layers.
   b. A difference between the surface and gradient wind due to the turbulent mixing of the boundary layer near the surface (due to surface friction, terrain, thermal activity, etc.) gives a marked difference in the surface wind direction and speed compared to the gradient wind. This can be classified as a marked vertical wind shear (CAT).
   c. Local surface winds, i.e., land/sea breezes, can give rise to marked differences between the surface and gradient wind.
   d. Terrain-generated winds can cause low-level CAT. Mountain/standing waves develop on the downwind side of a range of hills. The wind speed must increase with altitude but generally remains constant in direction. The air must be stable so the airflow is laminar or layered.

NB: The up and down currents associated with mountain waves can be very strong and can reach very high altitudes at great distances downwind from the mountain range.

   Rotor zone CAT can be extremely strong and occurs at low level within approximately 5 nm of the hill ridgeline and just below the ridge. Here one wavelength downwind, roll or rotor clouds can form that actually rotate about their own horizontal axis. Below these clouds, in clear air, rotor zone turbulence can be intense and on occasion violent.

NB: Isolated lenticular and/or roll clouds indicate the presence of mountain/standing wave and/or rotor zone clear air turbulence in the general area.

2. *Jetstreams.*   The most severe CAT can be found level with or just above the jet core in the warm air, but on the cold air/polar side of the jet.
3. *Frontal.*   At an active front where there is convergence and a sharp temperature difference between the warm and cold air, the isobars are usually sharply inclined at the frontal line. This produces a marked horizontal wind shear,

which can produce CAT at the front. In addition to this, as the frontal surface is at a very shallow angle, it is possible to hit vertical wind shear and CAT while descending or climbing through the frontal surface. A similar effect can be found on ridges and especially troughs (which are usually more marked), both at low and high levels.

4. *Thunderstorms.*   Serious CAT wind shear can be found at low level under and near thunderstorms, e.g., microbursts. CAT is also found at medium and high levels near cumulonimbus clouds (CBs) and thunderstorms, at the entry points from inflowing winds, and above the cloud tops. Active CBs should be avoided by 10 nm in visual flight; the top should not be flown over with less than 5000-ft clearance, and you should not fly underneath the anvil overhang. The recommended avoidance ranges when weather radar is in use are as follows:

- FL 0–250   10 nm
- FL 250–300   15 nm
- FL 300+   20 nm

5. *Wake turbulence.*

Clear air turbulence is one of the hardest forms of turbulence to detect. In fact, there is no specific equipment developed, i.e., radar deflection, to detect CAT. Therefore CAT can be detected only by using meteorological appraisal of the prevailing conditions, typically shown on met charts, and identifying the location of conditions that give rise to CAT. Because CAT is a form of wind shear, any reported wind and/or temperature change between two places and the presence of certain weather conditions (e.g., cumulonimbus clouds, lenticular clouds, fronts, jetstreams, etc.) are good indications of possible CAT.

NB: In particular, large, rapid fluctuations of the total air temperature (approximately ±10°C in a few seconds) and broken engine trails from a preceding aircraft are very good indications of CAT.

CAT I, II, and III ILS Limits

CAT I	DH not lower than 200 ft	RVR at least 550 m at touchdown or visibility 800 m RVR/VIS at midpoint N/A
CAT II	DH 200 ft or lower But not less than 100 ft	RVR at least 300 m at touchdown RVR at least 150 m at midpoint
CAT IIIa	DH 100 ft or lower But not less than 50 ft	RVR at least 200 m at touchdown RVR at least 150 m at midpoint
CAT IIIb	DH less than 50 ft (Down to 50 ft)	RVR at least 50 m at touchdown RVR at least 50 m at midpoint
CAT IIIc	DH of 0 ft	RVR of 0 m

N/A = not applicable.

NB: CAT IIIc has never been in commercial operation, as it would require an automatic taxiing guidance system. CAT IIIb, however, is widely used.

NB: Company, local authorities, or specific airfield limits may be greater than these limits. If this is the case, then the higher limits take precedence.

CAT II and CAT III are different categories of precision ILS approaches. They are based on the principle that the short-range visual reference required (RVR) would be present at the DH. They are collectively referred to as *low-visibility procedures* (LVPs). CAT II and CAT III approaches are usually terminated with the execution of an autoland.

The requirements to carry out a CAT II or CAT III in addition to the normal CAT I approach are as follows:

1. The aircraft's CAT II or CAT III systems are certified and operational.
2. The runway's ground CAT II or CAT III equipment is certified, operational, and protected (namely, low-visibility procedures are in force).
3. The crew are qualified and current.
4. Weather minima are above the approach ban, prior to commencing your final approach.

NB: If the reported RVR drops below the approach minima during the final approach (i.e., past 1000-ft approach landing height), the approach can be continued.

5. Alternate airfield weather is above CAT I minima.

A normal CAT I hold is marked by two close parallel yellow lines across the taxiway (old markings may still be white)—one continuous line and one broken line. It signifies a holding position beyond which no part of an aircraft or vehicle may project in the direction of the runway, without ATC permission. The CAT I hold is usually the last holding point prior to the runway.

NB: The pilot can determine if the hold affects him or her. If you encounter the continuous line first, then the hold affects you (you are on the taxiway holding side) and you need ATC clearance to cross. If you are moving in the reverse direction, where the broken line is encountered first (i.e., the runway side), the holding point does not affect you and you do not require a clearance to cross it.

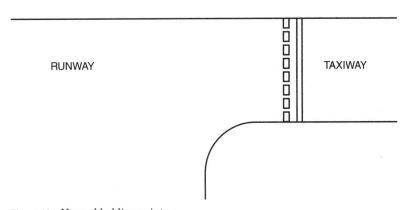

Figure 19   Normal holding point.

NB: Some runway hold markings may have two continuous lines and two broken parallel lines.

In addition, a runway holding point will usually have a red and white sign, on either side of the taxiway hold markings, denoting the hold category (e.g., CAT I Hold) and runway (e.g., Rwy 24/06).

A CAT II or CAT III holding point has a yellow ladder-type marking.

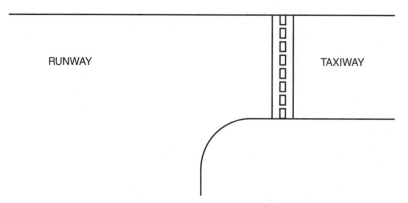

Figure 20   CAT II or III holding point markings.

A CAT II or III holding point is usually farther back from the runway than a CAT I hold (CAT III hold is farther back than a CAT II hold), yet still has the same restrictions as the CAT I hold when in force.

CAVOK   The weather term CAVOK is used to replace visibility, RVR, weather, and cloud in met reports and forecasts whenever the following conditions occur simultaneously:

1. Visibility equal to or greater than 10 km

2. No cloud below 5000 ft or below the highest minimum sector altitude (MSA), whichever is greater, and no cumulonimbus clouds at any altitude

3. No precipitation, thunderstorms, shallow fog, or low drifting snow

   NB: CAVOK does not mean clear blue skies.

Celsius   The Celsius scale is used for measuring temperature. The Celsius scale, measured in degrees Celsius (°C), divides the temperature between the boiling point and the freezing point of water into 100 degrees. That is, the boiling point of water is 100°C, and the freezing point of water is 0°C.

   NB: The Celsius scale is the most commonly used form of temperature measurement. The formula to convert from degrees Celsius to degrees Fahrenheit is

$$°F = \frac{9}{5}(°C + 32)$$

and the formula to convert from degrees Fahrenheit to degrees Celsius is

$$°C = \frac{5}{9}(°F - 32)$$

Center of Gravity   The center of gravity is the point through which the total weight of a body will act. (It is also called the c of g or cg.)

For an aircraft, the cg point is measured from a manufacturer-designated *datum* point and is expressed as either a distance along the aircraft's longitudinal axis or a percentage of the standard mean chord, known as the *arm*.

Any movement of the cg is due to a change in the actual or effective weight of the aircraft. The distribution of the aircraft's weight will change for three reasons and thereby causes the cg position to move. The three reasons for a change in the aircraft's weight are

1. *Fuel burn.* The most common reason for cg movement on a swept-wing aircraft is a decrease in weight as fuel (and to a lesser extent oil) is used in flight. It should be remembered that because of the sweep, the wing and the fuel tanks housed inside cover a distance along the aircraft's longitudinal axis. Therefore as fuel/weight is progressively reduced along this axis, the weight distribution pattern is changed across the aircraft's length.

A swept-wing aircraft will normally use fuel from the center tanks first. These tanks are normally a little forward of the cg; therefore, as this fuel is burned, not only does the aircraft's weight decrease but also the distribution of the aircraft's total weight shifts aft. Therefore this causes the cg to move aft. Fuel is then normally used from the wings' outer tanks; therefore fuel and its associated weight are progressively removed from the aft tanks, resulting in a change in the overall weight distribution, and the cg position moves progressively forward.

Heavy swept-wing jet aircraft have a requirement for a large cg range to facilitate the large cg movement due to the very large weight changes as a result of the fuel consumed during a single flight. Because of these weight changes an aircraft's weight can reduce by almost 50 percent in a single flight.

NB: On a piston-engine aircraft, because of its straight wing, fuel loading is not so critical.

2. *Passenger movement.* The extremely long passenger cabin and the indiscriminate movement of persons exaggerate the overall weight distribution and therefore give rise to the movement of the cg position. This also engenders the need for a large cg range.

3. *High speeds.* Theoretically high speeds cause a forward movement in the cg position. This is so because the greater the speed, the greater the lift created. To maintain a straight and level attitude, the aircraft adopts a more nose-down profile, which is accomplished by creating lift at the tailplane. This lift on the tailplane effectively reduces the weight of the tailplane section of the aircraft. Thus

$$\text{Arm} = \frac{\text{moment}}{\text{reduced weight}}$$

results in a longer tailplane to cg arm and the cg moving forward, albeit only slightly.

The greater the magnitude of the cg movement for any of these reasons, the greater the cg range required; therefore the greater the tailplane balancing force required.

Center-of-Gravity Moment This is the turning effect/force of a weight around the datum. It is the product of the weight and the arm.

$$\text{Moment} = \text{weight} \times \text{arm}$$

Center-of-Gravity Range (Limits)   The center-of-gravity range relates to the most forward and aft cg positions along the aircraft's longitudinal axis. Inside which the aircraft is permitted to fly because the horizontal tailplane can generate a sufficient lift force to balance the aircraft's lift/weight moment couple, so that it remains longitudinally stable and retains a manageable pitch control. (The cg range or envelope is listed in the aircraft's flight manual, and accordance is mandatory.)

It is very important to keep the cg *inside* its operating range (limits). The forward position of the cg is limited to

1. Ensure that the aircraft is not too nose-heavy, so that the horizontal tailplane has a sufficient turning moment available to overcome its natural longitudinal stability.
2. Ensure the aircraft's pitch control (rotation and flare) is not compromised, with high stick forces (tailplane turning moment), by restricting the aircraft's tailplane arm forward cg limit. [Remember that tailplane moment (stick force) = arm $\times$ weight.]

NB: This is particularly important at low speeds (i.e., takeoff and landing) when the elevator control surface is less effective.

3. Ensure that a minimum horizontal tailplane deflection, which produces a minimal download air force on the tailplane, is required to balance the lift/weight pitching moment. Therefore the stabilizer and/or the elevator is kept streamlined to the relative airflow, which results in
   a. Minimal drag; therefore performance is maintained.
   b. Elevator range being maintained; therefore the aircraft's pitch maneuverability is maintained.

The aft position of the cg is limited to

1. Ensure that the aircraft is not too tail-heavy, so that the horizontal tailplane has a sufficient turning moment available to make the aircraft longitudinally stable.
2. Ensure that sufficient pitch control stick forces (tailplane turning moment) are adequately "felt" through the control column, by guaranteeing the aircraft's tailplane arm to an aft cg limit. [Remember, moment (stick force)=arm $\times$ weight.]
3. Ensure that a minimum horizontal tailplane deflection, which produces a minimal upload air force on the tailplane, is required to balance the lift/weight pitching moment. Therefore the stabilizer and/or the elevator is kept streamlined to the relative airflow, which results in
   a. Minimal drag; therefore performance is maintained.
   b. Elevator range being maintained; therefore the aircraft's pitch maneuverability is maintained.

If the cg is *outside* its forward limit, the aircraft will be nose-heavy and the horizontal tailplane will have a long moment arm (tailpipe to cg point) that results in the following:

1. Longitudinal stability is increased, because the aircraft is nose-heavy.
2. The aircraft's pitch control (rotation and flare) is reduced/compromised because it experiences high stick forces due to the aircraft's long tailplane moment arm. [Remember, tailplane moment (stick force) = arm $\times$ weight.]

3. A large balancing download is necessary from the horizontal tailplane by deflecting the elevator or stabilizer. This results in the following:
   a. The wing angle of attack is increased, resulting in higher induced drag which reduces the aircraft's overall performance and range.
   b. Stalling speed is increased, due to the balancing download on the horizontal tailplane that increases the aircraft's effective weight.
   c. If the elevator is required for balance trim, less elevator is available for pitch control, and therefore the maneuverability of the aircraft to rotate at takeoff or to flare on landing is reduced.
   d. In-flight minimum speeds are also restricted due to the lack of elevator available to obtain the necessary high angles of attack required at low speeds.

Generally the aircraft is heavy and less responsive to handle in flight and requires larger and heavier control forces for takeoff and landing.

If the cg is outside its aft limit, the aircraft will be tail-heavy and the horizontal tailplane will have a short moment arm (tailplane to cg point), with these results:

1. The aircraft will be longitudinally unstable because it is too tail-heavy for the horizontal tailplane turning moment to balance.
2. The aircraft's pitch control (rotation and flare) is increased/more responsive, because it experiences light stick forces due to the aircraft's short tailplane arm. [Remember, tailplane moment (stick force) = arm $\times$ weight.] This lends itself to the possibility of overstressing the aircraft by applying excessive g forces.
3. A large balancing upload is necessary from the horizontal tailplane by deflecting the elevator or stabilizer. This has the following results:
   a. The wing angle of attack is decreased, resulting in lower induced drag, which increases the aircraft's overall performance and range.
   b. A lower stalling speed, due to the balancing upload on the horizontal tailplane, decreases the aircraft's effective weight.
   c. Also if the elevator is required for balance trim, less elevator is available for pitch control, and therefore the maneuverability of the aircraft to recover from a pitch-up stall attitude is reduced.
   d. In-flight maximum speeds are also restricted due to the lack of elevator available to obtain the necessary low angles of attack required at high speeds.

Generally the aircraft is effectively lighter and more responsive to handle in flight and requires smaller and lighter control forces for takeoff and landing.

In summary,

CG position	Forward	Aft
Stability	More	Less
Stick force	More	Less
Drag	More	Less

A jet aircraft needs a large cg range because its cg position can change dramatically with a large change in its weight during a flight. Therefore to accommodate a large cg movement, the aircraft has to have a powerful horizontal tailplane, to balance the large lift/weight pitching moments, so that the aircraft remains longitudinally stable and retains its pitch controllability.

Center-of-Gravity Effect on Stall Speed   A center of gravity forward of the center of pressure will cause a higher stall speed. This is so because a forward center of gravity would cause a natural nose-down attitude below the required en route cruise attitude for best performance. Therefore a downward force is induced by the stabilizer to obtain the aircraft's required attitude. However, this downward force is in effect a weight and so increases the aircraft's overall weight. Weight is a factor of the stall speed, and the heavier the aircraft, the higher the aircraft's stall speed.

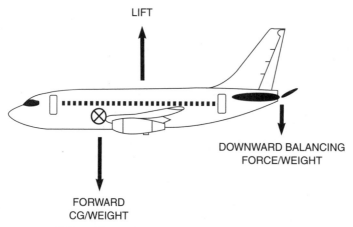

Figure 21   Effect of forward cg position on stall speed.

Conversely, a center of gravity aft of the center of pressure will cause a lower stall speed.

Center of Pressure   The center of pressure (cp) is a single point acting on the wing chord line at a right angle to the relative airflow, through which the wing's lifting force is produced. The position of the cp is not a fixed point, but depends upon the distribution of pressure along the chord, which itself is dependent on the angle of attack. So for a greater angle of attack, the point of highest suction (highest air pressure value) moves toward the leading edge. The distribution of pressure and cp point will thus be farther forward as the angle of attack increases and farther aft as the angle of attack decreases.

See Fig. 22 at the top of the next page.

Changes in the angle of attack are required to maintain straight and level flight in response to configuration and speed changes. An increase in speed or deployment of lift flaps will create an increase in lift. Therefore to maintain a level flight condition, a decrease in the angle of attack is required; otherwise the aircraft would climb. This decrease in the angle of attack results in the cp moving aft. For a decrease in speed or retraction of the lift flaps, the opposite is true, requiring an increased angle of attack to maintain level flight, resulting in the cp moving forward.

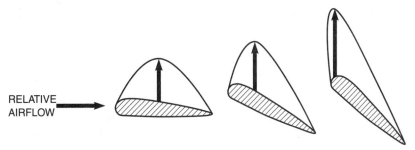

Figure 22   Center of pressure position/angle of attack.

Centrifugal Turning Moment   This is a force that tends to turn the blades to a fine pitch.

Chamber Line   The mean chamber line is a line from the leading edge to the trailing edge, equidistant on the upper and lower surfaces of an aerofoil.

Chimed Tires   A chimed tire has a special sidewall construction that takes the form of a ridge built onto the sidewall, which diverts runway water to the side, reducing the amount of water thrown up into the intake of rear-mounted engines.

Chord Line   The chord line is a straight line from the leading edge to the trailing edge of an aerofoil. The mean chord is the wing area divided by the wing span. (This is sometimes referred to as the *standard mean chord.*)

Circuit Breakers   Electric circuit breakers (CBs) are thermal devices placed in series with an electrical load. They "open-circuit" and thus cut off the electrical equipment when it experiences an abnormal (overload) current operating condition. Open-circuit conditions are indicated by a CB reset pushbutton being visible on a CB panel. Pressing the CB button in will reengage and complete the circuit.
   A non-trip-free circuit breaker can be held in under a fault condition as an emergency measure, to complete the circuit and engage the electrical load.
   A trip-free circuit breaker will not make any internal contact by pressing the reset button with an overload condition in the electric circuit.
   NB: Holding in the reset button under a fault condition cannot reset this type of CB.

Circuit—Direction   The standard direction of flight in a circuit is left-handed.
   NB: However, some runways might have a nonstandard right-hand circuit pattern. This is typical for aerodromes with neighboring noise-sensitive areas, high ground, or other restrictive airspace that precludes the use of a left-hand circuit.

Circumference of a Turn   *See* Turns.

Clear Air Turbulence (CAT)   *See* CAT—Clear Air Turbulence.

Clearway (Runway)   The runway clearway is the length of an obstacle-free area at the end of the runway in the direction of the takeoff, with a minimum dimension of 75 m either side of the extended runway centerline that is under the control of the licensed authority.

NB: The clearway surface is not defined and could be water. It is an area over which an aircraft may make a portion of its initial climb to a specified height, i.e., to the screen height of 35 ft.

Climb Gradient   The climb gradient is the ratio, in the same units and expressed as a percentage, of the change in height to the horizontal distance traveled. The climb gradients on performance charts are true gradients for the *all-up weight* (AUW) of the aircraft, which allows for temperature, aerodrome pressure altitude, and aircraft configuration; i.e., they are achieved from true rates of climb, not pressure.

Clouds   Cloud formation is only possible if you have the following properties:

1. Moisture present in the air

2. A lifting action to cause a parcel of air to rise

3. Adiabatic cooling of the rising air

NB: The four main lifting actions are convection, turbulence, frontal, and orographic.

If a parcel of air containing water vapor is lifted sufficiently, it will cool adiabatically and its capacity to hold water vapor will decrease; i.e., cooler air supports less water. Therefore its relative humidity increases, until the parcel of air cools to its dew point temperature, where its capacity to hold water vapor is equal to that which it is actually holding, and the parcel of air is said to be saturated (i.e., its relative humidity is 100 percent). Any further cooling will cause some of the water vapor to condense out of its vapor state as water droplets and to form clouds. Further, if the air is unable to support these water droplets, then they will fall as precipitation, in the form of rain, hail, or snow.

Furthermore the nature and extent of any cloud formation depend upon the nature of the surrounding atmosphere through which the parcel of air is ascending.

NB: It is important to realize that the dry adiabatic lapse rate (DALR) (3° per 1000 ft) and saturated adiabatic lapse rate (SALR) (1.5° per 1000 ft) are almost constant, whereas the environmental lapse rate (ELR) can be variable to the extent that it can be zero (isothermal) or even negative (inversion).

If the ELR is high, that is, if the temperature of the surrounding air reduces more quickly with height gained than the rising DALR/SALR parcel of air, then this parcel of air will always be warmer than the surrounding (ELR) air. Therefore it will be lighter, and so it will keep rising. This air is said to be unstable, and it produces cumiliform (i.e., heaped) clouds. That is, ELR > DALR > SALR.

See Fig. 23 at the top of the next page.

If the ELR is low, that is, if the temperature of the surrounding air reduces less quickly with height gained than the rising DALR/SALR parcel of air, then this parcel of air will always be cooler than the surrounding (ELR) air. Therefore it will be heavier, and so it will sink back to the ground. This air is said to be stable, and it produces only stratus (i.e., layered) clouds. That is, DALR > SALR > ELR.

See Fig. 24 at the bottom of the next page.

Finally, if the ELR is between the DALR and the SALR, a rising parcel of air could be stable or unstable, depending upon whether it is saturated. That is, rising air is stable if dry and unstable if saturated. This is known as *conditional instability*.

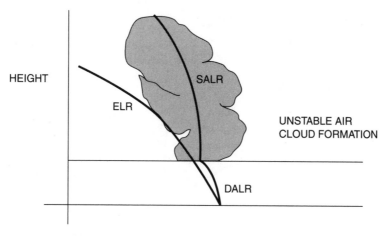

HEIGHT

ELR

SALR

UNSTABLE AIR
CLOUD FORMATION

DALR

TEMPERATURE

Figure 23   Unstable air/clouds.

See Fig. 25 at the top of the next page.
There are four main groups of clouds:

1. Curriform or fibrous

2. Cumuliform or heaped

3. Stratiform or layered

4. Nimbus or rain-bearing

These groups are further subdivided with the following prefixed names, according
to the level of their base above mean sea level.

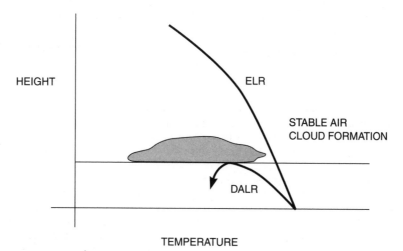

HEIGHT

ELR

STABLE AIR
CLOUD FORMATION

DALR

TEMPERATURE

Figure 24   Stable air/clouds.

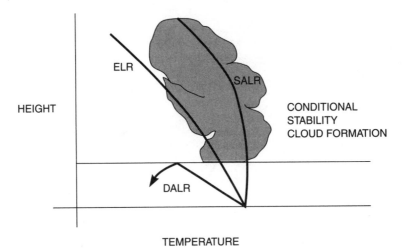

Figure 25   Conditional stability.

Cirro or high-level cloud	cloud base > 16,500–20,000 ft
Alto or medium-level cloud	cloud base > 6500 ft
No prefix for low-level clouds	cloud base < 6500 ft

NB: The state or extent of any cloud formation is dependent upon the stability of the air. Unfortunately, the classification of cloud types is not straightforward, because clouds may take on numerous different forms, many of which are continually changing. However, as pilots, we need to understand cloud classification, because meteorological forecasts use these cloud classifications to give a picture of the weather along a prescribed route. Therefore a typical classification of cloud types is as follows:

1. *High-level cloud.*   Cirrus (CI) clouds are detached white, fibrous, high-level clouds. Cirrocumulus (CC) clouds are regular, thin patches of high-level, small heaped clouds. Cirrostratus (CS) clouds are transparent, white, layered, "veiled" high-level clouds.

NB: The cirrus family of high clouds contains ice crystals rather than liquid water, due to the extremely low air temperature at this altitude. Therefore they do not produce any weather.

2. *Medium-level cloud.*   Altocumulus (AC) clouds are one or more medium-level layers of heaped clouds, which can be arranged as broken patches across the sky or merged together as a continuous layer. In an unstable atmosphere, the vertical development of AC may be sufficient to produce precipitation in the form of *virga* (rain that does not reach the ground) or light rain or snow showers. *Altostratus* (AS) is a medium-level uniform layer of cloud. Light rain or snow showers from thick altostratus clouds may reach the ground.

3. *Low-level cloud.*   Stratus (ST) is a gray layer of cloud with a fairly uniform base, which typically produces continuous fine drizzle or snow precipitation. *Stratocumulus* (SC) is a layer of small heaped clouds that are often aligned in waves that may or may not merge. They are usually formed by ground effect turbulence, or sometimes as cumulus dies down and spreads out in the evening before dispersing. Stratocumulus typically produces intermittent drizzle or snow precipitation, and thick stratocumulus over the sea in winter can lead to marked icing conditions.

Nimbostratus (Ns) is a thick layer of cloud, darker in appearance than either stratus or altostratus, with a variable base.

NB: Nimbostratus can also be a medium-level cloud. Nimbostratus clouds are associated with frontal lifting and typically produce quite consistent rain or snow precipitation.

Cumulus (CU) is a detached heaped cloud with sharp outlines, particularly as it develops. They either can be limited to small, "puffy" clouds, or may continue through to a considerable depth. Large cumulus clouds produce rain or snow showers, and possible turbulence and icing. *Cumulonimbus* (CB) is the ultimate in cumulus development. Their tops may reach the tropopause, i.e., in excess of 50,000 ft in the tropics. An anvil shape is common with the top of a CB, which is created by most of the cloud being stopped by the tropopause. Cumulonimbus clouds typically produce heavy rain, snow, and hail showers/thunderstorms. In addition, turbulence and severe icing are always expected with CB clouds.

The height of a cloud base can be determined by the difference between the dew point temperature and the ground temperature. The difference divided by the appropriate lapse rate/1000 ft will determine the height of the cloud base.

NB: The higher the moisture content present in the rising air, the higher the dew point temperature, the less the difference between the surface and the dew point temperatures, the lower the cloud base. Ergo the amount of moisture content in the air is a determining factor of the cloud base height.

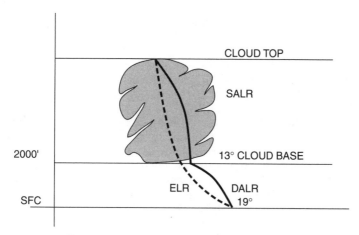

Figure 26 Cloud base.

$$\text{Cloud base height, ft} = \frac{19° \text{ (surface temp)} - 13° \text{ (dew point)}}{3 \text{ (DALR)}} \times 1000$$

It is worth noting that the relationship of the temperature in the cloud and the surrounding air (i.e., the ELR/DALR/SALR) determines not only the type of cloud formation but also the height of the cloud top. That is, when the surrounding temperature is the same as or warmer than the temperature inside the cloud, then the air in the cloud will become stable and will stop rising. This limits the height of the cloud top.

Cumulus clouds have flat bases due to the uniform decrease in temperature of the DALR. They have round and uneven tops because of the uneven decrease in the ELR temperature and different magnitudes in movement of the rising currents inside the cloud.

NB: If the ELR is high, that is, if the temperature of the surrounding air reduces more quickly with height gained than the rising DALR/SALR parcel of air, then this parcel of air will always be warmer than the surrounding (ELR) air. Therefore it will be lighter, and so it will keep rising. This air is said to be unstable, and it produces cumiliform (i.e., heaped) clouds.

If cumulus clouds were present in the morning, you would expect CBs (cumulonimbus) later in the day.

Clouds—Mountain (Lenticular)   *See* Mountain (Lenticular) Clouds.

Clouds—Tops   By using the weather radar system, you can calculate the height of the cloud tops by using the following formula:

Height of cloud tops relative to aircraft =

$$\frac{\text{range in feet} \times (\text{weather radar scanner tilt} - \frac{1}{2} \text{ beam width})}{60}$$

For tilt 1° up, beam width of 4°, and cloud 50 nm away,

$$\frac{50 \times 6080 \times (1 - 2)}{60} = -5067 \text{ ft (below aircraft's cruising altitude)}$$

Coefficient of Lift (CL)   The CL is the lifting ability of a particular wing. It is dependent on both the shape of the wing section (fixed design feature) and the angle of attack.

Coffin Corner   Coffin corner occurs at the aircraft's absolute ceiling, where the speeds at which Mach number buffet and prestall buffet occur are coincident, and although trained for, in practice they are difficult to distinguish between. Therefore a margin is imposed between an aircraft's operating and absolute ceiling.

Mach number and the slow-speed stall buffet are coincident at coffin corner because a stall is a function of IAS, and Mach number is a function of the *local speed of sound* (LSS) which itself is a function of temperature.

For a constant Mach number (which is the normal mode of speed management) the IAS decreases with altitude due to the decreasing LSS. To prevent the IAS decreasing to its stall speed, the Mach number must be increased, which results in an increasing IAS.

For a constant IAS the Mach number increases with altitude due to a decreasing LSS/temperature, to a point where the IAS exceeds the critical Mach number, denoted by Mcrit. To prevent the Mach number from exceeding the Mcrit, the IAS must be reduced, which results in a decreasing Mach number. Therefore there comes the point at the aircraft's absolute ceiling where the aircraft can go no higher. This is so because it is bounded on one side by the low-speed buffet and on the other side by the high-speed buffet, because the stall IAS and the Mcrit values are equal. This is coffin corner, and this effect restricts the attainable altitude by the aircraft.

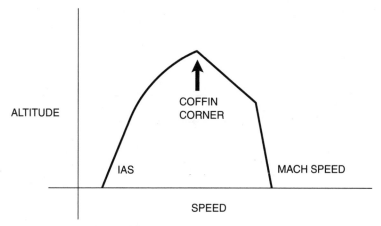

Figure 27   IAD/Mach speed versus altitude, coffin corner.

Col   A col is an area of almost constant air pressure, with a region of low pressure between two highs and/or a region of high pressure between two lows.

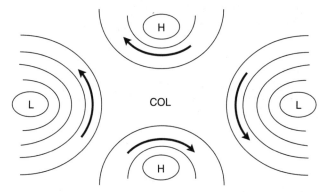

Figure 28   A col.

Winds in a col are extremely light and of random direction. They are not the familiar circular winds of either a depression or an anticyclone. Over land a col often produces showers or even thunderstorms in the summer due to high surface temperatures that can create thermal lifting, and fog or low cloud in the winter.

Cold Front   A cold front is the boundary produced between two air masses (i.e., cold air behind warm air) where the colder, denser air mass undercuts and replaces the warmer preceding air mass from the surface upward (slides under). The cold front position on a weather chart rather accurately displays the front's actual position, because the upper altitude position of the front is virtually overhead its surface position.

NB: A front is shown as a line on a weather chart, which represents the front's *surface* position. And *barbs* along the front line represent a *cold* front.

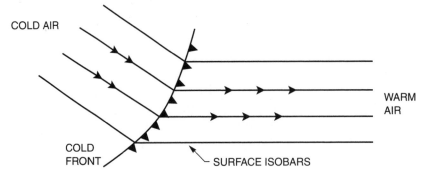

Figure 29 Cold front shown on a weather chart.

This is so because the slope of the cold front is relatively quite steep, approximately 1:50. Therefore there is little warning of the approach of a cold front, which typically contains cumulus and even cumulonimbus clouds with associated thunderstorms, covering a band of only 30 to 50 nm approximately.

A cold front is at the trailing edge of the warm (air mass) sector and generally behind the warm front in a depression (low-pressure system). It moves more quickly than the warm front ahead of it, with its cooler frontal air at altitude slightly lagging behind the cold air at the surface. The speed of a cold front is equal to the full geostrophic wind component. This can be determined by measuring the gap between the isobars along the front and comparing this distance directly against a geostrophic wind scale.

The direction of travel for a cold front is perpendicular to its surface position, provided the geostrophic component is greater than 10 kn.

NB: On a weather chart the barbs (that indicate it is a cold front) are *pointed* in the front's direction of movement.

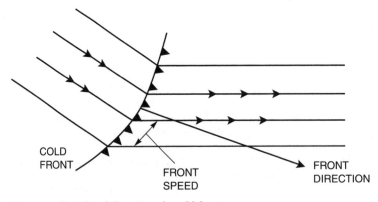

Figure 30 Speed and direction of a cold front.

However, if the geostrophic wind component is less than 10 kn, then this rule is unreliable and the front will tend to move from high pressure to low.

A cold front moves faster than a warm front because of the tendency of the warm air (in the warm front) to rise up and slide over the cold, more resistant, denser air (mass) in front of it. Therefore the force (geostrophic wind component) moving a warm front is used to move the warm air upward as well as horizontally forward. Hence the speed of a warm front is only two-thirds of the geostrophic wind component. However, the denser, cold air in the cold front meets very little resistance from the warm, less dense, sector air in front of it; and therefore its driving force is used solely to move it horizontally forward, at roughly the full geostrophic wind component.

The passage of a cold front has the following general characteristics and associated weather:

As the cold front approaches:

1. Cumulus and even cumulonimbus cloud cover is experienced. This is so because the passage of the cold front *undercuts* the warmer sector surface air ahead of it, forcing it to rise relatively sharply and to become *unstable*. This results in the following severe weather changes:
    a. Heavy rain
    b. Thunderstorms
    c. Turbulence (possibly severe)
    d. Wind shear
    All these can be hazardous to aviation.
2. Poor visibility is experienced. This is so because with the passage of a cold front, visibility is reduced due to the increase of heavy cloud through a large vertical band, and the heavy consistent rainfall.
3. The atmospheric pressure will usually fall as the cold front approaches.

As the cold front passes:

4. A sudden drop in air temperature is experienced. This is due to the arrival of the cold sector air mass.
5. Skies are clear with isolated cumulus clouds. This is due to the cold sector air mass (possibly polar maritime) that contains little moisture with a colder dew point temperature giving rise to a greater atmospheric stability, which only produces the occasional isolated cumulus cloud in an otherwise clear sky.
6. There is good visibility, except within the isolated cumulus clouds.
7. Wind veers in direction in the northern hemisphere and back in the southern hemisphere.
8. The atmospheric pressure will stop falling and may even rise.

See Fig. 31 at the top of the next page.

Collision Avoidance   Regardless of any ATC clearance, it is the duty of the commander of an aircraft to take all possible measures to see that he or she does not collide with any other aircraft or terrain.

An aircraft must not fly so close to any other aircraft or terrain that it creates a collision danger. An aircraft which is obliged to give way to another aircraft must avoid passing over, or under, or crossing ahead of, the other aircraft, unless it is passing well clear of the other aircraft.

NB: An aircraft with the right-of-way should maintain its course and speed.

A constant relative bearing of another aircraft, at the same altitude, means that there is risk of collision.

See Fig. 32 at the bottom of the next page.

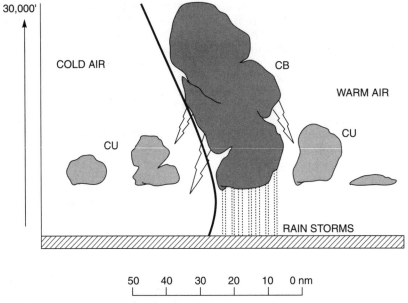

Figure 31   Cross section of a cold front.

Collision Avoidance—Rules of the Air   *See* Air Law—Rules of the Air.

Communication Failure   The basic procedure for a two-way communications failure, or a reception failure that is assumed to be a two-way failure, is as follows:

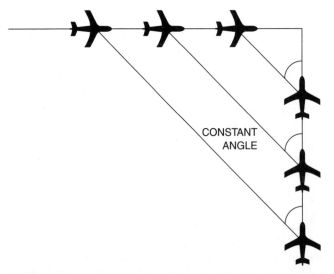

CONSTANT
ANGLE

Figure 32   A constant relative bearing means that a collision risk exists.

1. Continue your flight in accordance with your current flight plan. Maintain your last cruising level given by ATC; or if no level is assigned, then maintain the flight plan level. Select 7600 transponder code. Continue routine position reports, in case your transmissions are still being received by ATC.

2. Arrange your flight to arrive over the arrival holding point as close as possible to your last acknowledged estimated time of arrival (ETA) with ATC. Or, if you had no ATC-acknowledged ETA, then plan to arrive at the holding point at the calculated ETA computed from the last successful position report and flight plan data. Follow appropriate routes, i.e., STARS, etc.

3. Commence your descent, within 10 min, from over the holding fix at the last acknowledged estimated approach time (EAT) or the calculated ETA.

NB: If you are unable to start the descent within 10 min of the EAT, then do not attempt a landing, and proceed immediately to an alternate. Descend to the lowest level in the holding stack at not less than 500 ft/min, and plan to land within 30 min of the time your descent should have started.

4. If you are unable to complete the approach within 30 min of the time your descent should have started, or if you do not become visual at your MDA (H), then you should leave the area of the aerodrome and controlled airspace at your specified route and altitude to your alternate. Or, if no altitude or route is specified, fly at either the last assigned altitude or the minimum sector safe altitude, and avoid areas of dense traffic.

NB: If you have to carry out a missed approach within the elapsed 30-min period, then you can continue.

5. Inform ATC as soon as possible (ASAP) after landing at your destination or alternate airfields.

Combustion Cycle—Aero Piston Engine   The combustion cycle of an aero piston engine comprises *induction, compression, combustion (expansion),* and *exhaust.* The combustion of a piston engine occurs at a constant volume.

This cycle is known as the *Otto cycle* and is the basis of all four-stroke internal combustion engines. The cycle involves four strokes of the pistons (induction, compression, power, and exhaust) and occupies 2 revolutions (r) of the crankshaft.

Combustion Cycle—Jet/Gas Turbine Engine   The combustion cycle of a jet/gas turbine engine is *induction, compression, combustion, expansion,* and *exhaust.* In a jet/gas turbine engine, combustion occurs at a constant pressure whereas in the piston engine it occurs at a constant volume. The thermal efficiency of the cycle improves with increasing compression ratios, while the power output increases with the amount of fuel burned.

Compass Deviation   Deviation is the difference between the direction of magnetic north and compass north. Deviation is due to the effect of local magnetic fields, i.e., from metal and electrical equipment close to the compass, that distort the earth's magnetic field at a local level around the compass, causing the indicated compass direction to deviate from the magnetic direction; hence the term *deviation.*

Deviation is not a constant value for a given compass as it varies with the heading of the aircraft. Such deviation is predictable, and a *deviation card* stating the deviation experienced for a given compass heading is usually located next to the compass. Deviation is measured in degrees and, similar to compass variation, is described as east (+) or west (−).

Compass Direction   Compass direction (heading) is measured with reference to compass north. Compass north is the deviated direction away from magnetic north indicated by the compass needle.

To convert a compass heading to a true heading and vice versa, you have to apply both compass deviation and magnetic variation in the correct sequence:

Compass heading ± compass deviation =

magnetic heading ± magnetic variation = true heading

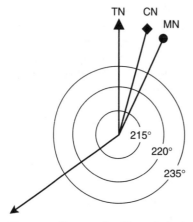

Compass heading	220°
Deviation	−5°
Magnetic heading	215°
Compass variation	+20°
True heading	235°

Figure 33   Compass heading calculation.

Compass Swinging   Compass swinging is a procedure used to check the accuracy of and to adjust an aircraft's magnetic compass. Compass swinging is accomplished at an area, usually on an aerodrome, where accurate magnetic directions are known. This allows the compass to be compared and adjusted as accurately as possible and any remaining errors to be entered on a deviation card. An engineer usually carries out this procedure, although some commanders are also certified to second-check this procedure.

A compass should be "swung" when any of the following occurs:

1. The compass is new.
2. Any electrical or magnetic influenced equipment in the vicinity of the compass has been altered.
3. It has passed through a severe magnetic storm.
4. There has been a considerable change in latitude.
5. There is an inspection of either the compass or nearby electrical or magnetic influenced equipment.
6. There is a doubt about the compass accuracy.

Component Arm   A component arm is the distance from the datum to the point at which the weight of a component acts (cg point). By convention an arm aft of the datum, which gives a nose-up moment, is positive; and an arm forward of the datum, which gives a nose-down moment, is negative. Therefore for a constant weight the longer the arm, the greater the moment.

The formula to calculate the length of the arm and thus the cg position is

$$\text{cg arm, ft} = \frac{\text{moment, lb}}{\text{weight, lb}}$$

Compressibility   This is the effect of air being compressed onto a surface (at its right angle to the relative airflow) resulting in an increase in density, and thus dynamic pressure rises above its expected value. It is directly associated with high speeds.

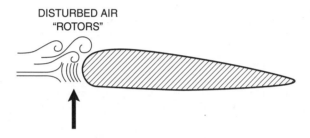

DISTURBED AIR
"ROTORS"

COMPRESSED AIR AT THE LEADING EDGE

Figure 34   Compressibility effect at the leading edge.

There are two main effects of compressibility:

1. There is compressibility error on dynamic pressure reading flight instruments; e.g., ASI shows an overread error which is greater as the aircraft's speed increases.
2. Compressed air is experienced on the leading edge of the wing, which disturbs the pressure pattern on the wing and causes the disturbed air, shock wave/drag effect at the critical Mach number.

Compression Ratio—Aero Piston Engine   In a piston engine this is the ratio of the total volume enclosed in a cylinder with the piston at bottom dead center (BDC) to the volume remaining at the end of the compression stroke with the piston at top dead center (TDC).

$$\text{Compression ratio} = \frac{\text{total volume}}{\text{clearance volume}}$$

However, the compression ratio is commonly considered as the ratio of the air pressures corresponding to these two volumes; but this is theoretically incorrect because a compression ratio in terms of air pressure does not reduce proportionally with a decrease in ambient air density. Therefore the compression ratio of a piston engine is a function of cylinder volume, because the combustion in a piston engine does occur at a volumetric constant.

Compression Ratio—Gas Turbine Engine The compression ratio in a gas turbine engine is a ratio measure of the change in air pressure between the inlet and outlet parts of either an individual compressor stage or the complete compressor section of the engine. Individual compressors, either centrifugal- or axial-flow types, are placed in series so that the power compression ratio accumulates. For example:

First compressor stage	Inlet pressure	15 psi	
	Outlet pressure	60 psi =	4:1
Second compressor stage	Inlet pressure	60 psi	
	Outlet pressure	180 psi =	3:1
	Overall compression ratio		12:1

Conductor *See* Electrical Conductor.

Constant-Speed Drive (CSD) Units—Electrical Constant-speed drive units (also known as generator drives) maintain the AC frequency output of an alternator, normally to 400 Hz.

NB: The number of cycles per second is the frequency of the AC supply. That is, a cycle is the complete reversal of direction of the electric ac, and 1 cycle per second is equivalent to 1 Hz.

The basic constant-speed drive unit (or generator drive) consists of an engine-driven hydraulic pump, which drives a hydraulic motor that itself drives the alternator. Most CSD units are capable of maintaining the alternator output frequency to within 5 percent of 400 Hz. The CSD unit can be disconnected from the engine input drive in the unlikely event of a malfunction. This allows both the drive unit and the alternator to become stationary, thus eliminating any chance of a malfunction affecting the engine.

The CSD unit can be disconnected at any time, but reconnection can only be done *manually* on the ground following the shutdown of the engine.

Constant-Speed Unit (CSU) A constant-speed unit, sometimes called a *governor unit,* controls the propeller's blade angle/pitch to maintain a rotational speed. The CSU controls the blade pitch variation, and the engine drives the propeller around at approximately a constant rpm. When it is required to accelerate the aircraft, the power control is opened by the throttle lever, supplying more fuel to the engine, causing momentarily an increase in rpm (overspeed). The CSU immediately detects this increase in rpm and converts it into an increase in pitch (coarser setting), which returns the rpm to its normal setting because the extra air load experienced by the propeller blade slows down the rpm. Moving the props to this coarser position increases the thrust produced, and the aircraft accelerates. As the forward airspeed increases, the angle of attack reduces, allowing the torque force to fall and therefore the rpm to rise. Again the CSU detects this change and further increases the blade angle (coarser) to provide efficient thrust. When the required forward speed is obtained, power is eased off, which causes a decrease in rpm; and the CSU detects this and reduces the blade angle to produce the required thrust at the same rpm. Thus the CSU is always adjusting the blade angle/pitch to compensate for changes in thrust requirements and for changes in forward speed by sensing changes in rpm, to maintain the maximum efficiency from the engine/propeller combination.

Contaminated (Wet) Runway—Landing *See* Landing—on a Contaminated (Wet) Runway.

Contaminated (Wet) Runway—Restrictions *See* Wet (Contamination) Runway—Restrictions.

Contaminated (Wet) Runway—Takeoff *See* Takeoff—on a Contaminated (Wet) Runway.

Control Surfaces The primary/main flight control surfaces are

Elevator, which controls the motion around the lateral axis, known as pitch/pitching

Ailerons, which control the motion around the longitudinal axis, known as roll/rolling

Rudder, which controls the motion about the normal/vertical axis, known as yaw/yawing

Control surfaces become more effective at higher speeds. Their effectiveness (the moment produced for a given control deflection) depends upon the size of the force produced by the control surface. The control force depends on the square of the airflow speed, and therefore the controls become more effective as speed increases. This results in a requirement for large control movements at low speeds and smaller control movements at high speeds to produce the same control force.

Control Surfaces—Hydraulically Operated On large, fast aircraft, especially modern jets, it is found that the control forces required to move a control surface are simply beyond the strength of the pilot, and are also too great to be controlled by pure aerodynamic designs, e.g., balance tabs. This is so because the shear sizes and weight of the control surface arm in question and the aerodynamic airflow lift forces (load) generated on the deflected control surface are too great.

NB: The control force required from the pilot increases with an increase of air load on the control surface, either as a result of an increase in control angle at a constant speed or with a given control angle an increase in speed, which is further compounded by Mach number effects.

For modern large/fast jet aircraft, the answer lay in the pure powered control surface, typically hydraulic powered systems, because they generate enough power to cope with the full air load force (i.e., not balanced) experienced on the control surface. Therefore the pilot merely signals the control angle required, and a hydraulic piston moves the surface to that position. The correct balance of the control surface is still required to be maintained within the structural strength of the surface, but in general the question of surface balance is no longer critical.

Control Surface Loads   If a control surface is deflected, the dynamic pressure/aerodynamic loads on it will increase and act as a lift force through its center-of-pressure (cp) point. When this is multiplied by the control surface arm, it gives the size of the moment trying to rotate the control surface back to its neutral position. This moment is known as the *hinge moment* or *air load force*.

Hinge moment (air load force) = lift force (air load) × arm

NB: The lift force is a design product of the size of the control surface, and the magnitude of the lift force experienced is dependent on (1) airspeed and (2) angle of deflection of the control surface. That is, the lift force increases dynamically in flight, either with an increased control angle of deflection at a constant speed, or at a constant angle of deflection with an increased speed. And the arm is a design product of the distance between the cp point and the hinge line. They produce a hinge moment/air load force that tries to return the surface to its neutral position.

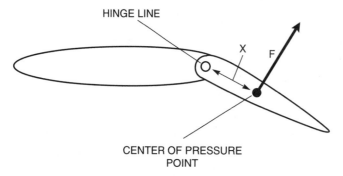

HINGE LINE

X   F

CENTER OF PRESSURE
POINT

Figure 35   Hinge moment/air loads. X = arm, F = lift force.

In flight the stick *control forces* reflect the hinge moment (air load force) which itself reflects the aerodynamic load/lift force and the arm. Therefore, the greater the lift force or the arm, the greater the hinge moment (air load force) and the greater the stick control force experienced by the pilot.

NB: Manual flying controls transmit the air load force (hinge moment) experienced on the control surface back through the cable system to the control column and rudder pedals. Powered flying control systems use an artificial feel system to transmit a proportion of the air load force experienced on the control surface back to the control column and rudder pedals.

This raises the issue of how the control forces required to move the different control surfaces are managed.

1. Light aircraft which fly slowly can rely on the pilot's physical strength alone to provide the control force to move the control surfaces.

2. On larger aircraft where the control forces are beyond the pilot's strength, the hinge moment (air load force) needs to be reduced. This is done by aerodynamic control balance designs, such as setback hinges, horn balances, weight balance, and trailing-edge balance tabs, that produce a force which opposes/balances (about the hinge line) the main control surfaces' lift force. This reduces the

overall hinge moment and associated stick control force to a level that is within the physical ability of the pilot. In this way the pilot supplies a proportion of the force to move the control surface, and the balance supplies the rest.

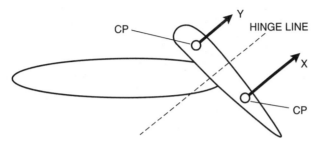

Figure 36 Hinge moment/air load balance. X = main hinge moment, Y = resistant hinge moment, $X - Y$ = resultant hinge moment Z, CP = center of pressure.

3. On even larger and faster aircraft, especially jets, the design of the pure aerodynamic control, although possible, was not practical or economically viable, namely, because the balancing of the opposing lift force and/or reduced arm that was required to reduce the overall hinge moment became a problem. Powered control surfaces (usually hydraulic) that could cope with the full hinge moment/air load force experienced on the control surface (i.e., not balanced) were the solution. However, there still remained a need for some degree of balancing, because the hinge moments had to be kept within certain structural limits. In general terms the balancing problem was no longer critical, namely because the hydraulic power that produced the control force was greater than the nonreduced hinge moment/air load.

Convergency Convergency represents the change of direction experienced along east-west tracks, except rhumb lines, as a result of the way direction is measured due to the effects of converging meridians at the poles.

To understand what convergency is, we need to understand the properties of navigating on a sphere. Direction in navigation is given as an angle from a datum, and the datum used to navigate is the north pole. Lines that form the shortest east-west distance between two points on the earth (Great Circle tracks) have a changing orientation to true north, and thereby direction, as a result of crossing converging meridians. The change of direction experienced between two points is known as *convergency*.

See Fig. 37 at the top of the next page.

Convergency is clearly dependent on latitude; it is zero at the equator, where the meridians are parallel, and a maximum at the poles, where the meridians converge. It is also dependent on how far you travel. A short distance will have only a small change of direction, while a long distance traveled will have a large change of direction.

To calculate convergency for a given east-west distance (change of longitude), the following formula is used:

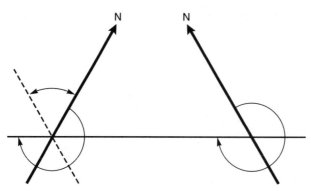

Figure 37   Convergency.

$$\text{Convergency} = \text{change of longitude} \times \sin(\text{mean latitude})$$

Conversion Angle   This is the angle of difference in direction between the rhumb line and the Great Circle track between two points; therefore it is used to calculate a rhumb line direction given a Great Circle track, or vice versa. The midpoint of a rhumb line and a Great Circle track are parallel. Therefore the direction of the Great Circle track's midpoint must be the direction of the rhumb line. Remember, the rhumb line has a constant direction.

Thus we can determine that the angle between the Great Circle track and the rhumb line track at either end is one-half of the convergency, and this is called the *conversion angle*. Conversion angle = ½(convergency).

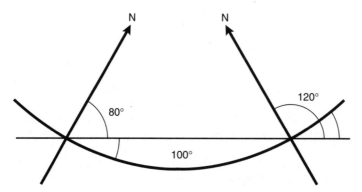

Figure 38   Conversion angle ($= \frac{1}{2} \times 40° = 20°$). Rhumb line direction = $80° + 20°$ or $120° - 20° = 100°$.

Coriolis Force   The Coriolis (also known as geostrophic) force is an *apparent* force that acts upon a parcel of air, which is moving over the rotating earth's surface. This means that the air does not flow simply from a high- to low-pressure system, but is deflected to the left or right, according to which hemisphere you are in. This is known as the *Coriolis effect*.

NB: In the northern hemisphere the Coriolis force deflects the airflow to the right (i.e., as a westerly wind), and in the southern hemisphere the airflow is deflected to the left.

The Coriolis force is a product of the earth's rotational properties. In that a parcel of air at the equator may be stationary over the earth's surface, but it is actually rotating with the earth at a speed of approximately 900 kn. If a parcel of air becomes displaced toward the pole, due to a pressure gradient force, the earth's rotational velocity is not lost, and the airflow still continues to move east (the direction of the earth's rotation) at approximately 900 kn. However, as you move away from the equator, the rotation speed of the earth's surface reduces with latitude, e.g., to 850 kn; therefore the speed of the airflow relative to the earth's surface has increased, and a westerly wind is born.

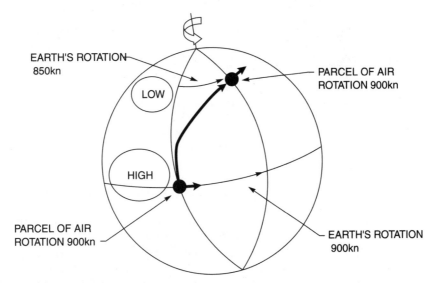

Figure 39 Coriolis force.

The Coriolis/geostrophic force is proportional to

1. The angular velocity of the earth a (approximately 900 kn).
2. The air density p (i.e., the lower the density of the air, the less resistant it is to being moved).
3. The wind speed V; i.e., the faster the airflow, the greater the Coriolis force. If there is no airflow, then there will be no Coriolis effect.
4. The latitude; i.e., the greater the distance from the equator (nearer the pole), the greater the change in latitude and the greater the change in the wind speed.

$$\text{Coriolis/geostrophic force} = 2aVp \sin (\text{lat})$$

Cost Index (CI) Cost index is a performance management function that optimizes the aircraft's speed for the minimum cost. Cost indices form part of a company's stored route and are inserted into the flight management computer (FMC). The CI takes into account specific route factors, such as the price of fuel at the departure and destination airports, so that the aircraft is flown at the correct speed to balance the fuel costs against the dry operating costs. An incorrect CI will always cost more money.

Critical Engine—Control Speeds (VMC)   In theory there is a higher critical control speed (VMC) for a failure of the critical engine, and a lower critical speed for a failure of the noncritical engine. However, in practice the critical speed relating to the critical engine is the only performance speed used.

Critical Engine—Crosswind Effect   A crosswind, depending on its direction, can help to either restore or aggravate the yawing moment of an aircraft with a failed critical engine. For instance, a failed critical number 1 engine will cause a yaw to the left. A crosswind component from the left will apply a restoring force to the aircraft's fuselage, while a crosswind from the right will aggravate further the yawing moment to the left, due to the sideways force experienced on the right side of the aircraft's fuselage which is from the right to left. Therefore a crosswind landing is even more important with a critical engine failure.

Critical Engine—on a Jet/Gas Turbine Aircraft   There is *no* critical gas turbine engine, because the engines are symmetrically positioned with opposing revolution direction. However, there is a governor engine, i.e., an engine that is the master that sets the rpm speed for the others.

Critical Engine—Propeller Aircraft   A critical engine is the engine that determines the critical control speed of the aircraft. It is especially vital on aircraft with propellers rotating in the same direction.

The number 1 engine on an aircraft with the propellers rotating in the same direction is the critical engine for two main reasons:

1. *Slipstream effect.*   If the propellers are rotating in the same direction (i.e., clockwise viewed from behind/anticlockwise viewed from in front), then only the number 1 engine will produce a sideways, slipstream, force on the fin. This has the effect of assisting the rudder side force needed to counteract the yawing moment for a number 2 engine failure. But for a number 1 engine failure the slipstream from the number 2 engine will produce a sideways force that aggravates the yawing moment, resulting in a more critical situation. Therefore a greater control force is required, resulting in a need for a higher critical speed to make the rudder more effective during a number 1 engine failure, which therefore determines that the number 1 engine is the critical engine.

See Fig. 40 at the top of the next page.

2. *Asymmetric blade effect.*   Propeller blades produce more thrust in the downward rotation than in the upgoing rotation. Therefore the point through which the thrust acts will be displaced toward the downgoing blade. Depending on the direction of rotation of the propellers, this either increases or decreases the thrust moment arm. And in aircraft with the propellers rotating in the same direction, failure of the engine with the shortest moment arm will produce the greatest yawing moment from the other live engine (i.e., normally anticlockwise rotation when viewed from in front).

See Fig. 41 at the bottom of the next page.

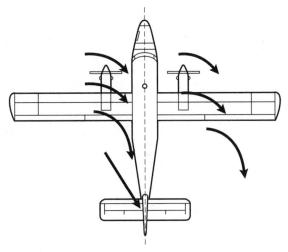

Figure 40 Slipstream effect of propellers.

Therefore the engine with the shortest thrust moment arm is the critical engine, and for anticlockwise rotating propellers this makes the number 1 engine the critical engine.

It can be seen that the aircraft's critical speed is based on the critical engine and is proportional to the length of the thrust moment arm; i.e., the shorter the thrust moment arm, the lower the critical speed (VMC).

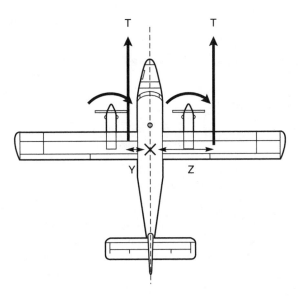

Figure 41 Asymmetric blade effect. T = thrust line, Y = engine 1 thrust moment arm, Z = engine 2 thrust moment arm.

The use of contra rotating propellers (i.e., engines with propellers rotating in opposite direction to each other), say number 1 engine rotating anticlockwise and number 2 engine rotating clockwise when viewed from in front, reduces the difficulties of asymmetric blade. If both propellers rotate inward, toward the fuselage, then (a) the thrust moment arm will be kept as short as possible and (b) slipstream effects counteract the yawing moment for both engine failures. Consequently the aircraft's critical speed is reduced.

Critical Point (CP)   The critical point, or equal-time point, is the en route track position where it is as quick (time) to go to your destination as it is to turn back. The critical point is calculated as a distance and time from the departure airfield, using the formula

$$\text{Distance to CP} = \frac{DH}{O + H} \qquad \text{given } D = \text{total sector distance}$$
$$H = \text{ground speed home}$$
$$O = \text{ground speed out}$$

$$\text{Time to CP} = \frac{\text{distance to CP}}{\text{ground speed out } O}$$

Critical points are normally required to ascertain the quickest flight time to an aerodrome, in the event the aircraft suffers an engine failure. Therefore the CP formula for engine failure scenarios is

$$\text{Distance to CP} = \frac{DH}{O + H} \qquad \text{given } D = \text{total sector distance}$$
$$H = \text{engine failure ground speed home}$$
$$O = \text{engine failure ground speed out}$$

NB: Engine failure ground speed is used to calculate distance because it represents travel on from the CP and travel home from the CP with a failed engine.

However, time to the engine failure critical point is calculated by using the same formula as the all-engine formula:

$$\text{Time to CP} = \frac{\text{distance to CP}}{\text{ground speed out } O}$$

NB: All-engine ground speed out is used because this time represents the ground speed travel out, prior to the engine failure.

The critical point moves into the wind. Given still air conditions, the critical point between two aerodromes is simply the halfway point. However, the effect of wind displaces the critical point to one side or the other of the midpoint. Flying into a headwind, the CP moves closer to the destination aerodrome; and flying with a tailwind, the CP moves closer to the departure aerodrome. This is so because the CP is the *equal-time point* to reach an airfield, and therefore ground speed is all-important, and ground speed is TAS × WV. (WV = wind velocity.)

Critical Speed   Critical speed is the lowest possible speed on a multiengine aircraft, at a constant-power setting and configuration, at which the pilot is able to maintain a constant heading after a failure of an off-center engine. VMC(G/A)/L are particular configuration and stage of flight critical speeds.

Crosswind Component *See* Wind Component.

Crosswind Landing *See* Landing—Crosswind.

Crosswind Takeoff *See* Takeoff—Crosswind.

Cruise—Below Optimum Altitudes (Jet Aircraft) A flight carried out below its optimum altitude uses more fuel but takes less time, when flying at a *constant Mach number,* to complete the trip.

NB: Optimum altitude is the aircraft's highest attainable altitude, i.e., the service ceiling, which for a jet would typically be an altitude above FL260, and so the Mach number (MN) speed flown is constant because the MN is the limiting speed.

NB: Higher FL = time increases and fuel consumption decreases (in still air and at constant MN)

Lower FL = time decreases and fuel consumption increases (in still air and at constant MN)

It uses more fuel because the engines are designed to achieve their best specific fuel consumption (SFC) at a high operating rpm, which can only be achieved at high (optimum) altitudes. Therefore when a flight is carried out below its optimum altitude, its engines cannot operate at their optimum rpm; therefore its SFC is higher, and consequently it uses more fuel to complete the trip. It takes less time, because when flying at a constant MN, which is normal practice, at high altitudes (i.e., above FL260), the LSS increases as the altitude becomes lower because LSS is a function of temperature.

Therefore, to maintain the same MN speed, the IAS and TAS must be increased; and therefore the ground speed increases (assuming a constant wind velocity at both altitudes). This results in the distance covered taking less time when a flight is carried out below the optimum altitude. For example,

$$\text{FL370/MN } 0.83 \ \frac{\text{TAS } 500}{\text{LSS } 600} \qquad \text{FL350/MN } 0.83 \ \frac{\text{TAS } 514}{\text{LSS } 620}$$

NB: Flights at lower altitudes, that is, less than FL260, *where IAS is the reference speed,* will have the opposite effect. This is so because the TAS decreases at lower altitudes for a constant IAS; therefore ground speed is reduced, and the trip will take a longer time to complete. However, a jet aircraft would normally fly at an altitude where the MN was the reference speed.

Cruise (Step) Climb A cruise (step) climb occurs when the aircraft in the cruise loses weight, due to fuel burn, which allows the aircraft to fly higher; so a cruise (step) climb is initiated to climb the aircraft to its new maximum altitude.

NB: An aircraft may have several step climbs, especially on long haul routes. A cruise climb is important because a jet aircraft's most efficient performance is gained at its highest possible altitude.

Cruise Climb Profile The cruise climb profile is a compromise between the best en route speed profile and the best climb profile most commonly used by commercial traffic. It provides faster en route performance, more comfortable aircraft attitude, and better aircraft control due to lower angle of attack and greater airflow over the control surfaces.

Cruise—High Altitude (Jet)   Jet aircraft climb as high as possible (i.e., the service ceiling) because the gas turbine (bypass) engines are at their most efficient when its compressors are operating at a high rpm, approximately 90 to 95 percent. This high rpm obtains the engine's optimum gas flow condition which achieves its best specific fuel consumption. This optimum high rpm speed can only be achieved at high altitudes, because only at high altitudes where the air density is low will the thrust produced be low enough to equal the required cruising thrust.

The primary reason for designing an engine's optimum operating condition at approximately 90 to 95 percent rpm is to make it coincident with the best operating conditions of the airframe, namely, the minimum cruise drag. Therefore at high altitudes there are two main consequences:

1. *Minimum cruise airframe drag* is experienced at high altitudes because drag varies only with equivalent air speed (EAS) (namely, as EAS decreases, drag decreases). At very high altitudes, i.e., above 26,000 ft, the Mach number speed becomes limiting, and therefore EAS and TAS are reduced for a constant MN with an increase in altitude. Therefore the lowest cruise EAS is at the highest attainable altitude (service ceiling), and because drag varies only with EAS, airframe drag is also at its lowest value at high altitudes. Consequently our thrust requirements are lowered at high altitudes because our thrust value must only be equal to our drag value.

2. *Best engine performance* is experienced at high altitudes, due to its ability to operate at its optimum high-rpm condition because of the low atmospheric air density. The engine's best operating conditions are a function of its internal aerodynamic design. This reflects the optimization of the engine to be generally at its best under the conditions where it will spend most of its working life, i.e., high-altitude, high-speed conditions at a comparatively high engine rpm setting.

High-altitude conditions optimize the engine's design by utilizing the reduction in the atmospheric air density as a reduced airflow mass into the engine for a given engine rpm speed with an increase in altitude. The fuel control system adjusts the fuel delivery to match the reduced mass airflow to maintain a constant mixture and so maintain a constant engine speed. This causes the thrust to fall for a given rpm speed and requires an increase in compressor rpm to maintain its thrust values with an increase in altitude until its optimum high rpm speed is reached.

In addition the required thrust is lower with an increase in altitude because the EAS and airframe cruise drag reduce with altitude. Therefore it follows that only at high altitudes will the thrust be low enough to equal the required thrust at its optimum normal cruising high engine rpm, which achieves its best SFC.

The advantages that a jet engine gains from flying at high altitudes are as follows:

1. Best SFC and thereby increased (maximum) endurance is achieved at high altitudes because of two effects.

NB: Endurance is the need to stay airborne for as long a time as possible for a given quantity of fuel. Therefore the lowest fuel consumption (SFC) in terms of pound weight of fuel per hour is required.

a. Minimum cruise drag is experienced at high altitudes, because the MN speed becomes limiting above approximately 26,000 ft; and for a constant MN (as is the normal operating practice) the TAS and EAS decrease with altitude, and drag varies only with EAS. As such the EAS is progressively reduced to a level closer to the aircraft's best endurance speed, as the altitude increases, which is obviously where drag is least.

NB: VIMD broadly speaking remains constant with altitude. And because minimum aircraft drag requires minimum thrust (i.e., thrust = drag) and given that thrust is a product of engine power and fuel consumption is a function of engine power used, the aircraft thereby has its lowest fuel consumption in terms of fuel used per hour. Hence it produces the maximum endurance flight time for a given quantity of fuel, when flying at its lowest cruise EAS (best endurance speed) which the highest attainable altitude will achieve for a constant MN.

b. In addition, an aircraft's fuel consumption decreases slightly at high altitudes because of the higher propulsive efficiency of the engine, which brings it closer to its best SFC operating condition. Therefore an aircraft's best SFC and maximum endurance are attained by flying at
(1) The best endurance speed (VIMD)
(2) Highest possible altitudes where the engine achieves its best propulsive efficiency

2. The higher the altitude, the greater the TAS for a constant IAS, which provides an increased (maximum) attainable range.

NB: Maximum attainable range is the greatest distance over the ground flown for a given quantity of fuel, or the maximum air miles per gallon of fuel.

Maximum range is also defined by an EAS, which is a slightly higher speed than the best endurance speed, because the benefits of the increased IAS, i.e., greater ground speed per distance covered, outweighs the associated increased drag and higher fuel consumption. The higher TAS for a constant IAS is simply due to the reduction of air density at higher altitude. Thus ground speed/ground distance covered and thereby range are increased (or the flight time is reduced for a given distance), at high altitudes, for a constant IAS that gives a higher TAS. So the rule of thumb for best range is: *The higher, the better.* Just how high depends upon other factors, such as winds at different levels and sector lengths.

NB: However normal operating practice for jet aircraft will be to have a limiting MN speed above approximately 26,000 ft. Therefore a constant MN is flown above 26,000 ft which would see TAS decrease with altitude, because IAS and LSS decrease to maintain a constant MN.

Therefore because TAS decreases at high altitude for a constant MN, this results in the ground speed being reduced slightly, and so an obtainable range will also be reduced slightly below its maximum range, and/or increases its flight time, for still-air conditions. To counter this, a slightly higher, long-range cruise MN (and thereby IAS) speed can be selected, which increases the TAS, for that altitude, although this would be detrimental to endurance.

In basic terms of best SFC and endurance and greater TAS and range, an aircraft should remain as high as possible for as long as possible. Even on a short sector you would still climb as high as possible, to gain the best SFC possible (which improves the higher the altitude). Additionally you gain a higher TAS, but not necessarily a higher ground speed.

Once an airframe and engine have been designed to be at their best at certain flight conditions, some deterioration must be accepted for significant departures from the optimum speed and altitude. At high altitudes, any lower-speed deviation could involve stability and control difficulties and a move away from the best SFC rpm. Any higher speeds would result in increased drag, reduced speed margins, and again a move away from the best SFC rpm. At low altitudes, the SFC will rise markedly because of the poor performance of the jet in an underthrottle condition. Any attempt to run the engine efficiently at a high rpm setting will produce so much thrust that the resulting speed will incur high drag penalties and easily exceed the aircraft's maximum permitted airspeed.

D

DALR (Dry Adiabatic Lapse Rate) DALR is the adiabatic temperature change for unsaturated air, as it rises. Unsaturated air is known as dry air, and its change in temperature is a rather regular drop of 3°C per 1000 ft of height/altitude gained.

DC (Direct Current)—Electric Power Direct-current electric power flows in only one direction around a circuit and has no appreciable variation in its amplitude. An aircraft's typical sources of DC electric power are

- DC generator
- Battery
- Ground DC supply
- Rectifier—changes AC to DC

NB: A transformer rectifier unit (TRU) changes AC to DC and changes its voltage level.

DC Generator A simple DC generator uses a rotating armature, inside a stationary magnetic field, that creates an emf/current in the armature in the same manner as a simple rotating armature alternator. The difference in the DC generator is that the slip rings and brushes of the simple AC alternator are replaced by a single split ring with two diametrically opposed brushes, thereby creating a simple DC generator. The brushes are arranged so that when an armature coil is passing through the vertical position in the magnetic flux, the two halves of the split ring are just on the point of changing contact from one brush to another. When used in this way, the split ring is called a *commutator.*

Decca Decca is a medium-range (300 to 500 nm by day and 200 to 250 nm by night) navigation system that uses ground-based beacons that operate in the low-frequency (LF) band using surface waves. It is capable of giving very accurate position information. It uses hyperbolic position lines; the system measures lines of equal difference in range between two beacons. Therefore, to fix your position you need a minimum of three stations. The first two stations would determine which hyperbolic position line you were on, and the third station would give you a cross-cut from a hyperbolic position line to obtain a minimum fix. Decca uses chains of four stations for added accuracy.

See Fig. 42 at the top of the next page.

Each decca chain of four stations is controlled by one master station with three slaves named green, red, and purple. The phases of the transmissions of the master and each of the slaves are compared so that there are three possible sets of hyperbolic lines to fix a position. This is achieved by using a fundamental frequency (approximately 14 kHz). The master station transmits at 6 times the fundamental frequency, the red slave station transmits at 8 times the fundamental frequency, the green at 9 times the fundamental frequency, and the purple at 5 times the fundamental frequency. This system allows the phases to be compared

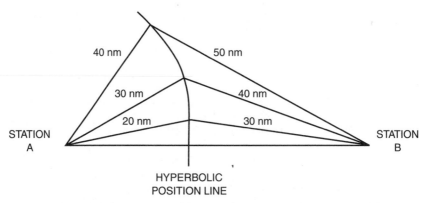

Figure 42 Hyperbolic lines measured from two beacons.

once all the stations transmitting frequencies have been reduced to the fundamental or comparison frequency (to compare phases, the frequencies must be the same). This system also allows the transmitting stations to be identified. Decca's onboard equipment can then run a *phase comparison* of the transmissions from a pair of beacons to determine the hyperbolic position lines as lines of equal phase = equal difference in range. Then decca fixes your position from a minimum of three stations, the first two will determine a hyperbolic position line, and the third station will give you a crosscut from a hyperbolic position line to obtain a fix. The fourth station crosscut provides added accuracy.

The accuracy of Decca is affected by the following:

1. *Height error.* There will be an increasing position error as altitude increases because the decca charts are drawn for ground level. However, height error is predictable and can be compensated for with correction tables.

2. *Decometer lag.* This is a delay in displaying the signal on the decometer (decca display instrument) giving a false reading, which commonly happens when moving across narrow lanes.

3. *Lane slip.* Lane identification signals interrupt the normal transmission and cause the decometer needle to stop moving. When the normal signal resumes, the decometer needle clicks back to its correct position. However, if more than one-half of a lane has been traveled before the signal resumes, the needle will move backward and fail to click up as another lane is transversed. This lane slip is overcome by installing a memory function, which keeps the decometer needle moving during lane identification.

4. *Night error.* Signals are refracted from the ionosphere at night and occasionally by day during the winter. This error reduces the accuracy of the system and thereby reduces its reliable range to 200 nm.

Decision Height (DH) Decision height is the wheel height above the runway elevation at which a go-around must be initiated by the pilot unless adequate visual reference has been established, and the position and approach path of the aircraft have been visually assessed as satisfactory to safely continue the approach and landing.

NB: DH is the height above the ground. That is, it is measured from the radio altimeter, or with the local QFE pressure setting of the barometric altimeter.

The DH minimum for an instrument-rated precision approach is calculated as follows:

1. Take the higher of
   - OCH for the aid and aircraft category
   - Precision approach system minimum, that is, ILS (CAT I), 200 ft; PAR, 200 ft; MLS, 200 ft.
2. Then add 50 ft for altimeter position error correction (PEC), especially for light aircraft.

NB: Many operators of advanced aircraft do not add 50 ft for PEC because their altimeter systems are extremely accurate and therefore do not suffer position error.

The approach may not be continued below CAT I DH (A) unless at least one of the following visual references for the intended runway is distinctly visible to and identifiable by the pilot.

- Element of the approach light system
- The threshold
- The threshold markings
- The threshold lights
- The threshold identification lights
- The visual glide slope indicator
- The touchdown zone lights
- The touchdown zone or touchdown zone markings
- The runway edge lights

The approach may not be continued below CAT II DH unless visual reference containing a segment of at least three consecutive lights, i.e.,

- The centerline of the approach lights   or
- The touchdown zone lights   or
- The runway centerline lights   or
- The runway edge lights

or a combination of these is attained and can be maintained.

The approach may not be continued below CAT IIIa DH unless visual reference containing a segment of at least three consecutive lights, i.e.,

- The centerline of the approach lights   or
- The touchdown zone lights   or
- The runway centerline lights or edge lights

or a combination of these is attained and can be maintained.

Deicing System   A deicing system, e.g., pneumatic leading-edge boots, is used where ice has built up on a surface and needs to be removed.

Density (Air)   *Density* is defined as a mass per unit volume of a substance. By definition, a block of gas is less dense than the same size block of liquid, which is less dense than the same size block of a solid.

Air density is easily changed by compression and expansion due to the influences of the following:

1. *Temperature.* Temperature is the main influence on air density, for a given pressure altitude. That is, a warmer temperature equals less dense air. An increase in the air temperature makes the air expand. The same numbers of air molecules are now spread over a larger volume, so there must be fewer molecules in the original volume. Therefore this original volume of air must be less dense.

2. *Pressure.* Pressure also influences the density of a column of air, over a range of altitudes. That is, the lower the altitude, the greater the air density. A column of air exerts a pressure force at any point as a result of the weight of air pressing down from above; that is, the effect of gravity. This effect progressively compacts the air molecules toward the earth's surface, which results in the air density being greatest at sea levels, and thereafter it decreases with a rise in altitude.

At high altitudes the air is also allowed to expand because of the increased concentric earth circumference, which also makes the air less dense.

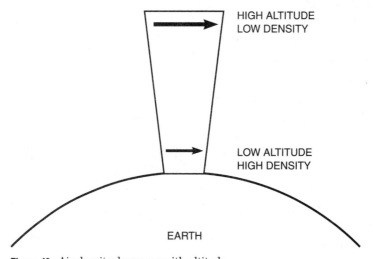

Figure 43   Air density decreases with altitude.

Here lies what on the surface seems to be a contradiction. Namely, at high altitudes, low air pressure creates a reduction in air density, but high altitudes have colder temperatures which would increase the air density. However, over a range of altitudes the effect of pressure on density is dominant, and therefore air density decreases with an increase in altitude. This is so because the influence of temperature is a function of a particular pressure altitude.

Therefore density is proportional to pressure over temperature.

3. *Humidity.* The last influence on air density is moisture or humidity. Moist air is less dense than dry air because lighter water molecules replace some of the air's heavier molecules.

A warmer than ISA temperature for a given altitude causes a decrease in air density which causes a decrease in the aircraft's engines and aerodynamic performance levels, because the aircraft's performance is based on air density.

Conversely, the opposite is true. A colder than ISA temperature for a given altitude causes an increase in air density which causes an increase in the aircraft's engine and aerodynamic performance levels.

Density (Air)—Errors *The altitude error* on the altimeter is due to air density differences. Let us examine this. Density for a particular pressure altitude is affected by temperature; thus an altitude error will exist when the actual temperature experienced is different from the ISA temperature. This occurs because the aircraft's altimeter is basically only a barometer, calibrated according to the ISA, which converts static pressure measurements to an altitude. The altimeter does not make any allowance for temperature deviation from the ISA, and temperature deviation causes the density altitude to differ from the pressure altitude. This is so because colder air will make a parcel of air denser and therefore heavier, and these additional weights push down the column of air with the effect that pressure levels are experienced at lower altitudes than the ISA. This results in the actual flight level being lower than the pressure level read by the altimeter. Or, in other words, the altimeter overreads.

Therefore, when you are flying from high to low temperatures, beware below. This density altitude error is significant when you consider terrain clearance.

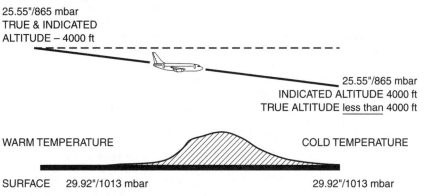

Figure 44 Altitude density error.

Conversely, warmer than ISA air makes the air less dense and lighter in weight, which allows the column of air to rise with the effect that the pressure levels are experienced at a higher altitude than the ISA. This results in the actual flight level being higher than the pressure level read by the altimeter.

See Fig. 45 at the top of the next page.

Airspeed error on the airspeed indicator (ASI) is due to air density differences. Let us examine this. The ASI is calibrated to ISA mean sea-level (MSL) conditions and measures dynamic pressure as the indicated airspeed (IAS). Because dynamic pressure is proportional, among other things to air density, the ASI will only read the correct airspeed at one air density, namely, ISA MSL conditions of 1225 g/m³. Density varies with temperature and/or pressure, and therefore any deviation from ISA MSL temperature and/or pressure conditions, such as experienced at higher altitudes, will effectively cause a density error in the ASI. Therefore a correction to the indicated airspeed needs to be made to find the true airspeed (TAS). This is important for navigation timings, etc.

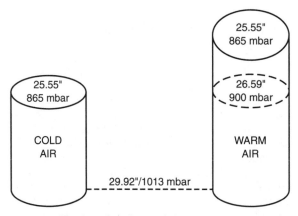

Figure 45 Heating of a column of air.

Density Altitude Density altitude is the altitude above the 1013-mbar/29.92-in datum at which the air density value experienced represents that of the level under consideration. That is, the altitude is measured against its air density value.

Density altitude allows for the consideration of ISA temperature deviation to pressure altitude. We know that the ISA temperature is +15°C at mean sea level (i.e., a pressure altitude of 0 ft) and decreases at a lapse rate of 2°C for every 1000 ft of altitude. Therefore any temperature deviation from this standard at any altitude will cause the air's density to differ from the pressure altitude, at a rate of 120 ft for 1°C deviation from the ISA temperature, and the corresponding temperature-corrected altitude is known as the density altitude.

For example, for a 3000-ft pressure altitude with a temperature ISA + 4°C,

$$4 \times 120 \text{ ft} = 480 \text{ ft}$$

$$3000 + 480 = 3480\text{-ft density altitude}$$

Remember: The warmer the temperature for a given altitude, the less dense the air; therefore warmer temperatures represent a higher density altitude, and vice versa.

Aircraft aerodynamics and engine performance, and terrain clearance are directly related to the corresponding density altitude.

Departure Departure is an east-west distance along a parallel line of latitude, other than at the equator. It requires the use of the following formula to calculate the variable east-west distance (nm) for a given change of longitude, at different latitudes.

$$\text{Departure, nm} = \text{change of longitude, min} \times \cos (\text{latitude})$$

The term *departure* is used because the farther you "depart" from the equator, namely, move north, the farther you depart from the 1 nm = 1 min value for a change of longitude distance.

Departure Profile Segments (Sectors) 1 to 4 The various segments and other terms relating to the takeoff flight path are as follows:

Reference zero is defined as the ground point at the end of the takeoff distance, below the net takeoff flight path (NTOFP) screen height. Screen height for a jet aircraft is 35 ft in dry conditions and a minimum of 15 ft in wet conditions. Therefore the reference zero point is also defined to this height. Reference zero defines the end of the TODR (A) and the start of the NTOFP, and it is the reference point to which the coordinates of the various points in the takeoff flight path, in terms of horizontal distance and vertical height, refer.

First (sector) segment extends from the reference point (35-ft height) to the point where the landing gear is retracted at a constant V_2 speed.

Second (sector) segment extends from the end of the first segment to a gross height of between at least 400 ft to a usual maximum of 1000 ft above ground level (AGL), at a constant V_2 speed.

Third (sector) segment assumes a level flight acceleration, during which the flaps are retracted in accordance with the recommended speed schedule.

Fourth and final segment extends from the third segment level off height to a net height of 1500 ft or more with flaps up and maximum continuous thrust.

Departure Separation The International Civil Aviation Organization (ICAO) departure spacing minima are as follows:

Leading aircraft	Following aircraft departing from the same position	Minimum separation at the time aircraft are airborne
Heavy	Medium	2 min
Medium	Light	2 min

Leading aircraft departed using full length of runway	Following aircraft departing from intermediate position on runway	Minimum separation at the time aircraft are airborne
Heavy	Medium	3 min
Medium	Light	3 min

Depression *See* Low-Pressure System.

Derated Thrust Takeoff Derating an engine reduces its *flat rating,* that is, to a lower maximum attainable thrust.

Descent Point The heavier the aircraft, the earlier its required descent point. The heavier the aircraft, the greater its momentum. Remember, momentum = mass × velocity. Therefore for a constant IAS or Mach number, that is, its VMO/MMO, the heavier aircraft will have to maintain a shallower rate of descent to check its momentum. The shallower the rate of descent, the greater its ground speed, and because an aircraft's descent is a function of rate of descent (ROD) (e.g., 700 ft/min), the aircraft will cover a greater distance over the ground per 100-ft descent, or per minute. Therefore, total descent is measured against distance over the ground, which is a function of ground speed, which is dependent on momentum, which is dependent on weight. Therefore the greater the aircraft's weight, the earlier its required descent point.

The lighter the aircraft, the greater its rate of descent (quicker). This is so because an aircraft is always restricted to a maximum speed, and during a descent the heavier aircraft has to maintain a lower rate of descent than a lighter aircraft (otherwise it would overspeed). (Remember, heavier aircraft have a greater momentum, and this weight-driven momentum will produce a greater speed in a vertical dive.) Therefore a lighter aircraft can descend later and more quickly than a heavier aircraft, because it can maintain a greater vertical descent profile without overspeeding.

Dew Dew is a water cover on the earth's surface that is formed when the following conditions exist:

- *Cloudless night* allows the earth's surface to lose heat by radiation that in turn cools down, by conduction, the air in contact with it, i.e., only a layer of air 1 or 2 in thick (remember air is a poor conductor of heat) to its dew point temperature. This causes its water vapor to condense out as a water liquid.

- *Moist air* with a high relative humidity only requires a slight cooling to reach its dew point temperature.

- *Light winds* of less than 2 kn.

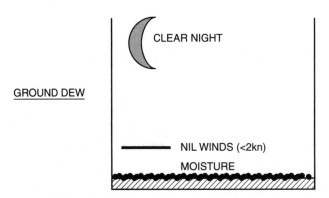

Figure 46 Dew conditions.

NB: The conditions for dew to form are the same as for radiation fog except for the lower wind or no wind.

Dew Point The dew point is the temperature at which a parcel of air becomes saturated. That is, its capacity to hold water vapor is equal to that which it is actually holding; or, in other words, its relative humidity is 100 percent.

NB: Dew point is also sometimes called the *saturated temperature.*

NB: The higher the moisture content in the air, the higher its dew point temperature.

DH (Decision Height) *See* Decision Height (DH).

Differences between a Jet and a Propeller Aircraft—Approach to Land   There are six main handling differences between the jet aircraft and the propeller aircraft on the landing approach. In all the differences, the jet aircraft is worse off than the propeller aircraft in maintaining the approach profile and in correcting errors on the approach. The six main differences are as follows:

1. *Momentum*

    The momentum of the jet aircraft is significantly greater than that of the propeller-driven aircraft because of its greater weight and velocity. (Momentum of a body is the product of the mass of the body and its velocity.)

    The matter of momentum is extremely important in the handling of fast/heavy jet aircraft, and the effectiveness of the aircraft's handling is of greatest importance on the landing approach, where its response to changes in its flight path is much slower and sudden changes are virtually impossible. Therefore jet aircraft require intelligent anticipation to stop unwanted momentum-fueled excursions, as opposed to the light piston aircraft days when it was acceptable to allow an error in our flight path to occur before applying a correction. Now because of the effects of momentum in heavy jet aircraft, we have to control our flight path more closely at a subconscious level to adjust to the required flight path to maintain the projected flight path in a much more preemptive manner.

    Jet aircraft have been designed to be able to exert greater control forces in both direction and magnitude to reduce the effects of momentum to an acceptable level. However, the jet aircraft's momentum is still a significant factor, especially on the landing approach, and the pilot should always be aware of it.

2. *Speed stability*

    Speed stability is the behavior of the speed after a speed disturbance at a fixed power setting. The behavior of an aircraft's speed after it has been disturbed is a consequence of the drag values experienced by the aircraft frame.

    The jet aircraft's speed stability is much poorer than that of the propeller-driven aircraft because of the following two main speed stability differences:

    a. The jet's recommended threshold speed (1.4V_S) tends to be in the *neutral* or *unstable* speed range.

    b. Thrust changes with speed which improve the propeller-driven aircraft's speed stability are absent from the jet aircraft.

    Speed stability is of greatest importance at low airspeeds on the approach. Here the jet aircraft's threshold speed is only just above VIMD and therefore only just in the stable speed region. This is a matter of concern because the jet aircraft could easily slip below VIMD and therefore into an unstable speed region during the most critical phase of flight. Here the aircraft is sliding up the back end of the drag curve, where thrust required increases with reducing speed. This is so because the drag values increase faster than the lift values, resulting in poor speed stability and a deteriorating lift/drag ratio which, if not balanced with increased thrust, will result in a tendency to progressively lose speed and to enter into a sinking flight path profile.

    However, the propeller-driven aircraft's recommended threshold speed is considerably greater than its minimum power required speed. Therefore the propeller aircraft is in the stable speed region with a large margin for error.

3. *Wing lift values*

    a. The jet aircraft's swept wing is designed to achieve a higher airspeed in the cruise configuration, but the consequence is that it has a poor coefficient of lift (CL), especially at low airspeeds on the approach.

Because the sweep of the jet's wing involves an apparently lower sensed airspeed it follows that if all the other parameters are kept constant, the swept wing will produce less lift than the straight wing on the propeller aircraft for a given true airspeed.

The swept-wing jet aircraft's lift is restored by increasing its angle of attack, which accounts for its rather nose-high attitude, which is especially evident on the approach.

  b. The straight-wing propeller aircraft has a rather large safety margin between its maximum CL angle of attack and its normal approach angle of attack, which lends itself to its rather stable approach configuration. However, because the jet aircraft's higher approach attitude is at or very close to its maximum CL angle of attack, it is only just stable. Any slight increase in angle of attack will produce a faster increasing drag penalty than the lift gained, resulting in a high sink rate of the aircraft.

4. *Engine response rate—acceleration and deceleration times*

On the approach where procedure margins are tight, it is important that the engine respond quickly to any input the pilot makes as she or he endeavors to maintain the aircraft's airspeeds and /or attitude on the approach. However, the jet engine has a poorer response time than the piston engine propeller installation on the approach, which is a critical consideration to the pilot.

  a. *Acceleration response time*

The piston engine propeller installation has a constant speeding ability that keeps the engine turning over at an rpm which is a compromise between the approach and balked landing power conditions, and varying the boost pressure alters the power. To increase power quickly, the boost pressure is increased, the propeller "coarsens off," and the demanded increase in thrust is quickly delivered in about 1 to 2 s. Therefore the propeller installation is extremely efficient at slow airspeeds, especially on the approach because of its quick response time to a demanded change in thrust.

The jet engine's best acceleration response time occurs when the engine is operating at its most efficient, i.e., at a high rpm, of approximately 90 to 95 percent. However, the jet engine operates at a lower rpm setting than this on the approach and therefore has a slower acceleration response time, known as the lag. The acceleration response of the engine from an approach rpm setting falls into two categories:

(1) Approach rpm below the rapid acceleration rpm

(2) Approach rpm at or above the rapid acceleration rpm. The engine's rapid acceleration rpm is the speed above which the engine can respond to a rapid acceleration demand to full thrust without causing distress to the engine, and it is usually around 75 to 80 percent. The acceleration response of the engine from a normal approach rpm that is at or above the rapid acceleration rpm to full thrust can be achieved in about 2 to 3 s. However, from an rpm setting below the rapid acceleration rpm speed, which can be easily experienced on an approach, this could take as long as 6 to 8 s on some jet engines.

NB: The engine's operating cycle is generally inefficient, and the engine's acceleration response time is limited to protect the engine from overfueling and causing it to overheat and surge, until it reaches the rapid acceleration rpm speed.

The acceleration time delay is of greatest importance on the approach, landing, and overshoot stages of flight, and it can be seen that it is vital to maintain the engine at or above the rapid acceleration rpm, in case a rapid acceleration to full thrust is required. Any delay in the delivery of the engine thrust could prove dangerous during the approach, particularly when trying to salvage an increasing sink rate, and especially so when you consider the cumulative effects of momentum and the CL maximum already being demanded by the wings. Therefore, don't let jet engines run down below their rapid acceleration rpm value.

b. *Deceleration response time*

When the throttles are closed to idle on a propeller aircraft, a reduction in power/thrust can be delivered just as quickly as it increases power/thrust. The propeller produces drag, or negative thrust, when the constant-speed unit cannot "fine off" the blade angle any further, and the higher the rpm, the higher the drag. The ability to produce drag from the propeller is useful, especially on the approach, because just as it is necessary to accelerate quickly, under certain conditions of flight, it is equally as necessary to be able to lose speed and height quickly. On a jet engine aircraft, however, when the thrust levers are closed to idle, the engine is still producing a forward thrust force. This is a disadvantage on the approach when it is necessary to lose speed and/or height quickly.

5. *Slipstream effects*

Both the propeller and the jet engine produce thrust by accelerating a large mass of air rearward. However, the displaced air then interacts with the aircraft's airframe in differing ways with different performance effects which are particularly relevant on the approach. The large mass of air accelerated rearward by the propeller passes over a comparatively large percentage of the wing area, which immediately produces extra lift. This is of particular importance on the approach, when a rapid response is often required to halt a sinking flight path, or to initiate a go-around, etc.

The large mass of air accelerated rearward by the jet engine is largely discharged clear astern and does not pass over the wing. Therefore the jet engine does not gain any significant lift bonus with an increase in power.

6. *Power on stall speed*

The main and dominant influence of the power on stall speed is the slipstream over the wing, which improves the propeller aircraft's stall speed but not the jet engine's stall speed.

The propeller slipstream over the wing increases the lift, which thereby reduces the aircraft's effective weight, which reduces the stall speed. Remember, the stall speed varies with weight. A power on stall speed can be significantly less than the power off stall speed on a propeller aircraft.

However, because the jet engine has no slipstream over the wings, it has no significant suppression of the power on stall speed, and this restricts the jet engine aircraft to higher airspeeds on the approach, which thereby makes the handling of the aircraft more sensitive.

Again when they are taken into context with the other jet aircrafts' limitations (i.e., sink rate momentum, slow engine acceleration response rate, etc.), the safety margins are even more restrictive for the jet aircraft when compared with the propeller aircraft.

NB: The jet aircraft does have a small reduction in the stall speed when the power is increased, because of an increase in the vertical component of the thrust, which effectively reduces the aircraft's weight. This also applies to the propeller-driven aircraft.

To summarize, the slower/lighter propeller aircraft is much more stable, responsive, and controllable on the approach than the jet aircraft. The pilot of the jet aircraft is much worse off in all the six crucial areas covered. The three engine-related areas all stem directly from the loss of the positive effects of the propellers, and the three aerodynamically related areas are worse for the jet aircraft because its design for greater cruise performance and size compromises the slow airspeed qualities that are common to its approach configuration.

These areas singularly or cumulatively make the jet aircraft's safe landing approach profile much more marginal than that of the piston engine/propeller aircraft, especially when you consider the requirements of the aircraft's handling qualities when salvaging an increasing sink rate.

Differences between a Jet and a Propeller Aircraft—Engine (Especially Performance)
The differences between a jet/gas turbine engine and a piston engine can be categorized into the following five main areas:

1. *Mechanical design, efficiency, and reliability differences*
   Obviously the designs of the gas turbine and piston engine are different. However, although they accommodate the same working cycle, how each achieves the working cycle and its properties differs. The jet is more reliable and mechanically efficient than the piston engine.

   The piston engine has high mechanical losses due to the heavy weight of the reciprocating parts (pistons, etc.) and is unreliable due to the highly stressed mechanism of the variable-pitch propeller and associated reduction gears of the piston engine. In contrast, the jet/gas turbine engine is mechanically efficient because of its minimal rotating parts, and it is very reliable because of its low mechanical stress and the simplicity of its overall design. Also the jet/gas turbine engine's reliability does not deteriorate as the engine's size is increased, unlike the piston engine whose reliability varies inversely with the number of cylinders. In general, the jet engine's life span is more than double that of the best piston engine.

2. *Working cycle/airflow differences*
   The working cycle of the jet/gas turbine engine is similar to that of the four-stroke piston engine. Both engine cycles show that in each instance there is induction, compression, combustion, and exhaust. For the piston engine these processes are intermittent, as only one stroke is utilized in the production of power, the others being involved in the charging, compression, and exhausting of the working fluid. In contrast, the jet/gas turbine engine's individual strokes occur continuously, thereby eliminating the three "idle" strokes and thus enabling more fuel to be burned in a shorter time. Hence it produces a greater power output for a given size of engine.

   This is achieved because of the gas turbine's compressor, combustion chamber, turbines, etc., being arranged in series; and this design allows for a constant airflow pressure to be maintained through the gas turbine's cycle. In essence, the gas turbine engine is a continuous power stroke engine.

   Due to the continuous action of the turbine engine, and because the combustion chamber is not an enclosed space, the pressure of the air does not rise during combustion, but the air volume does increase. This process is known as *heating at a constant pressure* and allows the gas turbine engine to use low-octane fuels and light fabricated combustion chambers because it does not have to withstand fluctuating pressures. However, the piston engine has an enclosed combustion chamber, and its four-stroke cycle causes the pressure of the air to rise during combustion, which requires heavy, durable combustion chambers and the use of high-octane fuels to withstand the high-pressure fluctuations. In essence, the gas turbine's combustion occurs at a constant pressure, and the piston engine's combustion occurs at a constant volume.

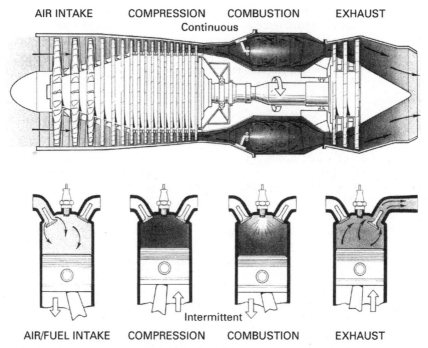

Figure 47   Gas turbine and piston engine working cycles. (*Reproduced with kind permission of Rolls-Royce Plc.*)

3. *Power output differences*

The jet/gas turbine engine can produce enormous power amounts, say, 20,000 to 80,000 lb, due to its internal aerodynamic efficiency which increases compression ratios, air mass, and cycle temperature that the aircraft requires for its high-altitude and high-speed operating conditions.

An important parameter of power is air mass flow in unit time. The piston engine is limited in the amount of air intake it can handle because of its limited cylinder size, and engine power is further limited because of propeller compressibility losses. This means it can only produce a fraction of the power produced by a jet engine, say, 1000 lb, under the same conditions.

Therefore the piston engine fails to produce the high power required in a single unit to meet its optimum required number of power plants per aircraft, whereas the jet succeeds. The power/thrust produced by the engine can be measured and compared in various performance-related ways as follows:

a. *Thrust to unit (power plant) weight ratio*

The piston engine produces a low thrust-to-unit-weight ratio whereas the jet/gas turbine produces an ever-improving high thrust-to-weight ratio. Obviously a better thrust-to-unit-weight ratio is desirable.

b. *Thrust to (power plant) frontal area*

The piston engine produces a low thrust output per square foot of frontal area, and the jet produces a much higher thrust output per square foot of frontal area. This is so because the development of the axial-flow compressor has resulted in smaller frontal area, as the development of the jet engine has been lengthwise and not radial. Therefore the jet/gas turbine has a smaller frontal area, which improves the aircraft's profile drag penalty.

The fan-type gas turbine engine, however, has a larger frontal area than the early jet engine, but the associated high bypass ratio substantially increases the engine's thrust which maintains the high thrust-to-frontal-area ratio.

c. *Propulsive efficiency differences*

The propeller's efficiency falls off rapidly above 0.5 Mach (due to compressibility losses), because the blade tip's revolution speed becomes sonic. In contrast, the jet/gas turbine's propulsive efficiency improves as speed increases, as the fan blades do not suffer from compressibility losses due to the shrouding of the engine's casing, and therefore they can rotate at a speed greater than the speed of sound. Therefore the jet/gas turbine engine's overall efficiency is greater than that of the piston engine.

d. *Cooling efficiency differences*

The jet/gas turbine engine is more efficient in terms of cooling than the piston engine, and this improves the power output efficiency of the engine. This is so because the piston engine's cooling can be expensive in terms of drag, because an inlet radial and an exit flap must be exposed to the airflow to allow a flow of cooling air through the engine. Additionally, the fuel/air ratio in a piston engine is increased under high-power conditions to provide fuel for cooling purposes in the combustion chamber, which is expensive in terms of fuel utilization. In contrast, the only power losses in the gas turbine engine arise from the use of compressed air for cooling instead of being used in the combustion and expansion processes.

4. *Specific fuel consumption differences*

The piston engine has a low SFC, in contrast to the early jet engine, which used twice as much fuel. However, this ratio is now improving with advancements in the gas turbine engine in response to demands for lower fuel consumption. (SFC = the quantity of fuel consumed, in pounds per hour, divided by the thrust of the engine, in pounds.) When the jet/gas turbine engine's fuel consumption is related to its better utilization (i.e., higher aircraft operating altitudes and speeds, and increased passenger capacity), the jet/gas turbine engine becomes more economical overall than the piston engine.

5. *Thrust-produced differences*

The thrust of an aircraft's propulsive system (engine) is the reaction to the force required to accelerate a mass of air through the system. Thrust is manifested as pressure forces on the propeller blades or on all the internal surfaces for gas turbine jet engines.

The major difference between the propeller and the gas turbine is seen in the relation between the values of mass and acceleration in the thrust equation: force = mass × acceleration.

The propeller only produces thrust by increasing the mass, because the acceleration through a conventional propeller is comparatively small and can only be increased within small limits.

In comparison, the jet/gas turbine engine produces thrust by giving a comparatively small mass of air a very large acceleration, because the mass airflow is limited by the engine's size, although high bypass fan engines have improved their mass airflow qualities.

To summarize, the difference between the gas turbine and piston engine, especially their performance, translates to a difference in their normal operating conditions.

The gas turbine engine is able to achieve high altitudes, high speeds, and high operating weights. Consequently this has improved the aircraft's range and endurance performance as well as its per-seat operating costs. Thus it has rendered the piston engine/propeller-driven aircraft obsolete on most routes. The propeller-driven aircraft is limited to low altitudes and speeds, which makes it efficient on a lot of short-sector, low-capacity routes, where the altitude and speeds flown are limited to within the propeller aircraft's performance range.

Difference between a Jet and a Propeller Aircraft—Speed Control   The speed control for both jet and propeller-driven aircraft is a function of drag against speed. The drag experienced by both aircraft has the same dynamic qualities, in that to balance drag against a speed, the thrust/power must be set to a corresponding required value. However, the value of drag against a speed is different between the jet and propeller-driven aircraft, and it is this difference that makes their speed control properties markedly different.

The jet aircraft's relatively flat drag curve over the low-speed range makes it difficult to select the correct thrust for the required speed, because of the minimal difference in thrust required values across a large speed range. Therefore it is easy to select an incorrect speed control thrust setting, and any slight difference, be it higher or lower, will set up a speed divergence which is also difficult to detect. A speed divergence due to poor speed control on jet aircraft can result in the aircraft's speed slipping into an unstable speed region where the aircraft suffers from poor speed stability.

In contrast, the propeller-driven aircraft's drag curve is markedly different across its low-speed range, and therefore it is much easier to determine and set the correct power, to balance the drag, and to deliver the desired speed. Also incorrect settings give rise to noticeable handling characteristics and more marked speed divergences which are very noticeable to the pilot, in stark contrast to the jet aircraft's qualities. Therefore the speed control of the propeller-driven aircraft is easier to manage.

Differences between a Jet and a Propeller Aircraft—Speed Stability   The jet aircraft's speed stability is much poorer than that of the propeller-driven aircraft because of the following two main differences:

1. Propeller-driven and jet aircraft have different relationships between the recommended threshold speed and the speed for minimum power, thrust/drag (VIMP/D). For the propeller-driven aircraft the recommended threshold speed is considerably greater than the minimum power required speed.

Therefore the aircraft is in the speed-stable area; i.e., less speed requires less power, etc.

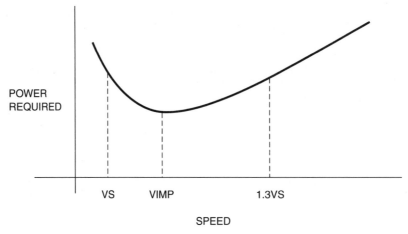

Figure 48 Location of significant speeds on the propeller-driven aircraft drag curves.

For the jet aircraft the recommended threshold speed (1.3VS) can actually be lower than the minimum thrust required/drag speed. Therefore the aircraft is already in the unstable speed area; i.e., less speed requires more power, etc.

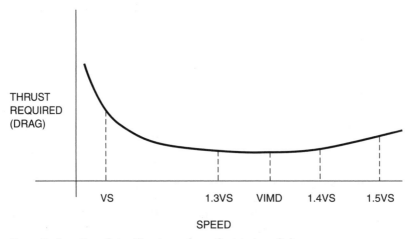

Figure 49 Location of significant speeds on the jet aircraft drag curves.

However, many modern jet aircraft have a recommended threshold speed (1.4 or 1.5VS) above the minimum thrust required/drag speed, albeit only slightly. But because of the relative flatness of the drag curve, the recommended speed remains in the *neutral* area. Here it is difficult to select an exact thrust for the required speed (speed control), and because of its close proximity to VIMD it is very easy for the aircraft's speed to slip below VIMD into the unstable speed area.

2. Thrust changes with speed, which helps to improve the speed stability on propeller-driven aircraft but does not improve it on jet aircraft. Speed stability is a

function of the airframe drag, which itself is a function of airspeed which is a function of thrust. Therefore speed stability can be said to be affected by the thrust/speed relationship of the aircraft.

On a propeller-driven aircraft, power tends to stay constant with a drag-induced change in airspeed, whereas its thrust (horizontal lift force) changes in a manner which improves its speed stability. That is, thrust tends to decrease with an increase in forward airspeed and to increase with a decrease in forward airspeed. These tendencies assist the aircraft to return to its original steady speed condition. Therefore a propeller-driven aircraft's drag curve measures speed against power required, with the minimum drag speed expressed as minimum power speed (VIMP). (See Fig. 48.)

On a jet aircraft, thrust tends to stay constant for a drag-induced change in airspeed, and this absence of thrust changes means the jet aircraft's speed stability qualities are reduced. Therefore the jet aircraft's drag curve measures speed against thrust, because thrust remains constant with a changing airspeed, and the minimum thrust speed is equal to and expressed as the minimum drag speed (VIMD). (See Fig. 49.)

Therefore a propeller-driven aircraft's speed stability is a function of

- Airframe drag
- Thrust changes with speed

And a jet aircraft's speed stability is *only* a function of airframe drag.

Differences between a Jet and a Propeller Aircraft—Stall Speed The stall speed of a piston engine propeller aircraft is generally a slower speed than that of the jet aircraft. Also the range of the stall speeds is much larger for the jet aircraft than for the piston engine propeller aircraft.

An aircraft's stall speed will vary with a number of parameters; but for any given aircraft the two most significant are weight and configuration. The stall speed is higher as the aircraft's weight increases, and will vary with configuration. That is, a reduction in the effective wing area and coefficient of lift (CL) because of the retraction of trailing-edge and leading-edge high-lift devices will increase the stall speed. Conversely, the opposite is true: With a reduction in weight and an increase in the CL due to a change in the configuration with the high-lift devices deployed, the stall speed will be reduced. This translates to the following two stall speed differences:

1. *Stall speed*

    The piston engine propeller aircraft typically has a lower weight than the jet aircraft and has a good CL at low speeds. Therefore the piston engine propeller aircraft has a low stall speed.

    By comparison the jet aircraft typically has a higher weight than the piston engine propeller aircraft and with a poorer CL at low speeds. Therefore the jet aircraft has a higher stall speed, in comparison to the piston engine propeller aircraft.

2. *Stall speed range*

    The variation or range of the stall speed for the jet aircraft is much greater than that for the piston engine propeller aircraft, because the jet aircraft has the ability to

    a. Experience a large weight change during a single flight *b.* Significantly increase its CL at low speeds by deployment of its sophisticated wing leading-edge high-lift devices

       The variation of the stall speed with different weights and configurations for a jet aircraft can be quite marked, as opposed to that for the piston engine propeller aircraft which has limited weight and configuration changes that result in a smaller stall speed range.

Because the stall speed is a *reference speed* for the calculation of any scheduled operating speed, say, V_2 or VAT, it follows that these speeds are also affected by weight and configuration changes. Therefore it becomes important for the pilot to know his or her aircraft's weight, so that the correct reference speeds are used, especially during the critical phases of flight such as the takeoff and landings. And because the jet aircraft has a large stall speed range, it is especially true for the jet pilot.

Differences between a Jet and a Propeller Aircraft—Wing Performance   The performance margin on a propeller aircraft's straight wing is greater than that of the jet aircraft's swept wing, especially when contaminated. The explanation for the improved relative performance on a propeller-driven aircraft is that the wing is unswept and there is a high-energy airflow (prop wash) over the upper surface of the wing which partly offsets some of the contaminant effects. On a jet, the available performance margin with a contaminated wing is practically zero, for two main reasons. First, there is no upper surface high-energy flow available. Second, the wing sweep decreases the amount of lift generated for a given angle of attack relative to the straight wing, so that higher angles of attack are required from the swept wings, which in turn results in greater performance sensitivity to wing contamination.

Differential GPS   The differential global positioning system (GPS) is a more accurate global positioning system than a normal GPS. It applies a correction factor called the *differential correction* to eliminate the two most significant errors in a normal GPS:

1. Selective availability (U.S. DoD downgrading of the normal GPS for civil users)

2. Ionospheric errors

GPS signals are received at a ground installation, which has been accurately surveyed. The ground installation then computes the difference between its known position and its position according to the GPS (known as *differential position error*). It then sends a differential correction factor to any aircraft within 70 nm using an ACARS link, which enables the aircraft's onboard GPS navigation computer to correct its own normal GPS-derived position into a refined *differential GPS position* which is accurate to within 1 to 3 m. Therefore differential GPS has the potential accuracy to be suitable for precision landing approaches, i.e., global landing systems (GLSs).

Differential Spoilers   The difference between differential and nondifferential spoilers lies in how they provide lateral roll control when already extended as speed brakes. Differential spoilers will extend farther on one side and retract on the other in response to a roll command; and when it is already fully extended as a speed brake, the spoilers will remain extended on one side and retract on the other side in response to a roll command.

Dihedral   This is the upward inclination of the wing from the root to the tip.

Direct Lift Control   This is provided by the elevator/stabilizer. The elevator and stabilizer are aerofoils, which by their position create an upward or downward balancing force that controls the direct lift force from the main aerofoils (wings), thus determining the attitude of the aircraft around the lateral axis.

Direct Reading Compass   *See* Magnetic Compass Instrument.

Directional Indicator (DI) Instrument   The directional indicator is a gyroscope that displays the aircraft's heading using a compass rose display. The directional indicator consists of the following:

1. A tied gyro.
2. The gyro rotates about the earth's horizontal axis.
3. Two gimbals.
4. Three planes of freedom.
   a. The gyro's spin axis
   b. Pitch and roll axis of the gimbals
5. The gyro axis is aligned (direction) to true north.

The gyro is mounted inside the gimbals, and the inner gimbal allows the gyro to rotate as a precessed force about the fore and aft axis of the aircraft. The outer gimbal, which is fixed to the instrument case, allows the gyro to turn about the horizontal plane to maintain its correct orientation. The indicator scale is fixed to the outer gimbal, and the aircraft effectively rotates about the outer gimbal as it turns to indicate the heading from a pointer attached to the aircraft, which is in front of the display scale on the outer gimbal.

The directional indicator gyro can be either vacuum- (air-) or electrically driven. The rotor is tied to its erect position; namely the aircraft's vertical, by streams of air striking the buckets on the rotor asymmetrically, when the rotor is not erect. This produces a sideways force, which precesses the rotor to reerect a vacuum- (air-) driven gyro.

Electrically driven directional indicators use the aircraft's electrical supply (usually DC battery power) to rotate the gyro about its spin axis and to keep the gyro erect. A slaving knob on the instrument face is used as an external force to align the DI (spin axis) with the magnetic compass heading.

The rigidity of the directional indicator gyroscope gives steadier heading information than the magnetic compass, which suffers from turning and acceleration errors.

Directional Indicator (DI) Instrument—Errors   The directional indicator suffers from the following:

1. *System failures*
   The vacuum- (air-) driven DI instrument will respond slowly to changes in heading or indicate incorrectly if the gyro is not up to speed as a result of a low air supply to the gyro.
   This is indicated by
   a. Suction gauge reading below normal
   b. Possible warning flag on the instrument face
   Electrically driven DI instruments will respond slowly to changes in heading or indicate incorrectly if the gyro is not up to speed as a result of a reduced

electric power supply to the gyro. This is indicated by a warning flag on the instrument face.

2. The DI suffers from a *total wander (drift) error,* which is the cumulative effect of all the gyro wander (drift) occurrences, that results in the DI being misaligned with magnetic north. Therefore displayed readings will be incorrect.

   a. Real wander is due to the instrument friction and imbalances.

   b. Apparent wander is due to the earth's rotation.

   c. Incorrect latitude nut correction. If the aircraft is at a latitude different from the latitude that the latitude nut has been set for, then an apparent wander error will exist.

   d. Transport wander is due to the aircraft's movement away from the global position that the gyro is aligned against.

That is,

Total wander (drift) = real wander ± apparent wander

$$\pm \text{ latitude nut correction} \pm \text{ transport wander}$$

The application of these individual errors (+*ve* or −*ve*) in the total wander formula is a consequence of the aircraft's relative position in either hemisphere.

Directional Stability This is the tendency for the aircraft to regain its direction (heading) after the aircraft has been directionally disturbed (e.g., an induced yaw) from its straight path. This is achieved naturally because the fin (vertical tailplane) is presented to the airflow at a greater angle of incidence, which generates a restoring aerodynamic force.

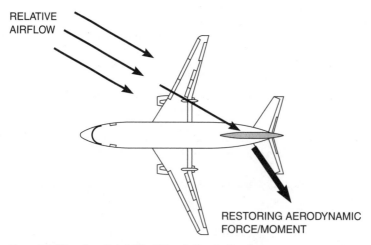

Figure 50 Directional stability following an induced yaw.

Disposable Load Disposable load is the weight of the payload and fuel.

$$\text{Disposable load} = \text{TOW} - \text{APS}$$

Diversion—Decision The most important question to ask in an emergency given two diversion aerodromes is, Which aerodrome is the quicker to get to?

Dive—Inverted   *See* Inverted Dive.

Dive—Spiral   *See* Spiral Dive.

Dive—Steep   *See* Steep Dive.

Diving (Deep-Sea)   You are not allowed to fly for 12 h after swimming/diving with compressed air. You are not allowed to fly for 24 h after swimming/diving with compressed air to a depth of 30 ft or more. These times are general guidelines, and some authorities/companies have greater time limits, that is, 24 and 48 h. The motivation for these restrictions is to protect against *decompression sickness.*

When diving with compressed air, you take in liquid nitrogen, which lingers in the body tissues. When you are flying at altitude, the lower air pressure allows this nitrogen to come out of its liquid solution as bubbles. These bubbles get into a person's joints, causing severe aching pain. This is known as the *bends.*

Flight crews have been known to experience decompression sickness at cabin altitudes as low as 6000 ft. Therefore it is very important that these guidelines be adhered to, so that the liquid nitrogen has time to dissolve out of the human system prior to flying.

DME—Arc Procedure   A DME arc is based on a navigation aid, e.g., a nondirectional beacon (NDB) or VOR, and is used to intercept a final approach track. It is flown by maintaining two parameters:

1. A constant DME reading from the aid on which the arc is based

2. The aid at a constant 90° off the aircraft's heading

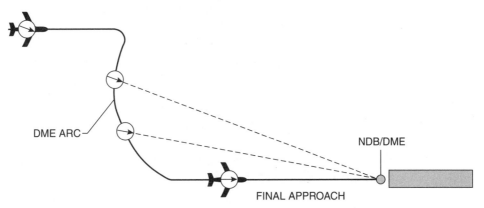

Figure 51   DME arc.

DME (Distance-Measuring Equipment)   DME is a form of secondary radar that gives a continuous distance readout, in nanometers, of the slant range to a ground station. DME operates in the UHF band, from 962 to 1213 MHz, and uses PON transmissions with line-of-sight propagation paths.

DME consists of an onboard aircraft interrogator and a ground beacon transponder (which is opposite to SSR, where the ground equipment is the interrogator and the aircraft equipment is the transponder). The aircraft's interrogator initiates the exchange by transmitting a stream of pulses to the ground station, which then retransmits them back to the aircraft. The time delay between sending and receiving these pulses is converted to a range/distance. Certain measures are built into the system to ensure that the transmitted and received pulses are correctly associated:

1. The aircraft identifies its own pulses, not other aircrafts' (because DME pulse trains are unique to each aircraft), by using a random pulse recurrence frequency (PRF), i.e., the number of pulses per second, also known as *jittering*. So the chances of two pulse trains being identical is effectively zero.

2. The aircraft also distinguishes between retransmitted pulses from the ground transponder and any reflected pulses of the ground surface, because the ground transponder retransmits at a different frequency, namely, 63 MHz apart from that of the interrogator. In total this system makes up 252 paired DME frequencies, which are called *channels,* in the UHF band.

3. The ground equipment will not be triggered by other non-DME UHF transmissions because it will only reply to pairs of pulses separated by 12 μs, which is unique to aircraft DME interrogator signals.

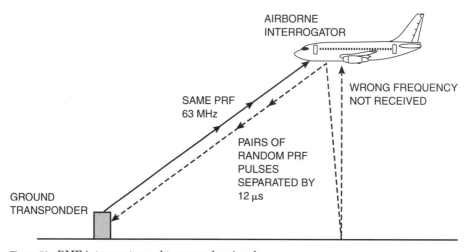

Figure 52   DME interrogator and transponder signals.

The maximum range of a DME beacon is a function of the height of the aircraft in question, and it can be calculated by using the formula

$$\text{Range (nm)} = \sqrt{1.5 \times \text{aircraft height (ft)}}$$

The accuracy of all DME installations, as required by ICAO, has to be within $\frac{1}{2}$ nm or 3 percent, whichever is greater, in slant measurement.

The pilot generally selects the DME automatically when she or he selects a VOR or ILS frequency, using the VHF NAV radio. *Note:* Most DME is paired with either a VOR or a localizer frequency. For a VOR/DME to be associated/paired with the same frequency and ident, they must be less than 100 ft (30 m) apart if used as a terminal aid, or less than 2000 ft (600 m) apart if used for any other purpose, such as en route airway aid. If a DME/VOR fails to meet these criteria, but the DME is within 6 nm of an en route VOR, they might be given similar idents. Also beacons that are not associated but are considered to be useful to each other may still be frequency-paired.

With an ILS the DME beacon is normally sited at or very near to the runway threshold and is usually paired with the localizer frequency so that it is automatically selected with the ILS. DME distance is usually displayed in the cockpit as a digital readout. In public transport aircraft there are normally two DME systems, and both are usually displayed on the HSI or the RMI. In all cases the DME has a memory function that allows the range to continue to count down at the same rate in the event that the signal is temporarily interrupted. However, after 8 to 10 s either an off flag will appear on the DME display or the range will not be displayed. Some airborne DME is also capable of computing the rate of change of DME distance, called the *rate of closure,* which is displayed on the DME cockpit. If it is assumed that the slant distance is equal to the horizontal ground distance, and that the aircraft is tracking either directly toward or away from the DME ground beacon, then the rate of closure display represents *ground speed.* Some DME indicators can further display "time to the beacon" in minutes, by computing ground speed with the DME distance.

Distance/range information is an important navigational indication, especially when used in tandem with bearing information, such as the following:

1. A paired VOR/DME will give a very good en route position fix and/or distance to run to a waypoint, e.g., the radial bearing from the VOR and the DME distance along the radial.
2. Many ILSs have their localizer frequency paired with a DME beacon sited close to the runway threshold. This gives very useful terminal procedure information:
   a. The distance to run to the runway threshold during an ILS or localizer approach.
   b. A DME arc procedure prior to intercepting the final approach track, by maintaining an approximate constant distance reading while keeping another bearing aid (e.g., NDB) approximately 90° abeam the aircraft nose.
3. By itself, DME range information will give a circle position line around a ground beacon.

The capacity of a DME's ground equipment is a function of the number of pulses per second it can handle. The capacity for a single DME ground station is normally 2700 pulses per second. Therefore because an aircraft interrogator transmits on average 150 pulses per second when it is searching for a DME ground transponder, this means that only 18 searching aircraft can search for the same facility at the same time. However, when the aircraft interrogator is locked onto the DME ground station, it only transmits about 24 pulses per second, which means that up to 112 locked-on aircraft can use the facility at the same time. In reality, a DME beacon has both searching and locked-on aircraft, and in this mixed manner an average of 100 aircraft are served by the DME ground transponder at a single time. If, however, the demand on a DME ground transponder is greater than its capacity, the DME station is said to be *saturated,* and it will only answer the strongest signals, not the nearest aircraft.

The only error of the DME is that it measures a slant range and not a ground range, known as *slant range error.* This is particularly evident when close to the beacon, i.e., when the ground distance is less than the height of the aircraft. In this area DME information should be disregarded; e.g., when you are passing directly overhead a DME beacon, the DME indicator will show the height of the aircraft in nanometers.

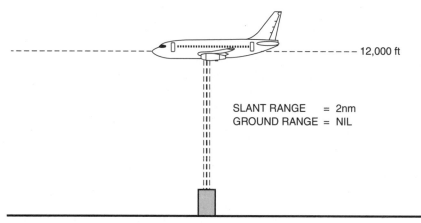

Figure 53   Slant error passing overhead a DME station.

However, away from the beacon the difference between slant and ground range is very small, and for most practical purposes the DME distance can be considered to be correct, as the following example illustrates.

$$(\text{Ground range})^2 = (\text{slant range})^2 - (\text{aircraft height})^2$$

$$= 50 \text{ nm}^2 - 2 \text{ nm}^2$$

$$= 49.96 \text{ nm}$$

$$\text{Slant range} = 50 \text{ nm}$$

Additional operational misuse is a pilot error and cannot be considered a DME system error, e.g., reading a DME display as a direct track to a beacon, when the aircraft is flying abeam the beacon.

Doppler Effect   The Doppler effect is the change of frequency between the transmitted and received signal, known as *Doppler shift,* due to the movement of the transmitter. This effect is present in radio waves and, in particular, radar.

With a static transmitter and a static receiver, the received frequency is the same as the transmitted frequency. If the transmitter is moving toward the receiver, more cycles are received every second; i.e., as the transmitter moves toward the receiver, the frequency increases. If the transmitter is moving away from the receiver, fewer cycles are received every second; i.e., as one moves away from the receiver, the frequency decreases.

The classic example of the Doppler effect is the change in pitch of a train's whistle as it passes a stationary observer.

Doppler System   Doppler is a self-contained onboard radio/radar navigation system, based on the Doppler effect principle, which operates in the 8.8- to 13.2-GHz frequency bands and is used to mathematically calculate the ground speed of the aircraft.

The first Doppler equipment developed was the single-beam system, which calculated ground speed only, while later Janus Doppler systems calculated the following:

■ Ground speed

- Drift

- Aircraft position

However, Doppler has been largely replaced today by satellite navigation and inertia navigation systems.

Doppler System—Janus Array   The Janus array system (Janus means "looking in two directions at the same time") is an improved development of the single-beam system. The Janus Doppler system measures

1. *Drift*

A Janus system uses a moving aerial that is initially aligned with the aircraft's nose and four beams that are not only depressed but also skewed away from the aerial centerline. To calculate drift, the beat frequency between the opposing pair of beams is compared; i.e., front left and rear right beams are compared with the front right and rear left beams. If there is any drift, one pair of beams will have a greater beat frequency than the other. The aerial is then rotated until the beat frequencies are the same, at which point it must be aligned with the aircraft's track. The angle between the aircraft's nose and the aerial represents the drift, which is displayed on a cockpit instrument.

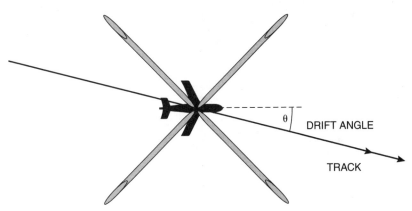

Figure 54   Janus aerial array and beams.

2. *Ground speed*

The Janus system measures the ground speed of the aircraft in a similar way to the single-beam system. However, it receives Doppler shift information from all four of its beams, and its computer uses the following formula to calculate the ground speed:

$$\text{Ground speed} = \frac{\text{beat frequency} \times \text{wavelength}}{4 \times \cos (\text{beam angle of depression}) \times \cos (\text{beam screw angle from aerial centerline})}$$

NB: The beat frequency between its opposing beams is twice the Doppler shift of a single beam as one faces forward and one rearward. Remember that a single beam experiences a double Doppler shift; therefore a pair of opposing beams will experience a quadruple Doppler shift. Ergo, the Doppler shift is divisible by 4, in the ground speed formula.

The Janus system also eliminates ground speed pitch and vertical speed errors.

3. *Aircraft position*

More sophisticated Doppler computers calculate the aircraft's position as well as display ground speed and drift, given certain manual inputs such as initial position and heading.

Doppler information is displayed in the cockpit on a gauge with drift and ground speed readings and a separate gauge with distance traveled information. Overall the accuracy of a Janus Doppler system is no worse than 1° left or right of track, and within 1 percent of distance traveled. However, all Doppler systems should be updated every 30 min with an independent, more accurate fix.

The errors and limitations associated with a Janus array Doppler system are as follows:

1. *Sea movement error.*   Because Doppler equipment gives ground speed relative to a surface, any movement of the surface will cause inaccuracies. This can happen over areas of water where there is significant tidal movement, i.e., sea movement error. However, this error is only transitory, while the Doppler beam is reflected off this type of surface.

2. *Reduced Doppler shift over water error.*   Because the sea is much smoother than land, less energy is reflected back from the edges of the beams, and the Doppler shift is reduced. The computer interprets this incorrectly as a ground speed reduction. However, because this error is predictable, it can be compensated for. A *land/sea bias* switch is fitted which, when selected to sea, boosts the ground speed reading by a set amount. It follows that if sea is selected over land, the ground speed will overread.

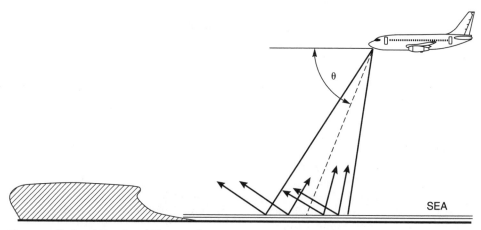

Figure 55   Reduced Doppler shift over calm water.

3. *Memory lock error.*   Over very calm seas, very little energy is reflected back to the aircraft that causes the Doppler to unlock, and therefore it will update neither ground speed nor drift. In these circumstances, most Doppler computers will continue to work on a dead reckoning position of the aircraft, based on the last wind calculated that is stored in the computer's memory. However, as with any dead reckoning procedure, it is prone to be inaccurate.

4. *Other errors.*   Thunderstorms can affect the transmitted signals, and the Doppler system can be limited in pitch and roll of the aircraft.

5. *Equipment errors.*   *Heading errors* will occur due to any stored errors of deviation and variation in the compass system that will affect drift, track, and ultimately the aircraft's position. *Latitude errors* will occur because the Doppler computer will count off 1 minute of latitude for each north and south mile traveled. But because 1 minute of latitude is only equal to 1 nm at 48°N&S, because the earth is not a precise sphere, the Doppler's calculated latitude position is likely to be slightly incorrect away from the 48°N&S latitude. *Altitude errors* occur because while 1 nm may be equal to 1 minute of arc at the earth's surface, at altitude it is less, and the computer makes no allowance for altitude/height difference. Therefore the position is likely to be slightly incorrect.

Doppler System—Single-Beam   The single-beam Doppler system is used to determine the ground speed of an aircraft. The system's forward-facing single beam transmits a radio wave/radar signals down toward the ground at a chosen angle, normally 67°, called the *depression angle*. The radio wave/radar signal is reflected off the surface and received back at the aircraft. The Doppler computer measures the difference between the transmitted and received frequencies, known as the Doppler shift. For example, a transmitted frequency of 9.0 GHz and a received frequency of 9.000006862 GHz equal a Doppler shift of 6862 Hz. Given the Doppler shift and the wavelength of the transmitted frequency, the Doppler computer can calculate the aircraft's ground speed from the formula

$$\text{Ground speed (m/s)} = \frac{\text{Doppler shift} \times \text{transmitted wavelength}}{2 \times \cos (\text{beam depression angle})}$$

NB: The Doppler shift × wavelength formula is divisible by 2 because the Doppler shift occurs twice—once at the reflected surface and once at the onboard receiver.

The depression angle is chosen to give a balance between 0°, which would give a maximum Doppler shift but no return signal, and 90°, which would give a maximum return signal but no Doppler shift as there is no vertical movement. High frequencies are used because they produce narrow beam widths and measurable frequency differences or shift.

See Fig. 56 at the top of the next page.

The disadvantage of a single-beam Doppler system is that any vertical speed of the aircraft will be shown as a Doppler shift, which the Doppler computer assumes is part of the ground speed. Also any aircraft pitch changes will cause a change in the depression angle, leading to an erroneous ground speed readout. These disadvantages severely restricted the reliability of single-beam Doppler systems.

The errors and limitations solely associated with a single-beam Doppler system are as follows:

1. *Pitch error.*   As the aircraft changes its pitch attitude, it creates a change in the depression angle of its single beam, which causes an erroneous ground speed indication. This error is virtually eliminated with the Janus array systems, as any change in Doppler shift to the front beams is canceled by the opposing change in the rear beams. Thus the beat frequency of the opposing pair of beams remains the same.

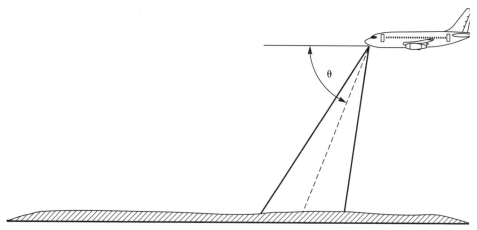

Figure 56 Doppler signal depression angle.

2. *Vertical speed error.* Any vertical speed of the aircraft will be shown as a Doppler shift of the single beam, which the computer assumes is part of the ground speed, causing an erroneous ground speed indication. This error is also eliminated with a Janus array because it affects equally the forward and rear beams, and therefore the beat frequency remains the same.

The errors and limitations associated with both Janus and single-beam Doppler systems are as follows:

1. *Sea movement error.* Because Doppler equipment gives ground speed relative to a surface, any movement of the surface will cause inaccuracies. This can happen over areas of water where there is significant tidal movement, thus the term *sea movement error.* However, this error is only transitory, while the Doppler beam is reflected off this type of surface.

2. *Reduced Doppler shift over water error.* Because the sea is much smoother than the land, less energy is reflected back from the edges of the beams and the Doppler shift is reduced. The computer interprets this incorrectly as a ground speed reduction. However, because this error is predictable, it can be compensated for. A land/sea bias switch is fitted which, when selected to sea, boosts the ground speed reading by a set amount. It follows that if sea is selected over land, the ground speed will overread.

3. *Memory lock error.* Over very calm seas, very little energy is reflected back to the aircraft that causes the Doppler to unlock, and therefore it will update neither ground speed nor drift. In these circumstances, most Doppler computers will continue to work on a dead reckoning position of the aircraft, based on the last wind calculated that is stored in the computer's memory. However, as with any dead reckoning procedure, it is prone to be inaccurate.

4. *Other errors.* Thunderstorms can affect the transmitted signals, and the Doppler system can be limited in pitch and roll of the aircraft.

5. *Equipment errors.* *Heading errors* will occur because any stored errors of deviation and variation in the compass system will affect drift, track, and ultimately the aircraft's position. *Latitude errors* will occur because the Doppler computer will count off 1 minute of latitude for each north and south mile traveled.

But because 1 min of latitude is only equal to 1 nm at the 48°N&S, because the earth is not a precise sphere, the Doppler's calculated latitude position is likely to be slightly incorrect away from the 48°N&S latitude. *Altitude errors* arise because while 1 nm may be equal to 1 minute of arc at the earth's surface, at altitude it is less, and the computer makes no allowance for altitude/height difference. Therefore the position is likely to be slightly incorrect.

Drag   Drag is the resistance to the motion of an object (aircraft) through the air. There are two major types of drag, *profile* and *induced,* which when combined create the total drag.

Profile drag is also known as *zero-lift drag* and comprises

- Form or pressure drag
- Skin friction drag
- Interference drag

Profile drag increases directly with speed, because the faster an aircraft moves through the air, the more air molecules (density) its surfaces encounter, and it is these molecules that resist the motion of the aircraft through air. This is known as profile drag and is greatest at high speeds.

Induced drag is caused by creating lift with high angles of attack that expose more of the aircraft's surface to the relative airflow, and it is associated with wing-tip vortices. A function of lift is speed, and therefore induced drag is indirectly related to speed, or rather the lack of speed. Thus induced drag is greatest at lower speeds, due to the high angles of attack required to maintain the necessary lift value. Induced drag reduces as speed increases, because the lower angles of incidence associated with higher speeds create smaller wing tip-trailing vortices that have a lower value of energy loss.

VIMD is the speed at which induced and profile drag values are equal. It is also the speed that has the lowest total drag penalty (VIMD = minimum drag speed). Therefore this speed also represents the best lift/drag ratio (best aerodynamic efficiency) that will provide the maximum endurance of the aircraft.

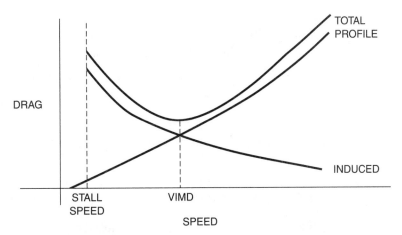

Figure 57   Total drag curve.

For higher aircraft weights, the induced drag value will be increased because the aircraft's lift [by increasing angle of attack (A of A)] must always be equal to its weight in straight and level flight conditions. VIMD is therefore also increased. However, profile drag will not vary with weight.

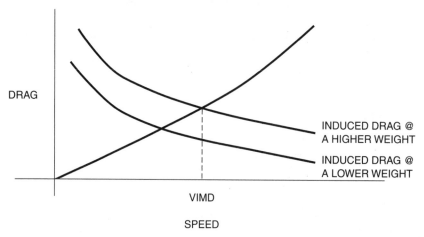

Figure 58 Increased induced drag with increased weight.

VIMD can be lowered by increasing profile drag with gear, air brake, and flap deployment, especially on the approach, that allows you to have a lower and more stable approach speed.

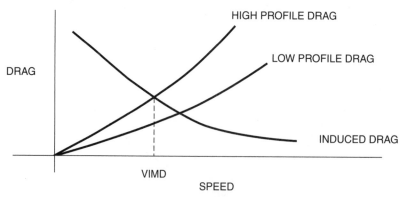

Figure 59 Reduced VIMD with increased profile drag.

Drag Curve A piston/propeller aircraft (for a piston engine/propeller aircraft, read straight-winged) has a well-defined steep profile drag curve at high speeds. This is so because the wing is not designed for high speeds; therefore, as speed increases, profile drag increases as a direct result. It also has a well-defined induced drag curve at low speeds. This is so because the straight-wing aircraft has a higher CL value, and with induced drag being proportional to lift, the lower the speed, the greater the angle of attack required to achieve the necessary lift, and therefore the greater the associated induced drag component.

It also has a well-defined bottom VIMD (minimum drag speed) point, which is due to the marked increase of profile and induced drag on either side of VIMD. Because this wing has a high CL, it is capable of a lower stall speed and also has a lower VIMD speed than that of a jet aircraft.

Flight below VIMD on a piston engine aircraft is very well defined by the steep increase of the drag curve, in flight as well as on paper. Speed is not stable below VIMD, and because of the steep increase of the curve below VIMD it is very noticeable when you are below VIMD. That is, below VIMD, a decrease in speed leads to an increase in drag that causes a further decrease in speed.

The drag curve on a jet aircraft is the same in composition as that of a piston aircraft in that it comprises induced drag, profile drag, and a VIMD speed, but its speed-to-drag relationship is different. This is so because the jet aircraft has swept wings, which are designed to achieve high cruise speeds, but as a consequence has poorer lift capabilities especially at low speeds. Therefore because profile drag is a function of speed and induced drag is proportional to lift, the drag values against speed are different on a jet/swept-winged aircraft. The three main differences are as follows:

1. Total drag curve is flatter.
    a. Profile drag is reduced especially against higher speeds (flatter drag curve) because the swept wing is designed to achieve higher speeds. This is done by reducing the resistance to higher speeds, the resistance being profile drag. Thus the design of a swept wing is to reduce profile drag, especially at high speeds. This makes a more even drag penalty across the speed range, resulting in a flatter drag curve on a jet aircraft.
    b. Induced drag is reduced (flatter drag curve) because the swept wing has very poor lift qualities, especially at low speeds. A swept-wing aircraft cannot maintain its lift value at low speeds by increasing the angle of attack because the value of induced drag experienced (which is proportional to lift) is greater than the lift produced; therefore it is unstable. So the aircraft maintains its required value of lift, especially on the approach, by using higher speeds, thereby replacing induced drag for profile drag, which the swept-wing jet aircraft is more aerodynamically efficient against.

    These factors combined give rise to a smaller total drag range against speed, which results in a flatter total drag curve.
2. The second difference is a consequence of the first, because of the relative flatness of the drag curve, especially around VIMD. The jet aircraft does not produce any noticeable changes in flying qualities, other than a vague lack of speed stability, unlike the piston engine aircraft where there is a marked speed drag difference. (Speed is unstable below VIMD, where an increase in thrust has a greater drag penalty for speed gained thereby with a net result of losing speed for a given increase in thrust.)
3. VIMD is a higher speed on a jet aircraft because the swept wing is more efficient against profile drag, and therefore the minimum drag speed is typically a higher value.

See Fig. 60 at the top of the next page.

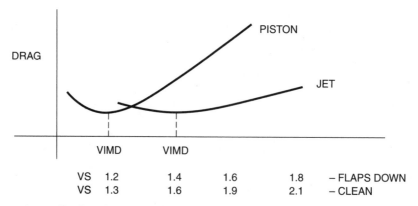

| VS | 1.2 | 1.4 | 1.6 | 1.8 | – FLAPS DOWN |
| VS | 1.3 | 1.6 | 1.9 | 2.1 | – CLEAN |

Figure 60 Total jet drag curve against piston drag curve.

A jet aircraft in the clean configuration, at a high altitude, with a speed of approximately 1.5 *VS* to 1.6 *VS* would be only just above VIMD. Therefore great care should be taken to avoid any decrease in airspeed, especially as the jet aircraft's flat drag curve means that the differences in the flying qualities are difficult to notice. If this should occur, then the jet will quietly "slip up the back end of the drag curve." Although the higher lift coefficient associated with lower speeds produces some extra lift, the drag increases faster than the lift; therefore the lift/drag ratio deteriorates, and the net result is a steeper flight path as the aircraft sinks. If you find yourself in this situation, then positive decisive action is required. Simply trying to rectify the situation with small thrust or attitude changes to increase speed is not sufficient. Therefore in this particular situation don't be afraid to give the aircraft a handful of thrust to recover.

The fact is that drag increases more rapidly than lift, causing a sinking flight path when below VIMD. The ease of slipping into this situation because of the flatness of the aircraft's drag curve and its unnoticeable handling qualities means that this is one of the most important qualities or aspects of flying a jet aircraft, which the pilot should be aware of.

Drag Devices *See* High-Drag Devices.

Drugs *See* Medication.

Dry Adiabatic Lapse Rate (DALR) *See* DALR (Dry Adiabatic Lapse Rate).

Dry V_1 Speed A dry (maximum) V_1 speed is the normal decision speed, on a runway classified as being dry, that following an engine failure allows the takeoff to be continued safely within the TODA or to be stopped safely within the EMDA.

Dutch Roll Dutch roll is an oscillatory instability associated with swept-wing jet aircraft. It is the combination of yawing and rolling motions. When the aircraft yaws, it will develop into a roll. The yaw itself is not too significant, but the roll is much more noticeable and unstable. This is so because the aircraft suffers from a continuous reversing rolling action.

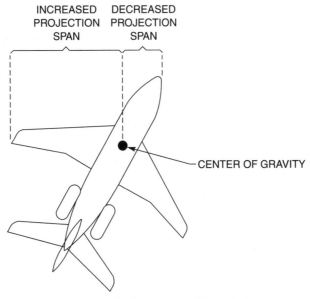

INCREASED DECREASED
PROJECTION PROJECTION
SPAN SPAN

CENTER OF GRAVITY

Figure 61   Change in effective aspect ratio/speed with yaw.

The cause of Dutch roll is the aircraft's swept wing. Roll occurs when a yaw is induced either by a natural disturbance or by a commanded or an uncommanded yaw input on a swept-wing aircraft. This causes the outer wing to travel faster and to become more "straight on" to the relative airflow (in effect decreasing the sweep angle of the wing and increasing its aspect ratio). Both of these phenomena will create greater lift. At the same time, the inner wing will travel more slowly and in effect become more swept relative to the airflow, and both of these phenomena will reduce its lift. Therefore a marked bank occurs to the point where the outer upward-moving wing stalls and losses all lift; therefore the wing drops, causing a yaw to the stalled wing, thus leading to the sequence being repeated in the opposite direction. This sequence will continue and produce the oscillatory instability around the longitudinal axis we know as Dutch roll. Pitch fluctuations only occur with an extreme degree of Dutch roll.

Depending on the design of the control surfaces, an aircraft's Dutch roll properties can be one of these:

Stable (positive)	Stable means that any oscillation will be naturally damped out.
Neutral	Neutral means that any oscillation will remain at the same magnitude and will not increase or decrease.
Unstable (negative)	Unstable means the magnitude of the oscillation will increase with each cycle by a variable amount.

The governing factor that determines whether a particular aircraft design is stable, neutral, or unstable in Dutch roll depends on its design stability. For an aircraft which has dominant directional qualities (vertical tailplane/fin), it will tend to be oscillatory-stable but spirally unstable. Let us examine this. As the aircraft yaws, the vertical tailplane becomes exposed to the relative airflow.

$$\frac{\text{Tailplane}}{\text{Fin area}} \times \text{air mass} = \text{restoring yaw force}$$

Restoring yaw force = oscillatory stable (but spirally unstable)

Therefore it is evident that the size of the fin/tailplane area determines the amount of the restoring force, which determines the oscillation stability of the aircraft. An aircraft that has dominant lateral qualities (wing span) will tend to be spirally stable but oscillatory-unstable.

Unfortunately, the designer has to trade one type of stability for the other, as both oscillatory and spiral stability cannot be accomplished in tandem. This is so because a feature which makes the aircraft stable in Dutch roll (i.e., a large vertical tailplane) will also make it unstable in spiral stability, and vice versa.

Dutch roll severity varies with an aircraft's configuration and altitude. Normally the Dutch roll will be worse at higher altitudes (due to the lower natural aerodynamic damping effect available) and usually at lower speeds (because of the lower airflow over the aircraft's fin surface, which results in a reduced aerodynamic damping ability). To minimize the possibility/severity of Dutch roll when the yaw damper is unserviceable, a typical pilot action would be to reduce altitude, reconfigure the aircraft, and maintain the cruise speed as recommended by the aircraft's checklist to gain the best natural aerodynamic damping possible.

Mechanically, *yaw dampers* prevent Dutch roll on swept-wing aircraft. A basic reason for the Dutch roll tendency of an aircraft (apart from the wing sweep, of course) is the lack of effective fin and rudder area to stop it. The smaller fin and rudder area is a design compromise that makes the aircraft, to a degree, spirally stable. Therefore the effectiveness of the fin area must be increased in some other way, to prevent Dutch roll. This is a known problem at the design stage, and the obvious solution is to apply rudder against the yaw to prevent the roll starting or building up. This is exactly what a yaw damper does. It is a gyro system sensitive to a change in yaw that feeds a signal into the rudder, which applies opposite rudder to the yaw before the roll occurs and therefore prevents Dutch roll. Yaw dampers are employed on unstable, neutral, or stable swept-winged aircraft alike. They are fully automatic with a limited range, e.g., 2° in either direction (port and starboard) which is about all that is required. The yaw damper is normally deemed of sufficient importance to warrant a backup hydraulic power system. To recover an aircraft suffering from Dutch roll, the pilot would apply the opposite aileron to the direction of the roll, assuming that the yaw dampers are not serviceable.

Although the root cause of Dutch roll is the yawing motion, application of a correcting rudder input by the pilot would normally worsen the situation. This is so because the yawing motion in the oscillatory cycle happens extremely quickly, and the pilot's reaction would not be quick enough to catch the yaw, which has already developed into a roll and dissipated. Therefore a rudder input to correct the initial yaw (which has since dissipated) would in fact aggravate the roll effect further into a sideslip. Aileron control is therefore employed because the roll cycle is long enough to allow the pilot to apply the correct opposite aileron control. A severe Dutch roll may require two or three aileron inputs to dampen the oscillation gradually.

Dynamic Pressure   *Dynamic pressure* is the pressure (over and above the static pressure) of the air molecules impacting onto a surface, caused by either the movement of a body, i.e., aircraft, through the air or by the air flowing over a stationary object. Dynamic pressure is sometimes referred to as *moving pressure*.

There are two main components that determine the amount of dynamic pressure experienced.

1. *The speed of the body relative to the air.* The faster the speed of the body, or the faster the wind speed over a stationary body, the greater the number of air molecules that impact upon the surface of a body per unit of time; therefore, the greater the dynamic pressure.

2. *The density of the air.* The density of the air varies under the influence of pressure and temperature. The density of a column of air, that is, 50,000 ft to MSL, is determined by the pressure exerted by the weight of the air pressing down from above (the effect of gravity). Therefore the air density varies over a range of altitudes, with the result that air density reduces with altitude. The density of the air for a particular pressure altitude is influenced by temperature; i.e., warmer temperatures create less dense air. Therefore less dynamic pressure is experienced as the air density decreases.

These two components are used to formulate the actual measurement of dynamic pressure:

$$\text{Dynamic pressure} = \tfrac{1}{2}\rho V^2 \text{ (TAS)}$$

where ρ represents the air density (which decreases with altitude and/or warmer temperatures) and V represents true air speed (TAS) (the speed of the body relative to the air). Dynamic pressure actually represents indicated airspeed, which itself is a function of aircraft performance. Therefore dynamic pressure ($\tfrac{1}{2}\rho V^2$) is a fundamental basis of aerodynamics.

An expression for dynamic pressure is

$$\text{Total pressure} - \text{static pressure} = \text{dynamic pressure}$$

E

EADI (Electronic Attitude Directional Indicator)   The EADI display includes the following:

1. Basic attitude information (pitch and roll) and a turn and slip indicator (yaw) received from an IRS

2. Additional attitude information, namely,
   - Flight director command bars
   - Pitch limit symbols, also known as eyebrows
   - Rising runway

3. Speed indicator
   - Speed tape (on the side of EADI)
   - Fast/slow speed indicator (speed trend)
   - Mach number and ground speed digital display

4. Navigation information
   - LNAV or localizer deviation indicator (bottom of EADI)
   - VNAV or glide slope deviation indicator (right-hand side of EADI)

5. Altitude, radio altimeter height, and decision height display

6. Autopilot, armed and engaged modes
   - Auto thrust
   - Pitch mode
   - Roll mode
   - Autopilot status

A rising runway symbol on the EADI normally becomes active at 200-ft radio altimeter, but this can vary, as it is a type-specific design feature. Zero feet is represented by the rising runway symbol reaching the base of the aircraft symbol.

Earthing   *See* Electrical Earthing.

Earth's Magnetic Field   *See* Magnetic Field—Earth.

EDR   *See* Emergency Distance Required (EDR).

Effective Pitch   This is the actual distance a propeller blade moves forward in 1 r.

EFIS (Electronic Flight Instrument System)   EFIS is a fully integrated computer-based digital navigation system, which utilizes color cathode-ray tube (CRT) types of electronic attitude directional indicator (EADI) and electronic horizontal situation indicator (EHSI).

An electronic flight instrument system has the following five main components:

1. Cathode ray tubes generate the displayed information, with light sensor units that modify the intensity of the CRT-generated display.
2. EFIS control panel provides a means to select the desired information.

3. Symbol generators receive inputs from the following:
   a. Flight management computer
   b. Weather radar
   c. Aircraft navigation and sensor systems to generate the proper visual displays
4. EADI
5. EHSI

The EFIS display has two distinct advantages over the older, mechanically driven instruments. First, it displays the same information in a clearer and more versatile manner. Second, it can bring together additional data from several different sources to present the pilot with the best possible attitude and navigation information for a particular stage of flight, on a dual or single display panel.

There is no standard color coding used by all the different EFIS manufacturers. However, in general, the following color scheme is the most common:

Green	Active or selected mode, changing conditions
White	Present situation and scales
Magenta	Command information and weather radar turbulence
Cyan	Nonactive background information
Red	Warning
Yellow	Caution
Black	Off

EGPWS (Enhanced Ground Proximity Warning System)   Initially EGPWS is only suitable for use on aircraft with digital avionics, although research is ongoing to develop an analog aircraft version. It provides a greater level of detection than a standard GPWS. For example, terrain mapping is a new feature on EGPWS. It can be shown on navigation displays by using the weather system. "Probable wind shear" aural and visual warnings can also be generated to warn of an impending possibility of encountering wind shear ahead.

EHSI (Electronic Horizontal Situation Indicator)   The EHSI typically has seven display modes:

1. Full VOR/ILS shows a 360° rose, oriented to the aircraft's magnetic heading.

2. Full NAV shows a 360° rose, oriented to the aircraft's magnetic track.

3. Expanded (arc) VOR/ILS shows a 70° head-up rose, oriented to the aircraft's magnetic heading.

4. Expanded (arc) NAV shows a 70° head-up rose, oriented to the aircraft's magnetic track.

5. Map mode shows a plan view of the flight's progress, oriented to the aircraft's magnetic track. This mode has a 70° expanded rose with the aircraft symbol at the bottom of the display.

6. CTR map mode is the same as the map mode except it has a 360° rose with the aircraft symbol in the center. Therefore it displays information behind the aircraft as well as in front.

7. Plan displays a static map of the FMC route oriented to true north, and it is commonly used to review, or step through, planned routes.

Weather radar can be typically overlaid on expanded VOR/ILS, expanded NAV, map mode, and center map modes.

Electrical Circuit   An electrical circuit consists of

1. Power source
2. Conductive material, to transfer the current
3. Item of electrical equipment
    a. Conductive material, to return the current to the power source
    b. Earth return, or the return to the power source via the aircraft's fuselage, called an earth return circuit

Electrical Conductor   Electrical conductors are materials, especially metals, whose atoms have free electrons that allow the transfer of electric energy (current).

NB: Circuit electrical conductors are usually copper metal wires in an insulator covering.

Most metals make good electrical conductors, especially copper, aluminum, silver, gold, as well as carbon.

Electrical Earthing   Electrical earthing literally means the connection of an electrical appliance or a particular electrical circuit point to the earth by a wire. The reason for doing this is safety. In the case of an appliance, earthing makes the person and the electrical appliance have the same electrical potential, i.e., earth potential or zero, so no current can pass from one to the other; therefore the person does not get a shock whenever he or she touches the appliance case. In the case of an electrical circuit, it is often useful to earth particular points to bring their potentials to a known value, i.e., zero, which enables the potential at other points to be worked out or specified. The metal chassis (fuselage) on which an electrical circuit is mounted serves as a useful earthing point for this purpose.

Electrical Insulator   Electrical insulators are materials, especially nonmetallic substances, whose electrons appear to be firmly *bound* to their nuclei; examples are glass, rubber, plastic, etc. This means there are few, if any, "free" electrons, and thus these materials are not conductors of electricity.

Nonmetallic substances make good electrical insulators, especially glass, mica, rubber, and plastic.

Electrical System—Aircraft (Basic Parameters)   The basic parameters of an aircraft's electrical system are that

1. There is no paralleling of the AC sources of power.
2. All generator bus sources have to be manually connected through the movement of a switch that will also disconnect any previous existing source.

Electricity   Electricity is the movement of electrons [electromotive force (emf) and current] that produces power. An atom has a nucleus containing at least one positively charged particle, called a *proton,* with at least one negatively charged particle, called an *electron,* in orbit around the nucleus. Particles with opposite charges attract one another, and it is this attraction which holds the electron in orbit around the nucleus. Normally the number of electrons in orbit is equal to the number of protons in the nucleus, and the atom in this state is said to be electrically balanced or neutrally charged. The atom varies only in the number of protons within the nucleus and the number of electrons in orbit around the nucleus. With some substances, mainly metals, some of these electrons are so loosely bound to their nuclei that it is relatively easy to remove them from their normal orbits. Such electrons are called *free electrons*. Freeing electrons creates electricity and is commonly accomplished either by exciting magnetic fields or by chemical action.

Electromotive force is developed by a source of electric energy, such as a generator or battery, and is measured in volts. The emf can be thought of as the electrical pressure/force of the freed electrons, which creates and maintains the flow of electrical current in a circuit. The difference in electrical *level,* called the *potential difference,* between two points in a circuit is also measured in volts (i.e., transformer). The two ideas of force and level are merely alternate ways of describing what causes the electrical charge to flow (current); therefore it is sensible to measure them in the same units. An electrical current is the flow of electric charge, carried by electrons, which can be either negative or positive. When electrons move from one conductive material to another, the material that gains the electrons becomes negative in charge while the material losing the electrons becomes positive in charge. So an electrical current can best be described as a flow of electrons. The flow of electrons (i.e., a current) depends upon the generation of an electrical force (emf) by an electric power source or from a positive difference in the voltage level between two points in a current. This causes the movement of electrons along the conductive wires and through the electrical equipment (e.g., lamps) or other circuit elements.

Electronic Attitude Directional Indicator (EADI) *See* EADI (Electronic Attitude Directional Indicator).

Electronic Horizontal Situation Indicator (EHSI) *See* EHSI (Electronic Horizontal Situation Indicator).

Elevator The conventional elevator is a hinged control surface at the rear of the horizontal tailplane (stabilizer) that is controlled by the pilot's control column. As the elevator control surface is deflected, the airflow and thereby the aerodynamic force around the elevator (horizontal tailplane) change. Moving the control column back deflects the elevator up, causing an increase in the airflow speed, thus reducing the static pressure, on the underside of the elevator control surface. In addition, the topside of the elevator becomes presented more face onto the relative airflow which causes an increase in the dynamic pressure experienced. These effects create an aerodynamic force on the elevator (horizontal tailplane) that rotates (pitches) the aircraft about its lateral axis.

That is, back control column movement moves the elevator control surface upward, producing a downward aerodynamic force that pitches the aircraft *up.* Ergo, the opposite is true. Forward control column movement moves the elevator control surface downward, producing an upward aerodynamic force that pitches the aircraft *down.*

Elevator Artificial Feel System Elevator artificial feel systems (normally duplicated) are employed on powered controls, especially the elevators. They meet the requirement of progressive feel against control surface deflection at constant speed, and against a constant angle at varying speed, based on our old friend $\frac{1}{2}\rho V^2$.

Whenever the feel on an elevator control is significantly reduced, great care must be exercised in its use. The control must be moved slowly and smoothly over minimum angles to avoid overstressing the control surface structure, but enough to maintain the flight path. Overstressing the control surface with a lack of feel is a significant problem, and for this reason turbulence should be avoided.

The best position for the cg (center of gravity) with a reduced or failed elevator feel system is *forward*. This can be accomplished by moving the passengers to the front of the cabin and/or by moving the fuel into forward tanks, if possible. This increases the aircraft's natural longitudinal stability and renders the pitch control less sensitive and feeling heavier. This makes the aircraft less responsive to small elevator movements, so the chances of the pilot's overstressing the elevator control surface are minimized, although not eliminated.

Elevator—Jam/Degraded   A degraded or jammed elevator will result in a less effective elevator maneuverability in pitch control. The best position for the cg with a degraded or jammed elevator is *aft*. This can be accomplished by moving the passengers to the rear of the cabin and/or by moving fuel to the outer wing tanks, if possible.

An aft cg position lessens the need for large pitch control demands, especially during the approach and landing flare.

If there is *no* elevator control (in other words, it is totally jammed), then the stabilizer trim can be used for pitch control. If the elevator control is reduced (degraded), then one should assess whether sufficient elevator range is still available to land safely, and it should be used to do so, if applicable. If not, then the condition should be regarded as having no elevator control.

With a degraded or jammed elevator, several actions can be taken to minimize the need for major pitch changes and/or to improve the handling and management of the aircraft:

1. If possible, move the cg rearward; this will reduce the need for large elevator angles, especially during the landing.
2. Plan a long final approach and make configuration changes, gear, and flaps earlier than usual, to allow more time to sort out the aircraft before the next change has to be made.
3. Restrict the flap angle for landing in order to reduce the flare demanded.

Elevator Reversal   This occurs at high speeds when the air loads/forces are large enough to cause a twisting moment on the deflected elevator surface to either a neutral or opposite position, which results in the sudden reversal of the aircraft's pitch attitude.

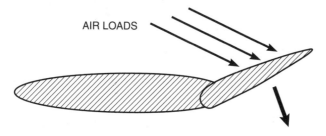

AIR LOADS

REVERSAL MOTION

Figure 62   Elevator twisting moment/reversal.

ELR (Environmental Lapse Rate) ELR is the rate of temperature change with height of the general surrounding atmosphere. The ISA assumes an ELR of 2° per 1000 ft of height/altitude gained. The actual ELR in a real atmosphere, however, may differ greatly from this; in fact, it can be zero (isothermal layer) or even a negative value (inversion).

Emergency Distance Available (EMDA) Emergency distance available [also known as the accelerate stop distance available (ASDA)] is the length of the takeoff run available, usually the physical length of the runway, plus the length of any stopway available.

$$\text{EMDA/ASDA} = \text{usable runway} + \text{stopway available}$$

Emergency Distance Required (EDR) Emergency distance required is the distance required to accelerate during the takeoff run on all engines to the critical speed V1, at which point an engine failure is assumed to have occurred and the pilot aborts the takeoff and brings the aircraft to a halt before the end of the runway or stopway, if present (i.e., RTO).

The whole emergency distance is factored by a safety margin, normally 10 percent.

NB: The use of reverse thrust in the EDR calculation differs from authority to authority, but usually it is not factored into the EDR calculation. The emergency distance (ED) is sometimes referred to as an *accelerated stop distance.*

The EDR must *not* exceed the EMDA.

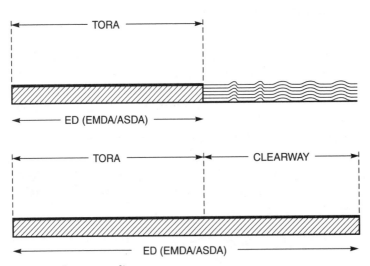

Figure 63 Emergency distance.

Emergency Frequency The emergency VHF radio frequencies are 121.50 and 243.00.

Empty Field Myopia Empty field myopia occurs when you are flying in cloud or pitch darkness; i.e., no, or very few, visual references make your eyes automatically focus to approximately 1-m distance ahead.

Engine Failure on Takeoff—Crosswind Effect   If you had an engine failure between V1 and *VR* and you had a maximum crosswind, it would be best to lose the upwind engine. This is so because the crosswind would then oppose the yawing moment of the downwind engine.

Engine—Fan   *See* Fan Engine.

Engine—Fire Detection and Protection System   *See* Fire Detection and Protection System.

Engine—Fire Drill   *See* Fire Drill—Engine.

Engine Instruments   The main primary engine instruments are usually

1. EPR gauge (thrust measurement)
2. N1 gauge (low compressor rpm)
3. EGT or TGT (engine temperature)
   Other possible primary engine instruments are
4. N2 gauge (intermediate compressor rpm)
5. Fuel flow (fuel flow indicator)
   Secondary engine instruments usually are
6. For oil—temperature, pressure, and quantity gauges
7. Engine vibration meter

Engine—Jet/Gas Turbine   *See* Jet/Gas Turbine Engine.

Engine—Piston   *See* Piston Engine.

Engine Power Monitored/Indicating Systems—Piston Engine   There are two main piston engine power monitoring/indication systems: manifold absolute pressure (MAP) and boost pressure.

1. *Manifold absolute pressure* is the U.S. indication/term for measuring the pressure in the induction system, based on 29.92 inHg for standard atmospheric conditions. For comparison, 1 lb of boost = 2 inHg.
2. *Boost pressure* is the turbocharged pressure in the induction system relative to sea-level standard pressure; it is calibrated in pounds per square inch (psi) above or below standard sea-level (SL) atmospheric pressure. The standard SL pressure reference datum is zero. For example, +4 lb indication = 14.7 + 4 = 18.7 psi.

Engine Pressure Ratio (EPR)   Engine pressure ratio (EPR) is the ratio of air pressure measurements taken from two or three different engine probes and displayed on the EPR gauge for the pilot to use as a parameter for setting engine thrust.

The EPR reading is the primary engine thrust instrument, with the temperature of the turbine stage governing the engine's maximum attainable thrust. Normally EPR on a gas turbine powered aircraft is a ratio measurement of the jet pipe pressure to compressor inlet ambient pressure, or sometimes the maximum compressor cycle pressure to the compressor inlet ambient pressure.

However, on a fan engine the EPR is normally a more complex ratio measurement of an integrated turbine discharge and fan outlet pressure to compressor inlet pressure. Therefore, the EPR is a measurement between any two or three engine parameters, and as such the parameters chosen are type-specific and have no great importance in themselves (other than the fact that they produce the EPR that reflects the engine thrust used, which itself is ultimately governed by the temperatures in the turbine section).

NB: Most modern gas turbine engines use an EPR reading as their primary engine indicator. The notable exception is the CFM 56 engine (commonly used on B737-3/4/500 and A320 aircraft) which uses an N1 gauge as its primary engine instrument. An N1 reading denotes the rpm speed of the engine fan (low-pressure compressor spool).

Engine Pressure Ratio (EPR)—Takeoff Roll   Let us assume that EPR is a measure of the jet pipe exhaust pressure (P7) against compressor inlet ambient pressure (P2). Prior to opening the throttle levers, the EPR reading will be very low, if not a 1:1 ratio. Now when the throttle levers are advanced, the EPR reading will initially decrease, because the P2 pressure increases against a constant or more slowly increasing P7 value; therefore the ratio decreases, before steadily increasing to its takeoff setting. The reason for this is twofold. First, the engine compressors, combustion, and turbine stages are in series, and this leads the engine to suffer from a slow response (or lag) to a throttle input, as the greater induced air into the engine takes time to move through the compressor, combustion, and turbine stages before it is expelled from the engine with a resultant/reaction forward force. Second, a consequence of the engine's slow response rate or lag is its effect on the EPR reading. The reading of the (P7) jet pipe exhaust pressure is taken from the rear of the engine, and the (P2) compressor inlet pressure reading is taken from the front of the engine. Therefore as the engine throttles are opened up, initially the compressor inlet air pressure will increase before the jet pipe exhaust air pressure increases proportionally, thereby creating an increased or initially more quickly increasing P2 value for a constant or more slowly increasing P7 value. This results in an initial decrease in the EPR reading. Then as the engine accelerates along its entire length, the engine turbine and compressor rpm speeds increase and the EPR reading increases steadily to its takeoff setting. For example,

$$\frac{P7\ (40\ psi)}{P2\ (38\ psi)} = 1.05\,EPR \qquad idle$$

$$\frac{P7\ (40\ psi)}{P2\ (45\ psi)} = 0.89\,EPR \qquad initial\ lag$$

$$\frac{P7\ (60\ psi)}{P2\ (48\ psi)} = 1.25\,EPR \qquad steady\ increase$$

$$\frac{P7\ (120\ psi)}{P2\ (60\ psi)} = 2.00\,EPR \qquad takeoff\ power$$

The EPR for takeoff has to be set by 40 to 80 kn (the exact speed is type-specific) for the following main reasons:

- So that the pilot is not chasing rpm needles on the takeoff roll.

■ To ensure an adequate aircraft acceleration, so that the performance calculated V_1 and V_R speeds are achieved by the takeoff run required (TORR) rotate point for the given aircraft weight and ambient conditions.

In essence, we calculate the takeoff point in terms of a speed (V_1, V_R), which relates to a specific point on the runway that ensures an adequate stopping distance, if so needed, for a given aircraft weight. For the aircraft's speed acceleration performance on the takeoff roll to meet or be better than our calculated TORR, the aircraft is dependent on its acceleration properties. Ensuring that the EPR is set by 80 kn means that the acceleration of the aircraft will bring the aircraft to its V_1 and V_R speed within its calculated takeoff run distance.

If the EPR is set before 80 kn, then the acceleration performance will be above that required, and the aircraft will achieve its V_1 speed before its calculated TORR point.

If the EPR is set exactly at 80 kn, then the acceleration performance will be as calculated and the aircraft will achieve its V_1 speed exactly at its calculated TORR point.

If the EPR is set at a speed greater than 80 kn, then the acceleration performance will be poorer than calculated, and the aircraft will only achieve its V_1 speed at a point past its TORR point. This reduces its emergency stopping distance, which if limiting could result in the aircraft being unable to stop before the end of the runway stopping distance—if you continued with the takeoff roll, which you obviously would not do.

Engine Relight Boundaries   *See* Relight Engine Boundaries.

Engine Surge   *See* Surge—Engine.

Engine—Windmill Start   *See* Windmill Engine Start.

Enhanced Ground Proximity Warning System   *See* EGPWS.

En Route Performance Limitations   The normal en route operating performance limitations for an aircraft are as follows:

1. *En route obstacle/terrain clearance with one or two engines inoperative.*

    An aircraft with one or two engines inoperative must be capable of maintaining a minimal vertical separation, say, 2000 ft, from obstacles or terrain, within its domain to any destination and alternative aerodromes.

NB: En route domain is typically 10 nm either side of the planned route or is reduced to 5 nm if the onboard navigation equipment and the en route navigation aids together facilitate accurate track keeping.

    If an aircraft is cruising at or near its all-engine maximum service ceiling and it loses an engine, it will drift down to its one-engine inoperative ceiling, because the lower total engine thrust produced is incapable of maintaining the aircraft at its all-engine altitude. It is given that thrust produces TAS that produces lift that balances weight. Initially the aircraft will have a relatively high rate of descent, but as it approaches its new ceiling, its descent gradient (profile) becomes much flatter. This is so for two reasons:

    a. As pressure altitude decreases, the air density increases, which means an aircraft is capable of producing a greater lift value and therefore more able to support the aircraft's weight.

b. During the drift down, descent, fuel is still being burned by the live engine(s), and therefore the actual aircraft weight will decrease. Remember, weight = lift ($\frac{1}{2}\rho V^2 S_{CL}$). Therefore the descent gradient becomes less marked in the descent. Obviously the sensible course of action on suffering an engine failure en route would be to turn to avoid any obstacle or to reduce the all up weight (AUW) of the aircraft. However, the regulations do not usually permit a jettisoning fuel schedule to reduce the aircraft's weight in such an eventuality. There are some exceptions, however.

The other alternative of altering the heading is not permitted because scheduling performance is a planned performance and it is impossible to make plans for an eventuality which may take place at any point along the route. Therefore only the safe procedure is scheduled, i.e., that you are unable to visually avoid the obstacle. Therefore the en route obstacle/terrain clearance must be assessed for the aircraft's one- or two-engine inoperative service ceiling. To do this, we have to consider the following:

(1) The aircraft weight required to achieve a one- (two-) engine inoperative 0 percent gradient at the minimum vertical separation limit, for example, 2000 feet over any en route obstacle.

(2) The drift-down distance from the worst engine failure point to the obstacle, to determine the maximum permissible weight at the flameout point, which achieves point (*a*).

Max. weight drift-down fuel burn weight =

one- (two-) engine inoperative 0% gradient weight

In turn this could limit the MTOW.

2. *Maximum range limit* of the aircraft.

Obviously the route must not exceed the maximum range of the aircraft, given its onboard fuel (full or not) and its cruising conditions (e.g., altitude, wind experience, AUW).

3. *ETOPS.*

At any point during the flight, the aircraft must be within its ETOPS (extended twin operations) range of a suitable aerodrome.

Environmental Lapse Rate (ELR) *See* ELR (Environmental Lapse Rate).

Equivalent Airspeed (EAS) This is RAS corrected for compressibility error. Compressibility error affects the ASI at high speeds because the air does not behave in the same way as it does at lower speeds. In fact, at high speeds the air comes to rest on the surface and becomes "compressed" resulting in an increase in air density, and so dynamic pressure begins to rise above its expected value. Normal calculations will then produce an excessive TAS.

Compressibility error is compensated for on navigation computers for TAS in excess of 300 kn and always produces a negative correction.

ETOPS (Extended Twin Operations) *Extended twin operations* (ETOPS) approval is an approval from the governing authority to allow an operator of a twin-engine aircraft type to conduct flights in which the aircraft is more than 60 min away from a suitable alternative aerodrome, in the event that the aircraft suffered an engine failure en route.

An adequate/suitable aerodrome for ETOPS diversions is one for which

1. Aircraft performance is suitable for the airfield.
2. Adequate emergency facilities are available at the aerodrome.
3. Adequate aerodrome lighting facilities are available for night flights.
4. A basic instrument approach is available for any expected instrument meteorological conditions.
5. Aerodrome is open.

ETOPS approval is based upon an airline's reliability (mechanical) record for an individual aircraft type and engines. Therefore a new aircraft type and/or airline is usually unable to get an ETOPS qualification until it has satisfactorily completed a stipulated period of non-ETOPS flights. When a state authority determines the qualifying period required for a new type or airline, it makes a judgment based on many factors, including these:

- For a completely new aircraft type, if other states have already granted ETOPS qualification for the aircraft type

- For a new airline operating that type, it already has other ETOPS granted types, and if it has a good engineering and reliability record for its other types and for its overall operations experience, e.g., flight operations department

Normal ETOPS categories vary between 60+ and 180 min and allows for most cross-Atlantic and cross-Pacific routes. However, in recent years (from 1999 onward) applications have been made for up to 207- and 240-min ETOPS, for more direct Pacific routings by transpacific airlines.

Exhaust Gas Temperature (EGT)   Exhaust gas temperature is an important engine parameter that indicates the temperatures being experienced by the turbine. The only real operating threat to the engine's life is excessive turbine temperatures. The maximum temperature at the turbine is critical because if the EGT limit is exceeded grossly on start-up (hot start), the excessive temperatures will damage the engine, especially the turbine blades. Also if the cruise EGT limit is slightly exceeded for a prolonged period, then this will shorten the engine's life.

Exhaust Smoke   Blue exhaust smoke indicates an oil burn in the cylinders. It is probably due to broken piston rings that allow oil seepage into the combustion chamber.

Black exhaust smoke indicates carbon granules burning in the cylinders. This occurs if the mixture is too rich, resulting in some of the fuel's not being burned and turning into carbon granules, which are then exhausted as black smoke.

White exhaust smoke indicates a high water content in the combustion chamber, which is exhausted as white "steam" smoke.

Extended V_2 Climb   An extended V_2 climb is one in which the aircraft's second-segment climb, i.e., at V_2 and takeoff flap, either

- Is continued to the highest possible level-off height, allowing for acceleration and flap retraction, if applicable, which can be reached with maximum takeoff thrust on all operating engines for its maximum time limit, normally 5 min   or

- Is continued to an unlimited height with maximum continuous thrust, instead of takeoff thrust, that meets the aircraft's minimum acceleration and climb gradient requirements in the takeoff flight path above 400 ft

An extended V_2 climb is used to clear distant, third-segment obstacles that are higher than the normal second- to third-segment level-off height, which is typically 400 to 1000 ft. An extended V_2 may only be used to clear the last obstacle in the flight path, so that a normal third-segment acceleration and final-segment profile can be achieved.

F

FADEC (Full Authority Digital Engine Control) FADEC automatically controls its engine functions, i.e., start procedures, engine monitoring, fuel flow, ignition system, and power levels required. FADEC computers can be found on the A320 aircraft's engine and on the EJ200 engine used on military aircraft.

Fahrenheit Fahrenheit is one scale used for measuring temperature. The Fahrenheit scale, measured in degrees Fahrenheit (°F), is based on water boiling at 212°F and freezing at 32°F. The formula to convert degrees Fahrenheit to degrees Celsius is

$$°C = \frac{5}{9} \, (°F - 32)$$

And the formula to convert degrees Celsius to degrees Fahrenheit is

$$°F = \frac{9}{5} \, (°C + 32)$$

Fail Operational Auto Land System A fail operational automatic pilot landing system, also known as a LAND 3 system, is one that employs three digital control computer channels (triplex system). It provides redundant operational capability; thus in the event of a single control channel failure below the alert height, the approach, flare, and landing can be completed by the remaining automatic systems, since the minimum required dual-channel system is still available. This allows the automatic landing system to work in a fail operational manner.

Fail Passive Auto Land System A fail passive automatic pilot landing system, also known as a LAND 2 system, is one that employs two digital control computer channels (duplex system). In the event of a single control channel failure, there is no significant out-of-trim condition or deviation of the flight path or attitude. However, the landing is not completed automatically, because a minimum dual-channel system is required for a fully automated landing; thus the pilot must assume control of the aircraft to complete the landing.

Fan Engine The fan engine can be regarded as an extension of the bypass engine principle, with the difference that it discharges its cold bypass airflow and hot engine core airflow separately.

The turbine-driven fan is in fact a low-pressure axial-flow compressor that provides additional thrust. Normally the fan is mounted on the front of the engine, and it is surrounded by ducting that controls the high supersonic airflow speeds experienced at the blade tips, preventing them from suffering from compressibility effect losses.

Either the fan is coupled to the front of a number of core compression stages (twin-spool engine), which restricts the width of the fan, and its bypass air is ducted overboard at the rear of the engine through long ducts; or it is mounted on a separate shaft driven by its own turbine (triple-spool engine) where the bypass airstream is ducted overboard directly behind the fan through short ducts (hence the term *ducted fan*). The CFM56-3 is a twin-spool fan engine, and the Rolls-Royce RB211 is a triple-spool fan engine.

The fan design reflects the specific requirements of the engine's airflow cycle, and it gives an initial compression to the intake air before it is split between the engine core and the bypass duct. The fan is capable of handling a larger airflow volume than the high-pressure compressor can. Therefore a fan engine will normally have a high bypass ratio, which means that the engine's resultant thrust properties are dominated by the large bypass air mass. Therefore the advantages of the bypass engine are increased further for the fan engine.

These are some of the main advantages of a fan engine:

1. Smaller engine size
   a. The engine is physically shorter because it has less depth to its compressor and turbine sections, because the majority of the engine's operating thrust is produced from the ducted bypass airflow, and the engine core produces only a small proportion of the engine's total thrust. Therefore the engine's core is correspondingly smaller.
   b. Because the engine is shorter in length, it is more rigid, it flexes less, and therefore it gives a better engine performance over the life span of the engine.
   c. It has a better thrust-to-weight ratio.
2. Better propulsive efficiency
   This is directly attributed to the bypassing of the N1 (fan) compressed air past the main engine core and delivery straight into the atmosphere via a bypass duct.
3. Better specific fuel consumption
4. Reduction in engine noise
   This is due to the bypassed air lessening the shear effect of the air exhausted through the engine core.
5. Contamination (i.e., bird strikes, heavy water) is centrifugally discharged through the bypass duct, therefore protecting the main engine core from damage and even a flameout from water contamination.

The advantage of a wide-chord fan engine is its ability to optimize the separate thrust properties of the bypass air mass and the engine core air mass from a single rotating blade row by utilizing the different speeds between the blade tip and center, without using inlet guide vanes. Thus the engine's weight and mechanical complexity are kept to an acceptable level. The chord is the length of the fan's blade from its center mounting to its tip.

The mass airflow generated by the fan is typically 6 times that required by the engine core, with the remaining five-sixths bypassing the engine core and being discharged into the atmosphere via the bypass ducts.

Bypass airflow on wide-chord ducted fan engine is generated at the outer section of the fan; and to optimize the efficiency of the bypass airflow, it needs to be raised to a pressure approximately 1.6 times that of the fan inlet pressure. This is achieved in the wide-chord fan by utilizing the very high tip speeds. The outer bypass section of the blade tips operates with a supersonic inlet air velocity of approximately 1.5 Mach, and this generates a very high airflow mass, which produces a high discharge air pressure of approximately 1.6 times that of the fan inlet air pressure.

At the center, or inner radius, of the fan chord, the airflow velocities required are of a lower value. This is so because the aerodynamic design of the next-stage core is more compatible with a lower level of supercharging, say, 1.3 to 1.4, which is akin to that of a conventional compressor stage.

So the advantage of the wide-chord (radius) fan engine is that it is capable of achieving the higher pressures required by the bypass air from the outer radius of the fan—and the more normal (lower) air pressures for the engine core section, by utilizing the speed difference between the blade tip and the center of the blade, without having to incorporate inlet guide vanes, etc., across the fan's chord.

Therefore, as a basic principle, the wider the fan chord, the higher the blade tip speed, the greater the generated airflow speed and mass, and the greater the discharged air pressure, resulting in an increased engine thrust.

A triple-spool turbofan engine, like the Rolls-Royce RB211, is a further development of the fan engine that has two distinct differences from the twin-spool fan engine:

1. The triple-spool turbofan engine has three independent compressor spools:
   - N1, low-pressure compressor spool or fan
   - N2, intermediate-pressure compressor
   - N3, high-pressure compressor spool
   Each is driven by its own turbine and connecting shafts.

2. The front turbofan, or N1 low-pressure compressor spool, is not connected to any other compression stages.

The turbofan on a triple-spool engine is further improved because it is not restricted to the size of other compressor spools (as it is on a twin-spool engine) and it is driven at its optimum speed by its own turbine. This allows it to have a larger frontal area that consists mainly of a giant ring of large blades, which act more as a shrouded prop than as a fan. It is responsible for producing an even larger bypass ratio (that is, 5:1) which generates approximately 75 percent of the engine's thrust in the form of bypass airflow delivered to the atmosphere via the engine's bypass ducts behind the fan. The one part of air that flows through the engine N2 and N3 compressors becomes highly compressed, of which one-third is used for combustion and two-thirds for internal engine cooling.

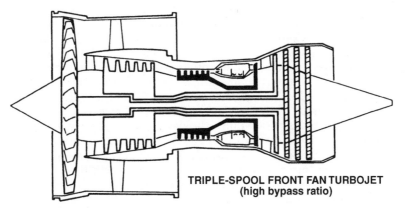

TRIPLE-SPOOL FRONT FAN TURBOJET
(high bypass ratio)

Figure 64   Triple-spool turbofan engine, the Rolls-Royce RB211. (*Reproduced with kind permission of Rolls-Royce Plc.*)

Advantages of the triple-spool fan engine, including the RB211, are twofold:

1. Particular to the triple-spool configuration, including the RB211:
   a. The N1 fan compressor can be built closer to its optimum design, namely, with a wider chord, because it is not restricted by any connection to booster compressors.
   b. The N1, N2, and N3 compressor sections all work closer to their optimum performance levels because they have their own independent turbine connecting shafts, especially the fan, N1 spool.
   c. The triple-spool configuration allows greater flexibility due to the aerodynamic matching at part load and lower inertia of the rotating parts.
   d. Engine thrust output is higher because of the improved fan section (point a) and the improved independent spool configuration (points b and c).
   e. Starting is easier because only one spool needs to be turned by the starter.
   f. The triple spool's modular assembly makes it easier to build and in particular to maintain. That is, if the N3 compressor suffers a fault, then the modular assembly of the engine allows for the N3 section of the engine alone to be removed for repair.
2. Advantages common to fan engines.

FANS (Future Air Navigation Systems)   FANS is the development of future navigation-based systems to meet the mutual demands of

1. U.S. and European airspace regulation changes, e.g.,
   a. Decommissioning of omega
   b. Development of a "free flight" airspace concept that incorporates the requirement for improved standards in
      (1) Navigation
      (2) Communication
      (3) Air traffic management (CNS/ATM)
      (4) Surveillance
2. The ever-increasing demand by the airlines for greater efficiency by reducing fuel consumption owing to more direct routes

FANS is not a single specific system but the ongoing development of several navigation-based systems that meet the requirements of both the airspace regulators and the airlines. A first step toward future airspace systems is to improve en route navigation capabilities, including communications. This has seen the United States develop a satellite global positioning system (GPS), and Europe to favor a new area navigation system, to be the basis of their new flight management systems. Both systems are fundamental parts of any air traffic management system, which are aimed at developing a free flight airspace. [Free flight is an air traffic management (ATM) system that allows pilots and airlines to set their own, more direct flight paths away from the restrictions of airways.]

Free flight plans also require the development of surveillance equipment that provides collision protection in the less regimented airspace. The development of onboard TCAS and ADS-B (Automated Dependent Surveillance—Broadcast system) satisfies this requirement for self-controlled collision protection. Therefore TCAS and ADS-B can be categorized as a future air navigation system.

FANS also includes

- Direct contact between airborne navigational computers and ground ATC/company computers, that will, for instance, allow a course deviation to be inserted by the pilot into the aircraft's navigation computer, which will be sent via a satellite link to ATC for acceptance, and approval to be received likewise from ATC. All this is done without any radio voice communication.

- Terminal navigation procedures, such as GLS (global landing system), which is a GPS-based landing system. A multimode receiver (MMR) avionics system was developed as a way for airlines to make the transition from today's ILS to GLS. The MMR system was created because of a requirement (especially in Europe) for aircraft to be immune from increasing radio interference. The MMR offers a way to update ILS units while providing an upgrade to GLS by way of the MLS (microwave landing system), if necessary.

The development of FANS can also include better utilization of present navigation equipment with head-up displays (HUDs) and flight deck instrument display upgrades on older aircraft types.

Fatigue Fatigue is caused by the lack of sleep and overwork/stress. Fatigue results in an increase in a pilot's response time as well as impaired judgment and an increase in the possibility of human error.

Feathering *Feathering a propeller* is the term given to describe the arrangement of the propeller blades turned beyond the coarse pitch until they are "edge on" (i.e., the chord line is parallel) to the airflow. This causes the propeller *not to windmill* or create extra drag, or hinder the extinguishing of an engine fire. Feathering the propeller is accomplished in the event of an in-flight engine failure or fire.

Fences *See* Vortex Generators/Fences.

Field Length Limits The most restrictive field length available from one of these

1. The all-engine operating runway length
2. The runway emergency distance length available
3. The one-engine inoperative runway length

limits the aircraft's MTOW (or MLW) so that it meets the required TOR/D (LDR) performance given the ambient aerodrome conditions of pressure altitude and temperature (density).

Final Approach Fix (FAF) The final approach fix denotes the start of the final approach segment of either a precision or a nonprecision approach. It is usually denoted as either a locator marker or a final approach distance.

NB: On some instrument precision approaches, the FAF is the same as the initial approach fix. For example, the IAF may be the locator outer marker when flying outbound, and the FAF is the same locator outer marker when flying inbound.

Where no final approach fix is shown, the final approach segment commences at the completion of the procedure turn and when the aircraft is within ±5°C of the final approach track, whereupon final descent may be commenced.

NB: The FAF should always be crossed at or above the specified intermediate height before final descent is initiated.

Fire The elements required for a fire are oxygen, combustible material (fuel), and ignition source (heat). The most practical way of eliminating a fire is to remove its oxygen supply.

Fire Bottles—Indications   The indications of thermal expansion and/or use of the fire bottles can normally be found on the side of the aircraft fuselage with separate disks, i.e., one for extinguishent release due to thermal expansion and one to indicate usage. If they are intact, they indicate that the extinguishent is still in the fire bottle.

NB: The color and location of such disks are aircraft type-specific.

Fire Detection and Protection System   A typical engine fire detection and protection system consists of the following:

1. Overheat and fire detection loop(s) with multiple sensors

NB: Normally a minimum of two separate systems (loops) per unit are used.

   Coupled with visual (lights) and aural indications for both overheat and fire-detected conditions, for
   a. Each engine
   b. Auxiliary power unit (APU)
   c. Wheel well (not on all aircraft types)
2. Fault monitoring system of overheat and fire detection systems. Coupled with visual indications for
   a. Engines
   b. APU
3. Fire extinguishers and firing circuits for
   a. All engines

NB: Usually a minimum of two extinguisher bottles are required, that can fire into each engine, so as to provide a second extinguishing supply to all engines. Coupled with visual indications of bottle discharge.

   b. APU

NB: Usually there is a separate single fire extinguisher bottle. Coupled with visual indications of bottle discharge.

4. Testing facility of
   a. Firing circuits for
      (1) Engine fire extinguisher(s)
      (2) APU fire extinguisher
   b. Fault monitoring system
   c. Overheat/fire detection loops for
      (1) Each engine
      (2) APU
5. Lavatory/cargo holds smoke detection system

Fire Drill—Engine   A typical engine fire drill is as follows:

- Thrust lever   CLOSE
- Auto throttle (if engaged)   DISENGAGE
- Start lever/switch   OFF or CUT OUT
- Engine fire warning switch   ENGAGE

If the fire or overheat warning light remains illuminated,

- Extinguisher number 1   DISCHARGE

If the fire or overheat warning light still remains illuminated,

■ Extinguisher number 2   DISCHARGE

If the fire or overheat warning light still remains illuminated, *land at the nearest airport.*

NB: Various type-specific system paper checks would follow.

Fire Extinguisher   Bromochlorodifluoromethane (BCF) should be used for electrical and flammable fires. Such fires are normally found in the cockpit. BCF is a lique-fied gas agent that vaporizes on deployment. It is stored in signal red, purple/brown, or green container.

Fixed-Pitch Propeller   The failing of the fixed-pitch propeller is that it only produces its maximum efficiency (i.e., best blade angle/pitch for rpm speed) at one single predetermined engine rpm, altitude, and forward airspeed condition. This is so because the forward airspeed affects the blade angle of attack (an increase in for-ward airspeed causes a decrease in blade angle of attack, which reduces the rear-ward air displacement and thereby decreases the thrust produced), and therefore propeller efficiency is reduced from its single predetermined condition.

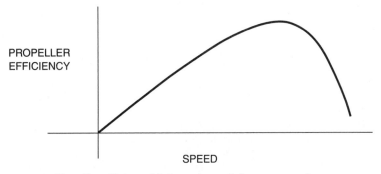

Figure 65   Propeller efficiency/blade angle or pitch versus speed.

Therefore, for a fixed-pitch propeller, a single setting must be made between the blade angle (and therefore the angle of attack and pitch) desirable for takeoff and that required for cruise conditions. And if a blade angle for high-speed cruise is chosen (coarse), then the aircraft has the disadvantage of poor propeller perfor-mance for takeoff, and vice versa.

Flaps   The primary use of flaps, especially on a jet aircraft, is to increase lift by extending the geometric chord line of the wing, which increases its chamber and area.

Lowering the flaps in flight will generally cause a change in the pitching moment. The direction and degree of the change in pitch are dependent on the rel-ative original position of the center of pressure (cp) and the center of gravity (cg). The factors that contribute to this are as follows:

1. The increase in lift created from the increased wing area and chamber will lead to a pitch-up moment, if the cp remains in front of the cg.

2. If the associated rearward movement of the cp is behind the cg, then this will produce a nose-down pitch.

3. The flaps will cause an increase in the downwash, and this will reduce the angle of attack of the tailplane, giving a nose-up moment.

4. The increase in drag caused by the flaps will cause a nose-up or nose-down moment, depending on whether the flaps are above or below the lateral axis.

The overall change and direction in the pitching moment will depend on which of these effects is predominant. Normally the increased lift created by extending the wing chord line when the flaps are extended is dominant, and will cause a nose-up pitching tendency as the cp normally remains in front of the cg.

The effect of raising the flaps in flight, if not compensated for by increasing speed and changing attitude, will result in a loss of lift. An aircraft can fly at a lower speed with the flaps down. Therefore it is important that the aircraft has reached a speed above its flaps-up stalling speed before the flaps are raised. The safety speed for each flap setting must therefore be known and the aircraft allowed to accelerate to that speed before the flaps are retracted to that setting; otherwise, the aircraft would lose height. If at any speed the flaps are raised without an increase in angle of attack, there will be a loss of lift, which will cause the aircraft to sink. The lift must be maintained by raising the nose of the aircraft to maintain the original coefficient of lift.

Flaps affect the takeoff ground run in the following manner:

1. When the flap is set *within* its takeoff range, a higher flap setting will reduce the takeoff ground run, for a given aircraft weight. The use of flaps increases the C_{LMAX} of the wing, due to the increased chord line for a low drag penalty, which reduces the stall speed (V_S) and consequently the V_R and V_2 speeds. This in turn provides good acceleration until it has sufficient kinetic energy that reduces the takeoff ground run.

Typically various flap settings from the first to the penultimate flap setting are available for takeoff (i.e., takeoff range). The higher the flap setting within this range, the less takeoff run required, because the drag is not significantly increased as the angle of attack is low. However, the drag increment is higher when the aircraft is in flight and out of the ground effect, because the aircraft's angle of attack is much higher.

NB: Initial- and second-segment climb performance will thus be reduced with a high takeoff flap setting.

2. When the flap is set outside its takeoff range, a high flap setting will result in a large drag penalty which will reduce the aircraft's acceleration, and therefore the takeoff run will be greatly increased before V_R is attained.

No setting or a very low flap setting, outside the takeoff range, on takeoff will result in a low coefficient of lift produced by the wing for a given speed. Thus a higher unstick (V_R) speed is required to create the required lift for flight. Therefore an increased takeoff run to attain the higher V_R speed is required.

Flat-Rated   The fan engine is flat-rated to give it the widest possible range of operation, keeping within its defined structural limits, especially in dense air.

NB: Flat rating guarantees a constant rate of thrust at a fixed temperature.

Flexible (Assumed) Temperature   *See* Assumed (Flexible) Temperature.

Flight Condition—Hazards   The commander of an aircraft must report to ATC, as soon as possible, any hazardous flight conditions encountered.

NB: Typical conditions worthy of reporting are moderate or severe turbulence, wind shear, volcanic ash, and rapidly deteriorating visibility.

Flight Deck   The commander of the aircraft is responsible for ensuring that

- One pilot is at the control at all times in flight.
- Both pilots are at the controls during takeoff and landing if the aircraft is required to carry two pilots. This usually means during the climb until 10,000 ft or transition, whichever is the later, and from the top of descent.

NB: Each pilot at the controls must wear a safety harness during takeoff and landing, where one is provided and required, and a lap safety belt at all other times.

Flight Director   The flight director (FD) system provides guidance information of the required aircraft maneuver in pitch and roll to gain, regain, or maintain the programmed flight path. The flight director display is found on the primary flight instrument (ADI) as either a *chevron* indicator or a pair of *vertical and horizontal bars.*

The chevron display consists of a magenta or yellow chevron shape that represents the aircraft's attitude and a hollow or white chevron shape that represents the required position. The aim is to center the magenta chevron inside the hollow/white larger chevron to attain the required aircraft maneuver. The aim of the vertical and horizontal bar flight director system is to center the bars in a central box to attain the required aircraft maneuver. Both the chevron and bar flight director systems are command displays; i.e., by flying toward the deflected bars they will center to gain, regain, or maintain the programmed flight path.

The FD system utilizes the autopilot (AP) sensors and the aircraft's FMC performance database to compute the aircraft's pitch and roll response to gain, regain, or maintain the programmed flight path, and the FCC position of the FD bars/chevron accordingly on the respective ADIs. The FD system has its own engagement switches, on or off, and as a system it is part of the APFDS, although it can be operated with or without AP/AT engagement, as long as pitch and roll operating modes have been selected. The FD system operates in most of the AP operating modes with a few usual exceptions that are normally type-specific; e.g., takeoff mode (TOGA) is a FD mode only (FDs have *no* landing flare capability).

Flight Director—Errors   The errors of a flight director system are primarily operating errors. First, an error commonly occurs with the interpretation of the FD indication by the pilot. A pilot interprets the flight director bars as the flight path, i.e., center and crossed flight bars as meaning the aircraft is on track. However, such a flight director indication can also—and often does—mean that the aircraft is on the correct intercept course. Therefore such errors of interpretation are associated with flight director systems although they are strictly a pilot error. Second, the FD system is part of the autopilot system, and because of this the FD bars are geared to the AP's quick response/reaction rate, which is far quicker than the ability of a human pilot. This therefore gives rise to problems when the pilot, especially on the approach, is manually flying the aircraft; i.e., a small localizer deviation will command a large correcting bank by the FD vertical bar to regain the localizer. This is a maneuver that the AP's quicker response rate can cope with, but the human pilot is too slow in his or her reactions when removing the bank and therefore cannot keep up with the FD bar. This results in the aircraft going through the localizer and then the aircraft will oscillate on either side of the centerline.

Therefore, this can be seen as a true FD system error when the system is used for manual flying maneuvers. However, awareness of the problem should allow the pilot to overcome this error by using raw data information in her or his maneuver decision making.

Flight Instruments—Gyro   The gyroscopic flight instruments are as follows:

1. *Directional indicator (DI),* which is a tied gyro, retains the space gyro's freedom of movement in all three planes, but has an external force controlling the direction of its spin axis.

2. *Artificial horizon (AH)* is an earth gyro. An earth gyro is a special form of tied gyro where its spin axis is tied by the earth's gravity to the earth's horizon.

3. *Turn and slip indicator* or *turn coordinator,* which is a rate gyro, only has a freedom of movement in two planes—the gyro's spin axis and 90° from the spin axis (precess).

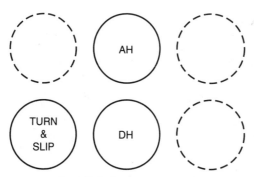

Figure 66   Gyro flight instrument panel.

Flight Instruments—Pressure   The pressure flight instruments are

1. Airspeed indicator (ASI)/Mach meter (MM)
2. Altimeter
3. Vertical speed indicator (VSI)

See Fig. 67 at the top of the next page.

Pressure instruments measure atmospheric pressure by utilizing the pitot static system, which is a combined sensor system that detects

■ The total pressure (static and dynamic pressure), also called the pitot pressure, which is measured by a pitot probe

■ Static pressure alone, which is measured by either the static port on a pitot probe or a separate static vent

The difference between the two will give a measurement of the dynamic pressure. That is, dynamic pressure = total pressure − static pressure. Dynamic and/or static pressure measurements are the basis of the flight instrument readings.

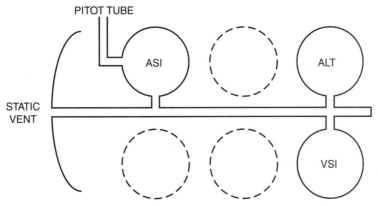

Figure 67 Pressure instruments and the pitot static system.

Flight Level A flight level is the measured pressure level above the 29.92-in/1013-mbar datum.

Flight Levels (IFR) The instrument flight rule (IFR) flight levels are based upon the quadrant and semicircle rule whenever an aircraft is above 3000 ft mean sea level (MSL) or above the appropriate transition level, whichever is higher, and the aircraft is in level flight. The following levels flown are based on the aircraft's *magnetic track*.

Flights at Levels below 24,500 ft (Quadrant Rule)

Magnetic track	Cruising level
0 to 89°	Odd 1000 ft
90 to 179°	Odd 1000 ft plus 500 ft
180 to 269°	Even 1000 ft
270 to 359°	Even 1000 ft plus 500 ft

Therefore vertical separation is only 500 ft. However, usually only 1000-ft levels are used.

Flights at Levels above 24,500 ft (Semicircular Rule)

Magnetic track	Cruising level	RVSM* levels
0 to 179°	250	
	270	
	290	290, 310
	330	330, 350
	370	370, 390
	410	410, 430

180 to 359°	260	
	280	300
	310	320, 340
	350	360, 380
	390	400, 420

*Reduced vertical separation minimum.

Therefore vertical separation is only 1000 ft up to FL290 and 2000 ft above FL290 in non-RVSM airspace (vertical separation is increased to 2000 ft above FL290 in non-RVSM airspace because of increased altimeter errors due to the lower air density experienced at these higher levels) but remains at only 1000 ft in RVSM airspace at all levels.

NB: Reduced Vertical Separation Minimum (RVSM) to 1000-ft separation above FL290 is granted to aircraft with advanced and more accurate altimeters, especially on crowded routes, e.g., cross-Atlantic routes.

Flight levels are based on the standard altimeter setting of 29.92 inches of mercury (inHg) or 1013.2 mbar (hPa).

The lowest usable en route flight level must be at least 500 ft above the absolute minimum altitude.

NB: Minimum altitude on an airway is at least 1000 ft above the highest obstacle within 15 nm of the airway centerline (hence you comply with the minimum height rule). Therefore terrain/obstacle clearance is at least 1500 ft at the lowest usable flight level.

Flight Management System (FMS)   The purpose of an FMS is to manage the aircraft's performance and route navigation, to achieve the optimum result. The managed data can then be used for either (1) advisory crew information only or (2) to direct the auto throttle and autopilot to steer the aircraft.

The FMS's three sources of data are

1. Stored databases

2. Pilot inputs via the CDU

3. Other aircraft systems, which are automatically fed into the FMS

These input sources are fed into the flight management computer (FMC) and are duplicated for both performance and navigation management functions.

See Fig. 68 at the top of the next page.

The following is based on the B737-300 FMS.

The flight management system combines data from various aircraft systems, which are fed into a central computer, the FMC, to manage the route navigation and the aircraft's performance. The main section of the FMS is the flight management computer system (FMCS), which is a dual system utilizing two flight management computers and two control display units (CDUs).

The CDUs are the single point of control used by the crew to access the FMC. Via the CDUs the crew can initiate and control a given flight plan's navigation and performance management and monitor its execution. It also provides a convenient access point to initiate the IRS units. Either CDU can input to both FMCs simultaneously, and each FMC will then compute and execute the command. The CDU

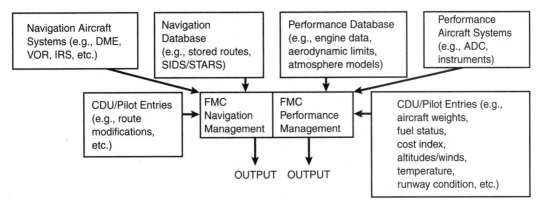

Figure 68 FMS input data sources.

consists of a monochrome CRT screen and a keyboard for data entry. The central part of the screen is used to display selected data. Below the data block is a "scratch pad" area where FMC messages can be shown and where data are held before being entered into the system.

Mode select keys are used to control the type of data available on the CDU, e.g., climb, route, and fix.

There are two main operational functions of the FMC/CDU:

1. *Route navigation management.* The FMC contains a navigational database, which includes a selection of companies' structured routes, and standard instrument departure and arrival procedures.

NB: The RTE key leads into the route selection procedure and the DEP ARR key into the SIDs and STARs.

Also the crew can manually amend company routes, or input a completely new flight plan route, via the CDU using its alphanumeric keys. Once the aircraft is airborne, the FMS screen shows the next active waypoint, and gives course and distance between waypoints with the speed and altitude at the waypoints. This can be used as advisory information only; or, when coupled to the auto flight system, can be used to automatically steer the aircraft along the flight plan route. The FMC's navigational function continuously computes the aircraft's position by using inputs from the aircraft's primary navigation systems, i.e., inertial reference systems, VOR, DME, and localizer information, which are fed into the FMC navigation management program.

When the VHF NAV is selected to auto, the system will tune its own DME frequencies according to the information in the navigational database in the following priority order:

1. DME/DME crosscut
2. DME/VOR
3. VOR/VOR
4. IRS—position information from the IRUs is used
   a. When out of VHF radio range
   b. When on the ground
   c. To solve any ambiguity from DME and VOR information

The FMS navigation information is shown on the (E)HSI flight instrument. On a EHSI, several different display modes can be selected.

2. *Aircraft performance management.*   A performance database is stored in the FMC and contains engine installation data, aircraft aerodynamic limits, and atmospheric models. In addition, specific operational data/limits are inputted by the crew via the CDU, and present conditions are automatically received from other aircraft performance systems, i.e., air data computer. The FMC uses both the database and the inputted data from the crew and the other aircraft performance systems to calculate the aircraft's performance parameters (limits) and predictions for its en route and takeoff stages of flight.

The following data entries are required from the crew for the aircraft's en route performance management:

- Either gross weight or zero fuel weight.
- Fuel reserves.
- Cost index.
- Cruise altitude.
- Transition altitude.
- ISA deviation.
- Winds at height. If winds are not entered, the route will be calculated with observed wind to the next waypoint and for zero wind subsequently.

In addition, the pilot may enter low- or high-speed restrictions for any phase of flight.

In tandem with the stored performance data and inputs from other aircraft performance systems, that is, ADC, the FMC calculates the aircraft's optimum speeds, power setting, and vertical profile performance for the aircraft's en route stage of flight. The FMC also provides takeoff performance to the pilot.

The following data entries are required from the crew for the aircraft's takeoff performance management:

- OAT
- Surface wind
- Runway direction and slope
- Runway condition (dry or wet)
- The cg position

   NB: The runway in use is known from the route selection page.

In tandem with the stored performance data, the FMC calculates the aircraft's correct

- V_1, V_R, and V_2 speeds
- Takeoff power

   NB: It can also provide reduced thrust settings for takeoff if performance is not limiting.

- Trim setting

The instruments that show FMS performance information are

1. Mach/airspeed indicators
2. Engine N1 rpm gauge
3. Attitude indicator (normally the flight director pitch bars on the ADI instrument)

The FMS drives cursors on the airspeed indicators and N1 gauges to show target speeds and thrust, respectively. The cursors can be manually overridden by pulling out control knobs.

Flight Plan   Flight plans are applications to air traffic control for flights

- Within controlled airspace *notified* as IFR only, irrespective of whether IMC or VMC conditions exist
- Within other controlled airspace, in instrument meteorological conditions (IMC) or at night, excluding special VFR (SVFR)
- Within certain special-rules airspace, irrespective of weather conditions
- Within upper information regions (UIRs), i.e., above FL245
- When the destination is more than a certain distance, for aircraft with a maximum total authorized weight exceeding 5700 kg
- When you cross flight information regions (FIRs) boundary, in most parts of the world
- During which it is intended to use the air traffic advisory service on an advisory route (ADR)

FMS   *See* Flight Management System (FMS).

Fog   Fog is simply a parcel of low-level air in contact with the ground that has small suspended water droplets, which have the effect of reducing horizontal visibility to less than 1000 m.

In general, fog is caused by a condensation process, due to a difference in temperature between the ground and the air next to it (radiation and advection fog); or between two interactive air masses (frontal fog), which as a result reduces the temperature of a parcel of air to its dew point temperature.

The most common different types of fog are as follows:

1. *Radiation fog*, which requires the following conditions:
    a. Cloudless night, that allows the earth's surface to lose heat by radiation, that in turn cools down, by conduction, the air in contact with it to its dew point temperature. This causes its water vapor to condense out in liquid form.

NB: Radiation fog only occurs over land and not at sea because the sea does not cool down diurnally.

    b. Moist air, with a high relative humidity, which only requires a slight cooling to reach its dew point temperature.
    c. Light winds, between 2 and 8 kn, that mixes the lower levels of the air, thereby creating a layer of low-level fog (mist).

NB: Without the light winds, dew would be formed.

The minimal mixing required can be provided by the slight turbulence caused as the ground starts to heat up just after sunrise. Therefore dew may form before sunrise, and then a fog may form later after sunrise. With stronger winds, i.e., greater than 8 kn, low-level stratus clouds may be formed.

See Fig. 69 at the top of the next page.

Radiation fog occurs inland, especially in valleys and low-lying areas. This is due in part to katabatic winds, common to this type of terrain, depositing cool

RADIATION FOG

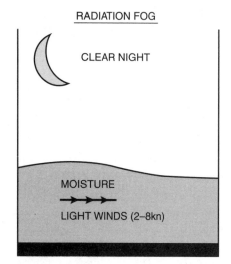

FOG LIFTS TO FORM LOW CLOUD

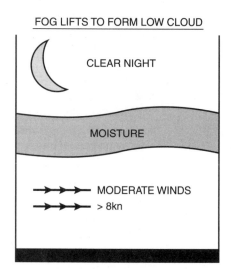

Figure 69   Radiation fog.

saturated air into the bottom of a valley or low-lying area, thus increasing the possibility of radiation fog's forming. As the earth's surface begins to warm up, usually sometime after sunrise, the air in contact with it will also warm up, causing the fog to gradually dissipate. The fog may rise to form a low layer of stratus clouds before finally clearing. If the fog that was formed overnight is particularly thick, it may act as a blanket, shut out the sun, and thereby restrict the heating of the earth's surface after the sun has risen. As a result, the air in which the fog exists will not be warmed from the earth's surface, and therefore the fog may last throughout the day. This is sometimes experienced in the autumn and winter seasons.

2. *Advection fog.*   The term *advection* means heat transfer by the horizontal flow of air. Fog formed in this manner is called *advection fog* and can occur quite suddenly, day or night, on land or sea, if the following conditions exist:

   a. A warm moist air mass flows across a significantly colder surface, which is cooled from below. If the temperature is reduced to its dew point temperature, then water vapor will condense out and fog will form; e.g., a warm moist maritime air over a cold land surface can lead to advection fog over land.

   b. Light to moderate winds encourage the mixing of the lower levels to give a layer of fog. However, stronger winds may cause the formation of stratus clouds rather than fog, although advection fog can persist in much stronger winds than radiation fog can.

3. *Sea fog.*   This is a form of advection fog and is formed by one of two ways:

   a. Tropical maritime air moves toward the pole over a colder air mass.

   b. An airflow off a warm land surface moves over a colder sea, which can typically occur in the summer season.

NB: Sea smoke occurs under the reverse conditions; i.e., very cold air, often in an inversion, passes over a warmer sea, which causes evaporating moisture to condense out into whispers of vapor. It only occurs with very cold air and therefore is only common in polar regions or from lakes in temperate regions, after very cold, clear nights.

4. Frontal fog usually forms in the cold air ahead of a warm occluded front as a prefrontal widespread fog. It forms due to the interaction of two air masses, in one of two ways:

    a. As a cloud that extends down to the surface during the passage of a front. It is mainly found over hills and consequently is called *hill fog.* Hill fog is formed as air is pushed up the side of a hill and cooled adiabatically to its dew point, to form low cloud/hill fog.

    b. As air becomes saturated by the evaporation of rainfall.

Fohn Wind Effect   The Fohn wind effect occurs when air is cooled as it is forced to rise over high ground, first at the DALR and then after the dew point/condensation level is reached, at the SALR. After the condensation level, clouds will form, usually on the hillside, and moisture will be lost, falling as either rain or dew. Then as the air descends over the far side of the hill, it has a lower water content and so the condensation level is higher. A longer period of warming at the greater DALR means the air on the far (downwind) side of the hills is warmer and dryer than it was on the upwind side of the hill.

    NB: As a rough guide, the air temperature can rise by 1.5°C per 1000 ft of hill, which of course is the difference between the DALR and the SALR.

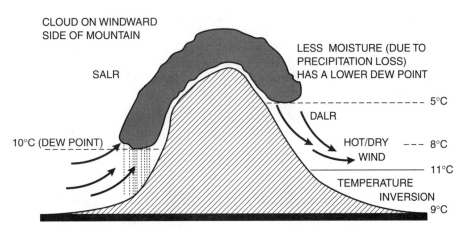

Figure 70   Fohn effect, cloud and wind.

    Therefore, the Fohn effect (i.e., descending air) can influence the wind direction and strength, and in some cases it even generates a wind that blows down the far/lee side (downwind) of the hill. This is known as the *Fohn wind.*

    In the mornings, this rather mild wind by itself could encounter a parcel/layer of denser, colder air sitting in the bottom of the valley. Such an encounter would create a temperature inversion, which could result in wind shear and/or CAT. When the cold air in the valley is finally displaced, the wind speed at the surface can increase from zero to gale force in a few moments. This Fohn wind effect can be seen in various places, including (1) Rocky Mountains in the United States

(Chinook winds), (2) Alps, and (3) east coast of Scotland. Obviously it is a danger to aviators and should be avoided if possible.

Forces The forces acting upon an aircraft in flight are drag/thrust and lift/weight.

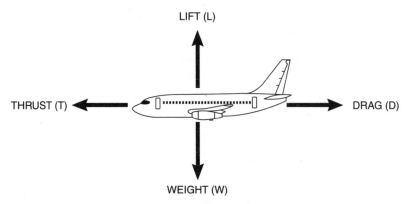

Figure 71 Forces acting upon an aircraft in flight.

When thrust and drag are in equilibrium, an aircraft will maintain a steady speed. For an aircraft to accelerate, thrust must exceed the value of drag. When lift and weight are in equilibrium, an aircraft will maintain a steady level attitude. For an aircraft to climb, lift must exceed the weight of the aircraft.

In a banked turn, weight is a constant but lift is lost due to the effective reduction in wing span. Therefore to maintain the altitude in a banked turn, the lift value needs to be restored by increasing speed and/or the angle of attack.

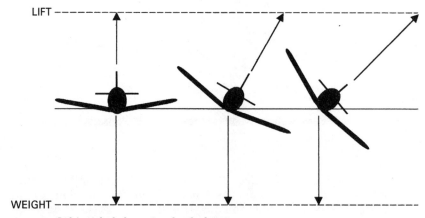

Figure 72 Lift/weight balance in a banked turn.

The forces acting upon an aircraft in a climb are

1. Thrust, which must equal the drag plus weight component along the flight path
2. Lift, which must equal the weight component perpendicular to the flight path

The forces acting upon an aircraft in a powered descent are

1. Drag, which must equal the thrust + weight component along the flight path. Therefore for a given drag value, the weight component required will be less than that with no thrust, and so the glide angle is less for power-on descent.

2. Lift, which must equal the weight component perpendicular to the flight path.

The forces acting upon an aircraft in a glide descent are

1. Drag, which must equal the weight component along the flight path

2. Lift, which must equal the weight perpendicular to the flight path

Fowler Flaps Fowler flaps are trailing-edge wing flaps (usually triple-slotted) used to increase the wing area and chamber, which increases the coefficient of lift maximum for low flap settings, e.g., 1 to 25°. High flap settings increase drag predominately more than lift, and therefore they are used to lose speed and/or height, most commonly during an approach to land.

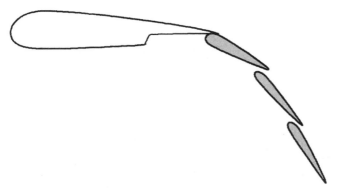

Figure 73 Fowler flaps.

Free-Flight Concept The free-flight concept is an air traffic management (ATM) system currently under development (since 1998, with trials in Alaska and Hawaii from 1999). It will allow pilots and airlines to set their own direct routes, altitudes, and speeds away from the restrictions of airways. The concept is based on two airspace zones, *protected* and *alert*. In principle, until the alert zone is breached, aircraft can be maneuvered with autonomous freedom, with the responsibility of collision avoidance in the alert zone moved to the pilot with the introduction of onboard TCAS and ADS-B systems.

The free-flight concept is the basis of the U.S. National Airspace System (NAS) which is planned to commence in 2005. It will enable an air traffic management (ATM) system to evolve from today's largely rigid procedural, analog and ground-based systems, into a flexible, collaborative system using current and emerging technologies (FANS) such as the following:

1. GPS navigation satellites, or area navigation system; for navigation purposes

2. ADS-B (Automated Dependent Surveillance—Broadcast)

3. TCAS (Traffic Collision Avoidance System) for airborne traffic collision protection

What all this means is that more direct point-to-point flights, which require less ATC and ground-based navigation system coverage, resulting in reduced flight times and therefore improved flight economics on most routes, will be possible. Also it will make the airspace less congested by reducing the number of aircraft using the same flight paths, especially over existing airway intersections, which is rightly portrayed as improving safety conditions as well as increasing the aircraft capacity of the airspace.

Frequency Meters (Gauge)   A frequency meter measures and indicates the frequency output of the selected AC power source.

NB: Aircraft AC power sources usually have a frequency of 400 Hz (1 Hz = 1 cycle/s).

Front   A front is a boundary between two different air masses. Air masses have different characteristics, depending upon their

1. Origin
2. Path over the earth's surface
3. Diverging or converging air

They usually give rise to a distinct division between adjacent air masses, which are known as fronts. There are two basic types of fronts, cold and warm. Fronts are normally found in low-pressure systems, because a depression is usually made up of two different air masses.

Front—Cold   *See* Cold Front.

Front—Occluded   *See* Occluded Front.

Front—Warm   *See* Warm Front.

Frontal Activity   *Frontal activity* describes the interaction between at least two air masses, as one air mass replaces another.

Frontal Depressions   Frontal depressions (low-pressure frontal system) develop when two different air masses meet (touch) but do not mix because one of them is warmer and less dense than the other.

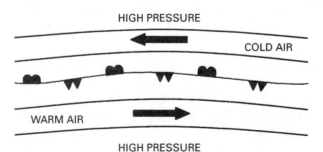

Figure 74   Stage 1 development of a frontal depression.

Therefore they remain adjacent to each other with a distinct boundary between them. This boundary area is called a *front*. For example, a polar front can be found between the cold polar air and the warmer air of the mid-latitudes.

This boundary becomes distorted into a wavelike shape because the warmer air mass bulges into the colder air mass.

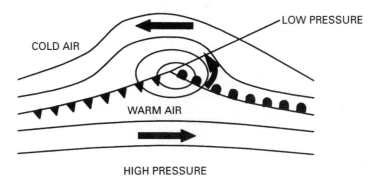

Figure 75 Stage 2 development of a frontal depression.

The leading edge of the bulge of warm air is a warm front that slides over the cold sector air, and its rear edge is a cold front where the cold air forces itself under the warmer air. The pressure near the tip of the wave falls sharply, and so a *depression* is formed. Any instability in the depression's airflow tends to produce a frontal depression with its usual anticlockwise flow of air (in the northern hemisphere).

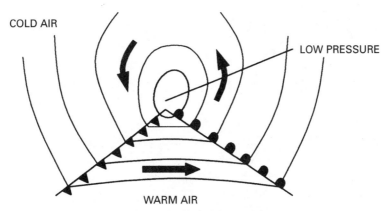

Figure 76 Stage 3 development of a frontal depression.

Once started, the circulation of air increases the penetration of the warm air into the cold air, making the depression more intense.

See Fig. 77 at the top of the next page.

A frontal depression is probably the most dynamic form of all low-pressure systems, and therefore it is commonly and incorrectly thought of as the only form of low-pressure system.

NB: It should be realized that frontal depressions are not the only form of low-pressure systems, but they are dynamic in that they have *frontal activity.*

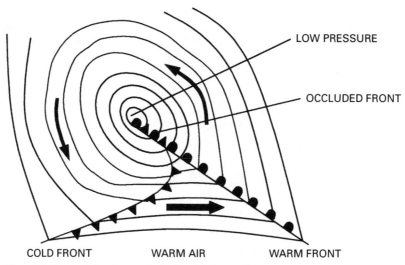

LOW PRESSURE

OCCLUDED FRONT

COLD FRONT   WARM AIR   WARM FRONT

Figure 77 Stage 4 development of a frontal depression.

A frontal depression (low pressure system) moves in the direction of the isobars in the warm sector. Its speed is taken from the geostrophic wind, which is measured as close to the center of a depression as possible.

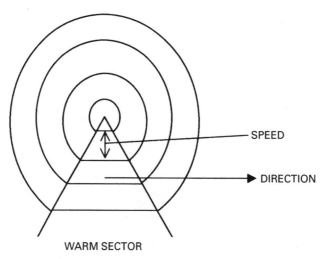

SPEED

DIRECTION

WARM SECTOR

Figure 78 Direction and speed of a frontal depression.

Frost   Frost is a frozen water cover on the earth's surface that is formed in the same manner as dew except the earth's surface has a subzero temperature that causes the water droplets that have condensed out of the air to freeze on the ground.

Fuel(s)   The fuels used for gas turbine civil aircraft are

1. Jet A1 (Avtar). This is a kerosene type of fuel with a normal specific gravity (SG) of 0.8 at 15°C. It has a medium flash point and calorific value, and a boiling range of 150 to 300°C and a waxing point of −50°C.

2. Jet A. This is similar to A1, but its freezing point is only −40°C.

   NB: Jet A is normally only available in the United States.

Fuel Contamination   The greatest risk of aviation fuel contamination comes from water. You can take steps to safeguard against water contamination of stored fuel:

1. Use water drains in fuel tanks.

   NB: Water is heavier than kerosene, and therefore it falls to the bottom of tanks, where it can be drained away.

2. Fuel heaters, typically used with gas turbine engines, are used to heat the fuel and evaporate the water prior to delivery to the engines.

3. Exclude atmosphere in the fuel tanks.

   NB: The main source of water contamination is the atmosphere that remains in the fuel tank.

   Therefore keeping the tank full of fuel removes any air and thereby prevents any water contamination.

Fuel Howgozit   Fuel howgozit is a comparison of the actual fuel remaining and the planned fuel remaining along the flight path. At any stage of flight an estimate of the fuel difference can be obtained, which is ideal to show any trends such as increasing fuel burn.

   NB: Fuel howgozit charts can also be used to plan the position of the PNR point.

Fuel Injection System   A fuel injection system delivers metered fuel directly into the induction manifold and then into the combustion chamber (or cylinder of a piston engine) without using a carburetor. Normally a fuel control unit (FCU) is used to deliver metered fuel to the fuel manifold unit (fuel distributor). From here, a separate fuel line carries fuel to the discharge nozzle in each combustion chamber (or cylinder head in a piston engine, or into the inlet port prior to the inlet valve). With fuel injection a correct mixture can be provided by its own separate fuel line.

   Advantages of a fuel injection system are

- Freedom from vaporization ice (fuel ice)
- More uniform delivery of the fuel/air mixture around the combustion chamber section or each cylinder
- Improved control of fuel/air ratio
- Fewer maintenance problems
- Instant acceleration of the engine after idling, i.e., instant response
- Increased engine efficiency

   Disadvantages of a fuel injection system are as follows:

- Starting an already hot fuel injection engine may be difficult due to vapor locking in the fuel lines.

- Having very fine fuel lines, fuel injection systems are more susceptible to contamination (i.e., dirt or water) in the fuel.
- Surplus fuel provided by the fuel injection system will pass through a return line, which is usually routed to only one of the fuel tanks. This may result in either the fuel being vented overboard (thus reducing fuel available) or asymmetric (uneven) fuel loading.

Fuel—Island Holding   *See* Island Holding Fuel.

Fuel—Measurement   The quantity of fuel onboard an aircraft is usually measured in terms of its mass, i.e., in either kilograms (kg) or pounds (lb). By using the SG, you can calculate the mass from a given volume of fuel, and vice versa.

Example: To convert fuel weights to volume and vice versa, using a SG of 0.8, do as follows:

$$\frac{\text{Weight, kg}}{\text{SG (0.8)}} = \text{volume, L}$$

$$\text{Volume, L} \times \text{SG (0.8)} = \text{weight, kg}$$

NB:

$$1 \text{ L} = 3.7853 \text{ U.S. gal} = 4.5459 \text{ Imperial gal}$$

$$1 \text{ kg} = 2.2046 \text{ lb}$$

Fuel is not measured in terms of a volume, because a volume of fuel will increase with a temperature rise and more importantly will decrease with a temperature drop. However, with a rise in temperature the SG decreases, and with a drop in temperature the SG increases, which ensures that the indicated mass (weight) of fuel remains the same. Obviously if fuel were uploaded in terms of a volume under high-temperature conditions and a change of temperature reduced the volume amount of fuel before or during the flight, then a situation could exist whereby the aircraft had insufficient fuel to complete the journey.

Fuel—Specific Gravity   *See* Specific Gravity.

Fuel—Trip Plan   A typical fuel profile for a flight would include sufficient fuel for the following:

1. Takeoff and climb at takeoff thrust.
2. Climb to the initial cruise altitude at maximum continuous thrust.
3. En route cruise including intermediate step climbs.
4. Descent to a diversion point (go-around point) overhead the destination aerodrome.
5. Contingency fuel.
6. Diversion to overhead an alternative aerodrome.
7. An instrument approach and landing.
8. An additional amount of holding fuel may be required if the destination is an island with no alternative or arrival will be at a major airport at its busiest period, e.g., London Heathrow between 07:00 and 09:30.

Fueling—Precautions   Before fueling commences, the following procedures should be carried out:

1. Fueling zones should be established. These zones should extend at least 6 m (20 ft) radially from the filling and venting points on the aircraft and the fueling equipment, regardless of whether it is in use. Within these zones the following restrictions should be enforced:

   a. No smoking is permitted.

   b. If the APU exhaust discharges into the zone, then it should not be started at any time during refueling, to protect against the APU exhaust igniting the fuel vapors.

   NB: If the APU is on prior to refueling, then it is safe to leave the APU running during refueling.

   c. Ground power units should be located as far away as practical from the fueling zones and not connected or disconnected while fueling is in progress, to protect against any electrical arching igniting fuel vapors.

   d. Fire extinguishers should be located so as to be readily accessible.

   During fueling these procedures should be carried out.

2. The aircraft should be earthed and bonded to the fueling equipment.

   NB: For over wing gravity refueling, the hose nozzle should be bonded to the aircraft structure before the tank filler cap is removed. For pressure refueling, the tank pressure relief valves, where fitted, should be checked, if possible; and the fuel hose bonding lead should be connected before the fueling nozzle is attached.

3. Passengers embarking and disembarking the aircraft during fueling operations should do so under supervision, and their route should avoid the refueling zones.

Fuse   An electric fuse is a piece of wire with a low melting point which is placed in series with the electrical load and melts, blows, or ruptures when a current with a higher value than its amperage rating is placed upon it. Thus the fuse protects the electrical load/equipment from excessive power surges.

There should be a minimum of 10 percent (or at least three) spare fuses of the total number of each rated fused installed.

Fusible (Tire) Plugs   Fusible plugs offer protection from tire blowouts caused by thermal expansion that is generated in the tire under extrahard braking conditions. These fusible plugs are fitted in tubeless wheel hubs by means of a fusible alloy that melts under excessive heat conditions and allows the plug to be blown out by the tire air pressure. This prevents excessive pressure buildup in the tire by allowing the air to leak away slowly.

The fusible plugs are made to three different temperatures and are color-coded for ease of identification.

Red   155°C
Green   177°C
Amber   199°C

G

Gas Turbine Engine *See* Jet/Gas Turbine Engine.

Generators *See* AC Generator and DC Generator.

Geometric Pitch This is the theoretical distance a propeller blade moves forward in 1 revolution (r).

Geostrophic Force *See* Coriolis Force.

Geostrophic Wind The geostrophic wind is the balanced flow of air from a wester- ly direction that is parallel to straight isobars with a low-pressure system to its left (i.e., Buys Ballot's Law) and at a strength directly proportional to the spacing of the isobars (i.e., pressure gradient). It is usually found at low/medium heights, from approximately 2000 ft and above.

The geostrophic wind is created when (1) the pressure gradient force and (2) the Coriolis (geostrophic) force are balanced.

The pressure gradient force gets the air moving toward the pole, and the Coriolis force turns it to the right (or left in the southern hemisphere), as a curve over the earth's surface, increasing its speed relative to the earth's surface. The pressure gradient force remains constant for a given pressure gradient, but as the air- flow/wind speed increases, the geostrophic (Coriolis) force, which always acts at right angles to the wind, also increases. It reaches an equilibrium where the geostrophic (Coriolis) force balances the pressure gradient force, which produces the balanced geostrophic wind.

See Fig. 79 at the top of the next page.

g Force Positive *g* force is the influence of the force of gravity on the normal ter- restrial environment beneath it. This is perceived as the normal weight of any body, including ourselves, in the terrestrial environment, that is, 1*g*. Acceleration in an aircraft can subject the body to forces of acceleration much greater than the normal, and these are denoted as multiples of the 1*g* (gravity) force. This is per- ceived as an increase in body weight, which results in making limbs harder to move and objects heavier to lift, say, at 2*g* or 3*g*. This increase in *g* force affects the blood, which is known as *hydrostatic force,* and increases with multiples of the *g* force experienced. It causes the blood to rush downward in the human body, which results in a variation in blood pressure levels in different parts of the body. Namely, the blood pressure in the head is reduced and the blood pressure in the legs is increased. This can result in a lack of blood flow to the eyes and brain that causes a graying out of vision and ultimately an unconscious state. The opposite effect occurs in the lower body where multiple *g* forces drive blood down into the legs, making it very difficult for blood to return and to recirculate. These effects occur at approximately 3.5*g*, but with the use of anti-*g* straining maneuvers these effects can be delayed until approximately 7*g* to 8*g*.

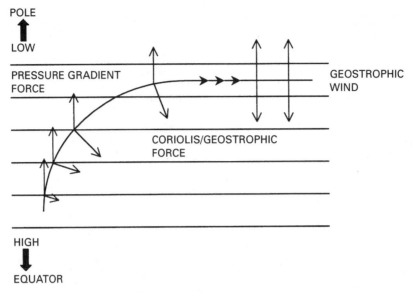

POLE

LOW

PRESSURE GRADIENT
FORCE

GEOSTROPHIC
WIND

CORIOLIS/GEOSTROPHIC
FORCE

HIGH

EQUATOR

Figure 79   Geostrophic wind—balanced forces.

Negative g force is the opposite of positive g force; i.e., it is the influence of the normal terrestrial environment above the force of gravity, or $-1g$. Negative g occurs during aircraft maneuvers such as inverted flight, outside loops, bunting, and some forms of spinning. The effects on the human body are opposite to those of positive g and much more uncomfortable. It causes blood to rush upward from the lower body, causing the heart to slow down and the face and eyes to experience pain as the blood pressure builds inside the head. This makes negative g much more dangerous than positive g. The maximum tolerance level is $-3g$ and then only for short periods.

Glide Range   A maximum glide range is achieved by producing a maximum lift/drag ratio, obtained by the aircraft being flown at its optimum angle of attack and corresponding VIMD speed.

The best glide range and time can be calculated as follows:

- Altitude divided by ROD, e.g.,

$$\frac{30,000 \text{ ft}}{1000 \text{ ft/min}} = 30 \text{ min}$$

- Time $\times$ speed per minute, e.g.,

$$30 \text{ min} \times 5 \text{ nm/min} = 150 \text{ nm}$$

NB: Weight does not affect the best glide distance.

The glide range does not vary with weight, provided the aircraft is flown at its optimum angle of attack and speed for that weight, because the glide range is proportional to the lift/drag ratio, which does not vary with weight.

Therefore, if a heavy aircraft were flown at the correct angle of attack and speed, it would glide the same distance as a lighter aircraft. However, the heavier aircraft will have a higher airspeed than the lighter aircraft, and although it will glide the same distance, it will take less time to do so.

Governor Unit   *See* Constant-Speed Unit.

GPS (Global Positioning System)   This is the U.S. Department of Defense (DoD) satellite system for worldwide navigation. The GPS consists of 24 satellites of which 21 are operational at any given time. This constellation of satellites is broken down into six orbital planes, each consisting of three or four satellites. Each circular orbital plane is at 55° to the equator, with the satellites at a height of 20,200 km (10,900 nm), and covers an orbit of the earth every 12 h. Four satellites will always be in line-of-sight range of an aircraft receiver at any position on the earth, at any given time. The orbiting satellites transmit accurately timed radio signals, and the receiver equipment uses the time delay between transmission and reception to calculate its distance from the satellites. The distance measured from two satellites will establish a latitude and longitude fix. The distance measured from a third satellite will confirm the fix, and the fourth satellite will give altitude information.

NB: GLONASS is the Global Orbiting Navigation Satellite System operated by the Commonwealth of Independent States (CIS), formerly the Soviet Union, and is similar to the U.S. Global Positioning System.

The advantages of the GPS are as follows:

1. Truly global

2. High-capacity use

3. High redundant satellite capacity system

4. Built-in confirmation of the aircraft's position from the third satellite

5. Cornerstone of future air navigation systems, including global landing systems, which offers the advantage of a curved approach path and missed-approach guidance information, at a lower installation cost compared to a normal ILS facility (The first GLS received FAA approval in 1997.)

6. Ability to integrate GPS into other flight management systems

7. Potential to be very accurate

8. Ability to fly great circle tracks accurately

9. Free (although rumors of imposed charges are often floated)

There are two main disadvantages of the civil use of the GPS:

1. *Selective availability*
   The potential accuracy of the system is deliberately downgraded for civil users by the U.S. DoD for security reasons. This is known as selective availability, but a better description is selective unavailability. This is achieved by making available to civil users only the coarse acquisition (C/A) modulation which achieves a lower accuracy level, i.e., to ±100 m on 95 percent of occasions. *Precise* (P) modulation has a greater accuracy of ±30 m but is available only to U.S. military users. This highlights the potential problems associated with using a system that is controlled by a third party. However, as part of a 1996 Presidential directive, President Clinton committed to discontinue selective availability in 2006; but in fact the White House decided to shut it off 6

years ahead of schedule on May 1, 2000. This move has allowed civil GPS users to enjoy the same high accuracy as that enjoyed by the U.S. military. It will also help the U.S. FAA in its commitment to adopting a GPS-based sole-means navigation system for civil aviation in the United States.

2. *System Errors*

a. *Clock bias* is an error in the range calculation between the satellite and the receiver that occurs because the clock in the receiver is less accurate than the atomic clock in the satellite, and therefore the synchronization of signals is impossible. Therefore a correction for clock bias has to be applied to obtain the true range.

b. *Satellite ephemeris* is an error caused by the satellite not being where it says it is. Control of position is so stringent that a difference between actual and believed satellite positions in the region of ± 0.5 m is recognized as a satellite ephemeris error.

c. *Ionosphere propagation* (or atmospheric propagation error) is an error due to the diurnal and seasonal changes that cause variations in the density of the charged particles in the ionosphere. This is a predictable error of up to ± 4 nm.

d. *Instrument receiver* error is caused by
   (1) Electrical noise
   (2) Computational errors
   (3) Matching the pseudorandom signals (pseudorange is the uncorrected clock bias range)

   However, this error is only expected to be in the region of 1 m.

e. *Satellite geometry* induces an error when there is a shallow angle of cut between the satellite and the receiver, which makes the fix liable to be less accurate than if the lines were to cut at an angle of 90°. Also if two or more satellites are close together, then the angle of cut will also be very shallow, thus introducing an error. The GPS term for this is *geometric dilution of precision* (GDOP).

f. *Signal jamming.* GPS satellites are susceptible to jamming, although this has been cleared up since the system was first invented.

Therefore the normal GPS is unsuitable for precision landing approaches given these poor accuracy levels.

See also Differential GPS.

GPWS   Ground Proximity Warning System (GPWS) is now mandatory equipment for public transport aircraft weighing greater than 5700 kg. It is essentially a central computer system that receives various data inputs on configuration (flap and gear), height/altitude (radio altimeter and barometric altitude), ILS glide slope deviation, approach minima and throttle position, etc. It then calculates these inputs to detect if any of the following dangerous and/or potentially dangerous circumstances exist:

1. Excessive barometric rate of descent

2. Excessive terrain closure rate

3. Height loss after takeoff

4. Flaps or gear not selected for landing

5. Too low on the ILS glide slope

6. Descending below approach minima

These circumstances form the six main working modes of the GPWS. Each mode has a precomputerized *active* range with distinct boundaries. At the *initial alert boundary,* a potential danger to the aircraft's safety exists, and at a *warning boundary,* a present danger to the aircraft's safety exists. A warning usually follows an alert if the condition persists. However, some modes (5 and 6) do not have a warning boundary, only an alert.

If the central computer determines that a boundary of any of the various modes has been exceeded, it will activate aural and visual alerts to warn the flight crew, using the following cockpit equipment.

- Speakers

- A pair of red warning lights with "Pull Up"

- A glide slope deviation inhibited amber advisory light

The initial action required for a GPWS alert is a *corrective response*; i.e., "too low flaps" require the corrective action of lowering the flaps accordingly. For a GPWS warning the *immediate* action required is a full wind shear go-around, i.e., maximum go-around thrust and climb attitude. The captain can decide to override a GPWS warning when

1. The aircraft is 1000 ft vertically clear of clouds.

2. The aircraft is 1 km horizontally clear of clouds.

3. There is 8½ km of clear visibility.

4. It is daytime.

5. It is obvious to the captain that the aircraft is not in danger.

When more than one warning or alert is triggered, the GPWS will give only the highest-priority warning or alert. The order of priority of GPWS modes (not including wind shear) is as follows:

Highest priority—"Whoop, Whoop, Pull Up"	Modes 1, 2, 3, 4
"Terrain, Terrain"	Mode 2
"Too Low Gear"	Mode 4*a*
"Too Low Flap"	Mode 4*b*
"Minima, Minima"	Mode 6
"Sink Rate, Sink Rate"	Mode 1
"Don't Sink, Don't Sink"	Mode 3
Lowest priority—"Glide Slope, Glide Slope"	Mode 5

The pilot's order of priority when given wind shear, GPWS, and TCAS warnings at the same time is

First	Wind shear
Second	GPWS
Third	TCAS

There are six *main* GPWS modes, of which some have further subdivisions.

1. *Mode 1: Excessive barometric rate of descent.* As the name suggests, this mode warns of a predetermined excessive rate of descent, as measured from the barometric altimeter. Inputs are received from

- Radio altimeter, to determine active range captures
- Barometric altimeter, to determine rate of descent

This mode is active from a height of 2450 to 50 ft. An aural alert "Sink Rate, Sink Rate" is generated whenever the barometric ROD is greater than 3 times the radio height (the threshold for the alert slackens off below 700 ft to avoid unnecessary warnings), followed by an aural warning "Whoop, Whoop, Pull Up," together with a visual warning of illuminated "Pull Up" warning lights, if the high ROD is maintained after the initial alert.

2. *Mode 2: Terrain closure.*   This mode warns of a rising ground terrain closure beneath the aircraft. Inputs are received from the radio altimeter, and Mode 2 is active from a type-specific height of either

- 1300 to 50 ft on the radio altimeter
- 1800 to 50 ft on the radio altimeter

An aural alert "Terrain, Terrain" is generated when the closure rate with terrain is in excess of 2000 ft/min within the mode active range. It is followed by an aural warning "Whoop, Whoop, Pull Up," together with a visual warning of illuminated "Pull Up" lights if the terrain closure is maintained after the initial alert.

NB: Mode 2 is subdivided into 2*a* and 2*b* depending on whether the landing flaps are selected or not.

3. *Mode 3: Sinking flight path after takeoff or go-around.*   Mode 3 warns of a sinking flight path after a takeoff or during a go-around. Inputs are received from the radio altimeter, and the mode is active from a height of 50 to 700 ft whenever the aircraft is in a typical takeoff or go-around configuration, i.e., with the flaps not in a landing position, the gear up, and the throttles well advanced. These preconditions prevent the mode 3 alert from going off on the approach.

An aural alert "Don't Sink, Don't Sink" is generated when an ROD of 40 ft or more is detected in the mode's active range. It is followed by an aural warning "Whoop, Whoop, Pull Up," together with a visual warning of illuminated "Pull Up" lights, if the sinking flight path is maintained after the initial alert.

NB: Sometimes this mode is inadvertently triggered by rising ground under the aircraft that effectively reduces the terrain clearance.

4. *Mode 4: Gear and flap not selected.*   This mode warns of closeness to the ground without the appropriate gear and flap selections. Inputs are received from the displayed position on the flap and gear indicator instrument. This mode is subdivided into two parts.

Mode 4a, gear selection, is active from a height of 500 to 50 ft on the radio altimeter. An aural alert "Too Low Gear" is activated whenever the gear has not been selected within the mode's active range. It is followed by an aural "Whoop, Whoop, Pull Up" warning, together with a visual warning of illuminated "Pull Up" lights if the condition persists after the initial alert.

Mode 4b, flap selection, is active from a height of 200 to 50 ft on the radio altimeter. An aural alert "Too Low Flaps" is activated whenever the flaps have not been selected to a landing configuration within the mode's active range. It is followed by an aural "Whoop, Whoop, Pull Up" warning, together with a visual warning of illuminated "Pull Up" lights if the condition persists after the initial alert.

5. *Mode 5: ILS glide slope deviation.*   This mode warns of a glide slope deviation. Inputs are received from the ILS glide slope signal and the aircraft's relative position, and the mode becomes active on glide slope capture. An aural alert

"Glide Slope, Glide Slope" is generated, and a visual "Pull Up" light illuminates whenever the glide slope deviation is greater than 2 dots (or 1.3 dots on some types) fly up command on the ILS. Mode 5 has *no* associated warning, as mode 3 becomes the priority active mode if this condition exists.

NB: Mode 5 can be canceled by pressing one of the "Pull Up" light or switch.

6. *Mode 6: Approach minima.*   This mode warns the crew that the minimum descent height has been passed. Inputs are received from the radio altimeter, at the minimum descent height, which is usually taken from the FMS, or the bugged height on the radio altimeter. An aural alert "Minima, Minima" is generated whenever you pass the minimum decision height. No warning boundary exists with this mode.

7. *Mode 7: Wind shear warning.*   This mode warns of a wind shear condition.

Wind shear warning is sometimes classified as a GPWS mode because it utilizes the GPWS flight crew visual and aural warning system. However, strictly speaking, it is not a ground proximity warning but a separate system. But because it is often perceived as a GPWS mode, we classify it as both an additional GPWS mode (mode 7) as well as a separate warning system.

Gradient Wind   The gradient wind is the resultant wind that blows around the curved isobars common to circular low- or high-pressure patterns.

NB: The geostrophic wind assumes straight, west-to-east isobars.

It is usually found at low/medium heights, from approximately 2000 ft and above.

Ergo a gradient wind has to have an additional force/effect, besides the Coriolis/geostrophic force and the pressure gradient force, to produce the resultant force to keep the air flowing (wind) in a circle parallel to the curved isobars.

The additional force needed to keep the wind going in a circle is called the *cyclostrophic force.* This additional (cyclostrophic) force acts as a resultant, or net, force on the geostrophic wind to pull the airflow into the turn, and it accelerates the airflow in the sense that its direction is changing.

NB: For a wind that is blowing anticlockwise (around a low in the northern hemisphere), the net force results from the pressure gradient force being greater than the Coriolis force, thereby pulling the air in toward the low. For a wind that is blowing clockwise (around a high in the northern hemisphere), the net force results from the Coriolis force being greater than the pressure gradient force, thereby pulling the air in toward the high. Since the Coriolis force increases with speed, it follows that the wind speed around a high will be greater than wind speed around a low for the same isobar spacing.

The gradient wind is such that if you stand with your back to the gradient wind, in the northern hemisphere, the low pressure will be to your left.

GRADU   GRADU relates to the gradual change of the original weather at approximately a constant rate throughout the period or during a specified part thereof, to a different and new weather condition.

Great Circle Track   A great circle track is a line of shortest distance between two points on a sphere (or a flat surface) with a constantly changing track direction as a result of convergence.

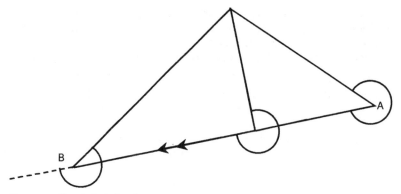

Figure 80   Great circle track.

Great circle tracks are always found on the pole side of the rhumb line. Meridians and the equator are both great circle tracks and rhumb lines.

Gross Flight Path/Performance   *See* Net and Gross Flight Path/Performance.

Ground Effect   Ground effect is the cushioning of the aircraft on the air between it and the ground when the aircraft is close to the ground, such as in the flare to land.
NB: In effect, the cushioning effect produces more lift.
Ground effect is caused by rising warm air off the ground, by a reduction in the amount of downwash behind the wings, and by the tendency of the aircraft not to slow (decreased drag) resulting from the reduction of wing tip vortices.
In summary, the two main results of ground effect are

- An increased *lifting ability* of the wing (i.e., increased C_L)
- A reduction in drag, i.e., a lower formation of vortices and less induced drag

Both of these cause a floating effect near the ground.
There are two further points to note about ground effect:

- The disturbance in airflow may cause the airspeed indicator and the altimeter to read inaccurately when flying near the ground.
- The disturbance of normal airflow as compared to free air may cause a change in the stability characteristics of the aircraft near the ground.

Ground effect will, if allowed, cause the aircraft to float along the runway past its touchdown point. This results in an increased landing distance.

Ground Proximity Warning System   *See* GPWS.

Ground Speed   Ground speed is the actual speed of an aircraft relative to the ground. It is a combination of the TAS (the speed of an aircraft through an air mass) and W/V (wind velocity represents the movement of an air mass relative to the ground).
Therefore ground speed can be greater than the TAS if it has a tailwind, less than the TAS if it has a headwind, or equal to the TAS in still-air conditions. Ground speed is used for navigation and flight planning purposes.

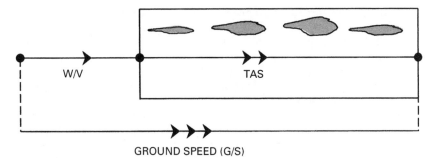

GROUND SPEED (G/S)

Figure 81 Ground speed.

Gyroscope   A gyroscope is a body (usually a rotor/wheel) rotating freely in one or more directions that processes the gyroscopic properties of rigidity and precession.
The properties of a gyroscope are

■ *Rigidity,* which is the tendency of a body to maintain the same absolute direction in space despite whatever happens around it, i.e., to maintain upright while the gyroscope is rotating and to fall over when it stops rotating. The rigidity of a system depends upon its angular momentum, where

Angular momentum = moment of inertia × speed of rotation

Therefore, the greater the moment of inertia and/or speed of rotation (faster the spin), the greater the gyro's rigidity.

■ *Precession,* which is the tendency of a force on a rotating body to act at a displaced point on the body, namely, at 90° in the direction of its rotation. That is, where a force is applied (applied force) to the bottom of a gyroscope that is spinning around a horizontal axis, the applied force is precessed by 90°, the *precessed force*. The greater the rigidity of a body, the greater the precession force required.

A gyroscope measures the *force* experienced on its rotor body during a maneuver of the aircraft. The rotor is usually suspended in a system of frames called gimbals, which are arranged at right angles to each other and are used as a conduit to

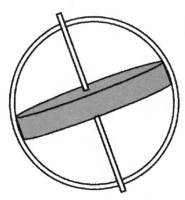

Figure 82 Gyroscopic precession.

transfer the force experienced on the rotor to a displayed measurement on the instrument face.

The gyro's rigidity allows it to remain stable in space while the aircraft moves around it. Any force applied to the gyro, as a result of the aircraft changing direction or attitude around the gyro, will *precess* the gyro. This precess is translated to a movement of the gimbals that are attached to a direction scale (on the DI) or an attitude horizon bar (on the AH) which measures and displays the movement of the aircraft on the instrument face. One gimbal is required for each axis around which you wish to measure a precessed movement. The spin axis of the gyro and the axis of precessed movement measured by the gimbal represent a gyro's *planes of freedom,* and a gyro is classified in terms of its planes of freedom. In total a gyro can have three planes of freedom—the spin axis of the gyro and two other possible planes at right angles to the gyro's spin axis and to each other. A gyro's rigidity is all-important and is itself a product of the spin (drive) speed of the rotor, which can be either vacuum (air) or electrically driven.

A *vacuum (air) driven system* draws high-speed atmospheric air through a nozzle and directs it at the gyro rotor blades or buckets, to produce a force on the rotor to spin (drive) it around its axis at high speeds, i.e., 10,000 rpm, and therefore it is very rigid. A vacuum pump that draws air in (suction) is generally preferred, and the suction rate is shown on a suction gauge in the cockpit. A normal working reading is in the range of 3 to 5 inches of mercury (inHg) below atmospheric pressure. If the vacuum reading is too low, then the airflow is low and the rotor will not be up to speed; therefore the gyro/rotor will be unstable, or will only respond slowly to any precessed force. If the airflow is too high, then the gyro/rotor may spin too fast and could be damaged. Many gyroscopes are operated by a vacuum system, which is usually provided by an engine driven vacuum pump.

An *electrically driven gyroscope* can be driven using an electrical supply (usually dc battery power) to spin (drive) the gyro about its axis. It self-erects and is very reliable and accurate. If the electric power supply to the gyro is interrupted, then the instrument will indicate incorrectly and a red "Power Failure" warning flag will appear on the instrument.

Finally the gyro has to be set up, first, by its orientation in space. Specifying the direction of the axis about which the gyro rotates does this. Specify either vertical gyro, which has its spin axis in the vertical, or horizontal gyro, which has its spin axis in the horizontal. The gyro has to be set up, second, by its alignment, e.g., aligned to true north. These two properties define a gyro; i.e., a horizontal gyro with its axis aligned to true north. A gyro's correct alignment and orientation are vital, because if the direction of the spin axis is not correct, then any measurement of the gyro will be starting from an incorrect alignment and therefore the gyro will produce an incorrect measurement.

Gyroscope—Apparent Wander   *See* Apparent Wander.

Gyroscope—Caging System   *See* Caging System.

Gyroscope—Real Wander   *See* Real Wander.

Gyro—Transport Wander   *See* Transport Wander.

Gyro—Wander   *See* Wander.

Haze Haze is reported when suspended solid particles, e.g., dust, smoke, and sand, reduce the horizontal visibility.

NB: If the visibility in haze falls to below 1000 m, the cause of the haze will be reported, e.g., smoke haze.

Head-Up Display (HUD) HUD is displayed on a glass panel or the cockpit window at eye level ahead of the pilot, and normally it consists of EADI information, i.e., speed, attitude, flight director bars, etc.

Head-Up Guidance (HUG) This is displayed on a glass panel or the cockpit window at eye level ahead of the pilot, and normally it consists of EADI information and additional navigation guidance raw material, i.e., localizer and glide slope.

NB: HUG allows CAT III ILS landings on runways only certified for CAT I landings.

Headwind Component *See* Wind Component.

Heat Heat is a form of energy that is measured in calories. One calorie (1 cal) represents the energy required to raise the temperature of 1 g of water 1°C. The energy required to raise the temperature of 2 g of water by 1°C would be 2 cal, etc.

The sun in the form of solar radiation provides the earth's heat and light. The wavelengths of solar radiation are such that only about one-half of the radiation that hits the upper atmosphere finally reaches the earth's surface.

NB: The rest is either reflected back into space or absorbed into the upper layers of the atmosphere.

This shortwave radiation is absorbed by the earth's surface, causing its temperature to increase. This is called *insulation.* Energy is then reradiated out from the ground as long-wave radiation (in the 10-μm band). It is this radiation that heats up the lower atmosphere, near the ground, with the result that any parcel of air warmer than the surrounding air will rise.

The heat held in the earth's surface can be transferred to the atmosphere by

1. *Radiation.* The sun emits shortwave radiation that is then reradiated (off the earth's surface) as electromagnetic long-wave energy that is absorbed by and heats CO_2 and water vapor in the lower atmosphere. The amount of energy radiated will depend upon the surface type and temperature.

2. *Conduction.* Conduction is the transfer of heat between two bodies in contact with each other, such as the earth's surface and the atmosphere. The flow of heat energy increases as the temperature difference between two bodies increases and is always in the direction of the lower temperature. A parcel of colder air in contact with the ground is poorly heated by conduction, and does not transfer this heat energy to neighboring parcels of air, because both the earth's surface and the atmosphere are poor conductors of heat. (See item 3, convection.)

This is very significant in the production of weather systems in the lower layer of the atmosphere to which this phenomenon is restricted.

3. *Convection.* Whenever air is heated, by either radiation or conduction, it will expand, become less dense, and therefore rise. This is known as *thermals.* Such a body in motion (i.e., vertical movement due to heating) has its heat energy transferred with it. This is known as *convection.* The vertical movement of a parcel of air leads to a horizontal movement of air, with its own heat and moisture content, across the earth's surface which replaces the air risen by convection (e.g., land/sea breezes). This is known as *advection.* If there are areas with different surface temperatures, then patterns of vertical and horizontal airflow circulation can occur on a large global scale (e.g., regional Hadley cells) or on a smaller localized scale (e.g., land/sea coastal breezes). Such movements help to distribute the heat through the atmosphere and contribute considerably to the complexity of weather patterns.

These three processes of heat transference contribute considerably to the complexity of weather patterns.

During the day, the cloud cover stops the incoming solar radiation from penetrating the earth's surface, by reflecting it back into space. This causes the heating of the earth's surface and thereby the lower layers of the atmosphere next to the ground to be reduced and results in lower temperatures and less atmospheric convective movement.

During the night, the cloud cover creates the opposite effect, by trapping the heat energy in the lower atmosphere, on and near the ground, and reflecting it back to the surface, in effect recirculating the heat. This results in the lower atmosphere, beneath the clouds, maintaining a higher temperature because it experiences less cooling than on a clear night.

It is important to clearly distinguish between heat and temperature. Heat is a form of energy which, once absorbed by a substance, agitates its molecules. Temperature is a measure of the degree of molecular agitation in the substance, which is represented as the hotness of the body. Therefore measuring the temperature of a body is not the same as measuring the heat energy within a body, because the amount of heat energy corresponding to a given temperature depends upon the capacity of the substance to hold the heat energy. This is known as *specific heat capacity.* However, in a general sense the temperature of a body gives a rough indication of the heat energy.

Hectopascal *See* Millibar.

Height In aviation terms, height is the measured distance above the ground, that is, QFE.

HF Communications High-frequency (HF) radio transmissions are used for long-distance communications, between two specific distance points only, unlike VHF. It uses predictable sky wave propagation paths, which are refracted off the earth's ionosphere over great distances.

NB: The vertical angle of the sky wave is called the *critical angle.* Higher frequencies require an increased critical angle from the vertical, so that the stronger wavelength can be bounced off the ionosphere at a glancing angle. This results in an increase in range. Range is usually referred to as the *skip distance,* i.e., the distance from the transmission point to the point where the first returning wave is received. A surface wave is the HF signal that follows a surface propagation path

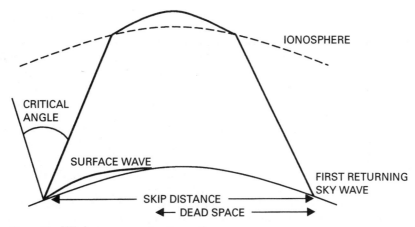

Figure 83 HF sky wave propagation path.

near to the transmitter, on which a communication can also be carried. *Dead space* is the area in between the end of the surface wave and the return of the sky wave, inside which no signal is present and therefore no HF communication is possible.

The HF band used in aviation starts from 2 to 22 MHz. Different frequencies have different range capabilities; i.e., the higher the frequency, the greater its range. Therefore a frequency is chosen by the pilot to meet the range between the transmitting and reception points, and the ground station personnel will monitor a range of frequencies, because they are unsure of the exact distance to the aircraft. Unfortunately the HF selection is not just a product of the ground distance between two points. It also has to account for the ionosphere's prevailing refractive properties, because the HF system utilizes the ionosphere to refract or bounce the sky wave back to the earth's surface. However, the ionosphere, especially its height, is changeable, and this affects the range capabilities for a particular HF band. Therefore, to calculate the exact frequency required, we need to understand the ionosphere's refractive properties.

The ionosphere is a product of the sun, which generates ion particles that make up the ionosphere's layers. The quantity of ion particles generated determines the ionosphere's density and height, which in turn determine its refractive properties. First, the ionosphere's height changes (i.e., has a lower height with increased exposure to the sun), and therefore a sky wave's vertical distance and thus horizontal distance traveled change. Second, the ionosphere's density changes. This is important because the ionosphere has to be dense enough to refract the sky wave; otherwise, the sky wave will just pass through the ionosphere. To match the different density levels in the changeable ionosphere, the sky wave's angle of collision with the ionosphere is altered so that the full force of the sky wave is not exposed to the ionosphere and the HF sky wave is refracted back to the earth's surface. The higher the frequency, the stronger the sky wave, the greater its vertical angle, and therefore the greater its range. Thus the effect of the prevailing density and height levels of the ionosphere on a sky wave has to be calculated to a basic range, to obtain the required refracted propagation path to the desired reception point, to determine the exact frequency required.

This frequency is further refined as a *maximum usable frequency* (MUF) that is slightly lower than the optimum frequency.

NB: The optimum frequency is the frequency at which the first returning sky wave just hits the receiver. However, the optimum frequency cannot be used because of the ionosphere's constantly changing conditions, which cause slight variation in the skip distance that moves the receiver in and out of the dead space, resulting in the optimum frequency signal being constantly interrupted.

Therefore a MUF is preferred because it keeps the receiver in constant usable contact with the sky wave. However, the frequency is kept as high as possible to maintain the signal condition (clarity), because ionospheric attenuation decreases with an increase in frequency. This is due to the fact that higher frequencies have a greater vertical critical angle, which has a lower penetration of the ionosphere.

NB: Ionospheric attenuation is the loss of signal strength due to the sky wave's energy being absorbed by the electrically charged atoms in the ionosphere.

Therefore the clarity of the signal's reception is maximized by keeping the frequency as high as possible to minimize static from ionospheric attenuation.

A HF sky wave can have many returns, known as *multihop*; i.e., after the first return point it is refracted off the earth's surface and back up to the ionosphere and then refracted back to the earth's surface, and so on. Therefore it is possible to receive and transmit a HF signal to a station thousands of miles away when you cannot transit or receive to a HF station much closer to your position. For example, an aircraft positioned over the north Atlantic can be outside the first return point for the HF station at Shannon, Ireland, but the aircraft is able to pick up Athens, Greece.

HF radio transmissions have the ICAO designator J3E:

J = single sideband with suppressed carrier

3 = analog information

E = voice

The following factors affect the range of a HF communication:

1. *Transmitter power.* The greater the power, the greater the possible range, especially for multihop signals.

2. *Frequency.* The higher the frequency, the greater the critical angle from the vertical to ensure its refraction from the ionosphere, and therefore the shallower the sky wave's path and the greater its range.

3. *Time of day.* The ionosphere is affected by the sun. At different times of the day the properties of the ionosphere are different, and therefore the refraction of sky waves varies. For example, frequency at night typically covers twice the distance it does during the day. HF agencies produce propagation charts to guide pilots in their frequency selection.

4. *Season.* The season also affects the ionospheric properties in a similar manner to the time of the day, albeit to a lesser extent. High frequencies in the winter typically cover a greater distance than in the summer.

5. *Location.* In some parts of the world, especially over the polar ice caps, prediction of both frequency and range is impossible, as a result of surface attenuation on the sky waves as they hit the ground. Surface attenuation increases as frequency increases.

6. *Disturbance of the ionosphere.* Solar activity, usually unpredictable, causes changes in the ionosphere that result in a change in density and/or height of the usual ionospheric layers. This changes the range of sky wave propagation paths.

HF Communications at Night (Winter) By night, one-half of the high frequency of the daytime produces the same range (skip distance). This is so because of the variation of the ionosphere. By day, especially in the summer, the sun generates ion particles that make up the ionosphere's D layer at a height of approximately 75 km. This layer is of sufficient density to refract HF sky waves. However, at night, or during winter days when the exposure to the sun is less, the D layer disappears, and therefore the HF sky waves are refracted by the ionosphere's E layer at a height of approximately 125 km. This increases the range (skip distance) of a HF wave because of its greater vertical distance to the higher E layer before it is refracted. Therefore because of the higher ionosphere at night, you need a lower frequency to reach the same receiver distance; typically one-half that of the daytime frequency is needed, because the signal is refracted more.

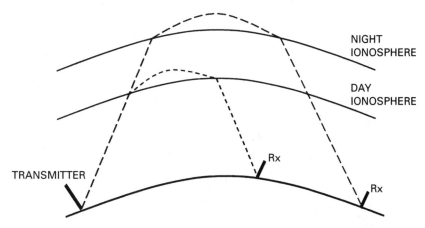

Figure 84 Day/night variation of ionospheric sky wave refraction. Rx = reception.

High-Altitude Cruise (Jet) *See* Cruise—High-Altitude (Jet).

High-Drag Devices These devices increase the drag penalty on an aircraft. Examples of high-drag devices are

1. Trailing-edge flaps (in high-drag/low-lift position)
2. Spoilers—in fight detent used as a speed brake and on the ground as lift dumpers
3. Landing gear
4. Reverse thrust (ground use only)
5. Braking parachute

High-Lift Devices These increase the lift force produced by the wings. Examples of high-lift devices are

1. Trailing-edge flaps (Fowler flaps), which increase lift at lower angles of deflection
2. Leading-edge flaps (Krueger flaps) and slats, which increase lift by creating a longer wing chord line, chamber, and area.

3. Slots (boundary layer control), which prevent/delay the separation of the airflow boundary layer and therefore obtain an increase in the maximum coefficient of lift

High-Pressure System   A high-pressure system, also known as an *anticyclone,* has
the following characteristics:

1. *Pressure gradient*
   In a high-pressure system, the surface barometric air pressure rises as you
   move toward its center; or in other words, its pressure drops as you move away
   from its center.

   NB: A high is depicted on a weather chart by a series of concentric isobars joining places of equal sea-level pressure, with the highest pressure value at its center.

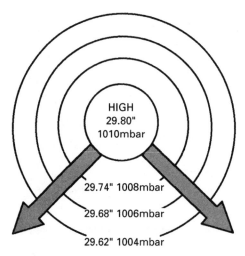

Figure 85   Pressure gradient of a high.

The pressure gradient (which is depicted by the closeness of the isobars)
relates to the wind speed, i.e.,

Weak pressure gradient = wide isobar spacing = weak wind

2. *Airflow pattern*
   The surface pressure at the center of an anticyclone is higher than that of the
   surrounding horizontal areas, i.e., a high-pressure system. However, the pressure gradient at its upper levels is reversed, i.e., creates a low-pressure system,
   where the pressure at the center is lower than that in the surrounding horizontal areas.

   NB: It is important to realize and remember that a surface high-pressure system is associated with a low-pressure system higher up.

   However, the center of a depression has a lower pressure at the surface than
   above it. Because the airflow/wind is essentially pressure-driven, these phenomena create the airflow pattern, and the airflow pattern contributes to the pressure

differentials within the air mass. The surface high-pressure system originates when the convergence of the upper-level air at the center of the system becomes greater than that of the air below it; i.e., converging air at the upper layers is added faster than the diverging air at the lower layers. Therefore the air pressure is greater at height, causing it to sink/descend (become stable) in the center of an anticyclone. At the surface it flows outward (this is known as divergence), because the air pressure at the center of the anticyclone is greater than at the surrounding areas; this is our surface high-pressure system.

The airflow pattern of an anticyclone can be categorized as follows:

a. Convergence (inflow) at the upper layers

b. Subsiding air at the center

c. Divergence (outflow) at the lower layers

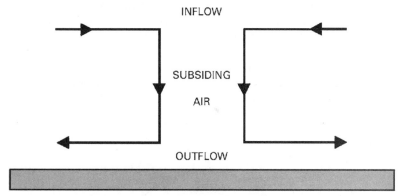

Figure 86 Airflow pattern in an anticyclone.

3. *Wind direction*

   The wind circulates clockwise around a high-pressure system in the northern hemisphere and anticlockwise in the southern hemisphere.

   NB: Flying toward a high in the northern hemisphere, an aircraft will experience left (port) drift.

   However, it should be remembered that above a surface high there is an upper low system; therefore the wind direction will reverse at height in a high-pressure system.

4. *Air mass*

   A high-pressure system will usually be made up of only one air mass and therefore will have no defined fronts.

5. *Movement*

   High-pressure systems are generally greater in extent, but with a weaker pressure gradient (change of pressure with distance) and are therefore slower-moving, more persistent, and longer-lasting than a low-pressure system. High-pressure systems tend to be over the land in the winter and over the sea in the summer, and they are classified as either warm or cold [e.g., Siberian high cold (over land in the winter) or Azores high warm (over sea in the summer)].

6. *Weather*

In general the weather associated with a high-pressure system is as follows:

a. There are clear upper skies with no or few low-level clouds and little precipitation. This is so because the descending *stable* air (subsiding air is very stable) in a high-pressure system is warmed as it descends. Therefore the dew point temperature is increased, and the relative humidity is reduced. Hence cloud formation and any associated precipitation will tend to disperse.

b. Winds are light because a high-pressure system has a weak pressure gradient, and its wind strength is represented by its pressure gradient.

c. Poor visibility is possible at low levels. This is so because the descending air

(1) May bring airborne particles from the upper levels down to the lower levels that pilots fly in.

(2) May warm sufficiently to create an inversion layer, as a result of the upper descending air warming to a greater temperature than the lower surface turbulent layer. This phenomenon may cause stratiform clouds to form and/or trap airborne particles, dust, smoke, and moisture (fog) beneath the inversion layer.

In the northern hemisphere you would experience a drift to the left (port) when flying from a low- to a high-pressure system. This is so because the wind circulates clockwise around a high in the northern hemisphere, and therefore flying toward the center of a high-pressure system, you would have a wind from the right (starboard) causing you to drift to the left (port).

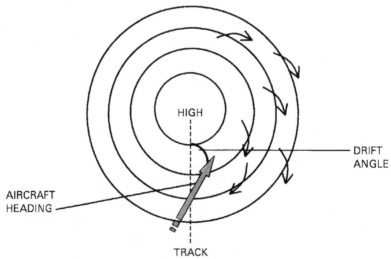

Figure 87   Left/port drift—high-pressure area ahead.

In the southern hemisphere, when you fly from a low- to a high-pressure system the opposite is true—you would experience a drift to the right (starboard). This is so because in the southern hemisphere the wind circulates clockwise around a low, and anticlockwise around a high, and therefore the airflow/wind experienced will be coming from the left, causing you to drift to the right (starboard).

Hinge/Horn Balance   A *setback hinge* is another form of aerodynamic control balance on a control surface. The design of the control surface sets the hinge line back into the control surface, thereby reducing the center of pressure (cp) to hinge line arm, which reduces the control surface hinge moment.

Remember:

$$\text{Hinge moment (air loads)} = \text{lift force} \times \text{arm}$$

Thereby the stick control force that reflects the overall hinge moment experienced on the control surface is reduced to a manageable level for the pilot.

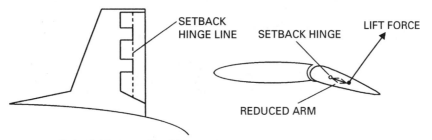

Figure 88   Setback hinge—reduced hinge moment.

Another form of aerodynamic control balance on a control surface is a horn balance. A horn balance is a *protruding* control surface that produces a balancing lift force in the opposite direction from the main lift force, which reduces the overall hinge moment/air load force (e.g., this is common on elevator/stabilizers). Thus the stick control force that reflects the overall hinge moment experienced on the control surface is reduced to a manageable level for the pilot.

NB: A horn balance and setback hinge are typically used in tandem on a control surface.

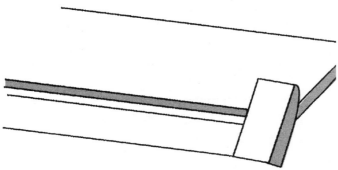

Figure 89   Horn balance.

Hold(ing)   A holding procedure is a predetermined maneuver which keeps an aircraft within a specified airspace while it awaits further clearance.

A holding procedure/pattern is generally a racetrack shape, with the following five basic elements:

1. A holding fix, normally based on an NDB or VOR radio navigation aid such as one of the following:
   a. Overhead aid
   b. A distance on a radial or bearing
   c. A cross fix of two radials/bearings from two different radio navigation aids
2. An inbound holding radial (VOR) or bearing (NDB)

   NB: The outbound leg usually has no tracking aid.

3. The location to the holding radial or bearing which is relative to the eight main points of the compass
4. Direction of the turns
5. Timing

   NB: A sixth element is altitude or height, whereby minimum and/or maximum height or altitude limits may be assigned to a holding procedure.

A holding pattern is designed with a specific protected area around it. The protected area is complex to calculate, involving many formulas:

- Outbound time or distance
- Speed of the aircraft
- Bank angle and rate of turn
- Fixing accuracy
- Wind velocity
- Temperature
- Timing tolerance
- Heading tolerance
- Entry procedures

It is often assumed that the protected area around a holding pattern is box-shaped; this is not so. However, if the aircraft is flown within these limits, it will keep the aircraft clear of protected airspace, terrain, etc.

The standard holding pattern direction comprises right-hand turns.

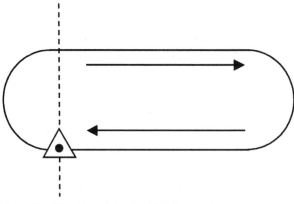

Figure 90   Direction of standard holding pattern.

Therefore the nonstandard holding pattern direction comprises left-hand turns.

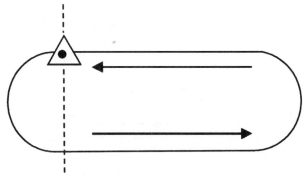

Figure 91   Nonstandard holding pattern direction.

There are three entry procedures into a holding pattern. The three sector regions have been devised based upon the direction of the inbound holding track and an imaginary line angled at 70° to the inbound holding track, through the fix. The three entry procedures based upon the sector of entry are as follows:

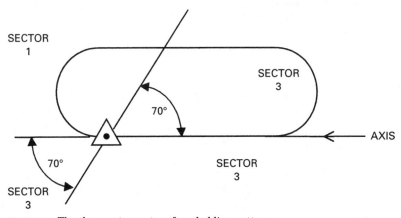

Figure 92   The three entry sectors for a holding pattern.

However, although the sectors are determined by track lines from the fix, the sector from which the aircraft joins the holding pattern, and therefore the entry procedure employed, is determined by the aircraft's *heading* to the fix.

NB: Be aware that you may be approaching a holding fix with a track in one sector but a heading in another sector. It may feel strange, but the sector that relates to the aircraft's heading determines the entry procedure employed.

Sector 1 procedure: parallel entry.   Fly to the fix and turn onto an outbound heading to fly parallel to the inbound track, on the nonholding side for a period of 1 min plus or minus 1 s/kn wind correction. Then turn in the direction of the holding side through more than 180° to intercept the inbound track to the fix. On reaching the fix, turn to follow the holding pattern.

SECTOR 1

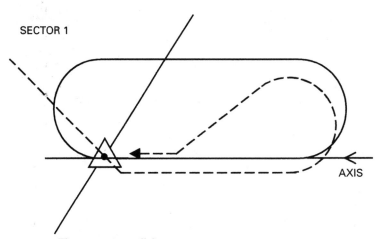

Figure 93   The sector 1 parallel entry.

Sector 2 procedure: teardrop entry.   Fly to the fix and turn onto a heading to fly a track on the holding side, at 30° offset to the reciprocal of the inbound track for a period of 1 min plus or minus 1 s/kn wind correction. Then turn in the direction of the holding pattern to intercept the inbound track to the fix. On reaching the fix, turn and follow the holding pattern.

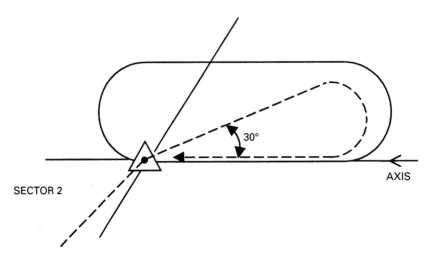

Figure 94   The sector 2 teardrop entry.

Sector 3 procedure: direct entry.   Fly to the fix and turn to follow the holding pattern. On the face of it, the sector 3 direct entry procedure is the easiest to carry out. But a little thought will show that when one is joining from the extremities of the sector area, it is necessary to apply some finesse to the procedure. If a full 180° or greater turn is required over the fix when joining to take up the outbound heading, then commence turning immediately when you reach overhead the fix.

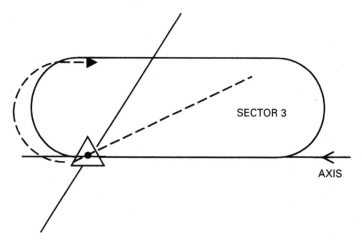

Figure 95   The sector 3 direct entry, >180° turn.

If, however, the turn onto the outbound heading is less than 180° but greater than 70°, then hold your heading for an appropriate time past the fix, approximately 5 to 15 s, before commencing a rate 1 turn onto an outbound track. For example, for a turn of 170°, hold your heading for 5 s. For a turn of 70°, hold your heading for 15 s.

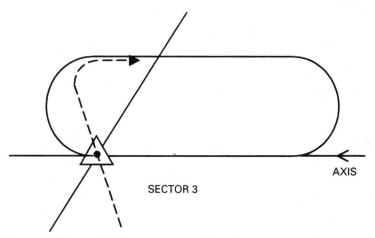

Figure 96   The sector 3 direct entry turn is less than 180° but greater than 70°.

If, however, the turn onto the outbound heading is close to 180°, then you are already very close to or on the inbound track of the holding pattern; i.e., your joining track is close to the hold's inbound track. Therefore, simply turn onto a normal outbound track from over the fix, with only small finesse adjustments, as if you are already in the holding pattern.

The aim of flying a holding pattern is to arrive back overhead the fix in an approximate total time of 4 min, or 3 min when timed from abeam the fix (for holds at or below 14,000 ft) in an *approximate* racetrack shape, which establishes the aircraft on the inbound track prior to arriving back overhead the hold's fix.

There are many different methods that achieve the aim of flying a hold. The following description is one method that is offered as an example only. There are many other correct methods, and the reader should be aware of the variations when contemplating this question. In the opinion of the author, flying a hold, especially an NDB hold, is one of the most difficult procedures to fly. And the best way to plan a hold is to be aware of the requirements to come and to manage the procedure one step at a time. Therefore I offer the following description, step by step.

1. *The start.* The hold commences overhead the fix at the end of the joining procedure (and not when starting the joining procedure), at which point the RT call is "Call sign...entering the hold" (as opposed to a "Call sign...joining the hold" when overhead the fix at the start of the joining procedure).

2. *Turn outbound.* Over the fix, immediately commence an accurate rate 1 turn onto the outbound heading.

NB: All turns in a holding pattern should be standard rate 1 turns, that is, 3° per second.

3. *Outbound track/leg.* There is usually no radio aid tracking for the outbound track, which is simply the reciprocal of the inbound track. However, the outbound leg is probably the most important part of the maneuver. If it is flown accurately, it will position the aircraft for a successful turn onto the critical inbound track.

In still air the ground track of the holding pattern will be a straightforward racetrack. However, with a crosswind condition, the outbound heading should be plus or minus single drift until you reach the abeam of the fix. At the abeam fix point, adjust your heading to plus or minus 2 times the drift if the crosswind is less than 60° off the outbound track, and to 3 times the drift if the crosswind is more than 60° off the outbound track. In both cases a maximum of 30° wind correction angle/drift should be applied to your heading. This simple system compensates for the drift experienced in the turns as well as the outbound leg; and in tandem with good rate 1 turns, you should approximately achieve the inbound track.

Start timing the outbound leg from abeam the fix or when on the outbound heading, whichever comes later. If you have a headwind on the outbound leg, this will tend to keep the outbound turn inside the abeam fix point; therefore you should achieve the outbound heading prior to reaching the abeam fix point. Conversely, if you have a tailwind on the outbound leg, this will tend to push the outbound turn past the abeam fix point; therefore you are likely to achieve the outbound heading after the abeam fix point. The outbound track timing should be 1 min at or below 14,000-ft and 1½ min above 14,000-ft altitude. For tailwind or headwind components, a reasonable correction is to reduce the outbound leg by 1 s/kn of tailwind and to increase the outbound leg by 1 s/kn of headwind.

Heading and timing are the two basic rules for the modification of a holding pattern. In addition to applying these, useful information can be gained at the end of the outbound leg when the *gate* position can verify whether the aircraft is in the correct position.

At the end of the timed outbound leg, if the aircraft is in the correct position, the QDM to the fix, known as the gate, should be about 30° from the inbound track in still air. In crosswind conditions the 30° gate angle should be modified by single drift. Namely, an *open gate* is when the outbound leg has been altered outward, and the correct gate angle should be 30° plus the single drift. Conversely, a *closed gate* is when the outbound leg has been altered inward, and the correct gate angle should be 30° minus single drift. However, the gate reading must *not* be used instead of timing as an indication of when to commence the inbound turn; it is just a useful guide to an impending overshoot or undershoot of the inbound track.

4. *Turn inbound.* Once the outbound time has elapsed, immediately commence an accurate rate 1 turn onto the inbound heading. The turn should be increased or decreased to achieve the inbound track. There are two indications to the progress of the inbound turn. The first is the gate position at the start of the turn. Namely, if your gate angle is greater than expected, then you are likely to undershoot the inbound track. Therefore decrease your rate of turn. However, if the gate angle is less than expected, then you are likely to overshoot the inbound track. Therefore increase your rate of turn, keeping within limits. For the second indication of progress, when you are turning within 60° of the inbound track, the ADF needle should indicate 10° to the inbound track (if the aircraft suffers from magnetic dip error, then the needle will indicate the inbound track at this point) if the turn is to roll out on the inbound track. If, however, the needle indicates an angle to the inbound track greater than 10°, then this indicates that you will undershoot the inbound track. Therefore you should roll wings level and fly straight until the needle indicates 10°; then you should continue the turn onto the inbound track. If, however, the needle indicates an angle to the inbound track of less than 10°, then this indicates that you will overshoot the inbound track. Therefore you should increase your rate of turn, keeping within bank limits, and be prepared to turn back through to intercept the inbound track.

5. *Inbound track / leg.* Roll out and *establish* to within 5° of the inbound track, and maintain the track over the fix. If you are not within 5° of the track, make bold ADF adjustments initially but be aware of the increasing sensitivity as you approach the beacon.

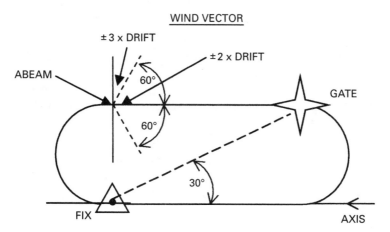

Figure 97 Hold adjustments.

NB: The main change to this method, that is commonly used, is to start the timing over the fix, thereby making the timing of the hold 4 min long.

Normally the timing in a hold should be commenced from abeam the fix at the start of the outbound leg or on attaining the outbound heading, whichever comes later.

NB: Care should be exercised about when the outbound timing is commenced, in compliance with EAT (expected approach time) or OCT (obstacle clearance time), and when the timing is modified for the effect of the known wind.

The outbound timing should be 1 min for a standard holding pattern up to and including 14,000 ft and $1^1/_2$ min above 14,000 ft.

NB: The pilot should make corrections to the timings (and headings) to correct for the effects of the known winds, during the entry and when flying the holding pattern. For tailwind or headwind components, a reasonable correction is to reduce the outbound leg by 1 s/kn of tailwind and to increase the outbound leg by 1 s/kn of headwind.

A complete holding pattern (at or below 14,000 ft) will, if flown perfectly, be timed at 3 min from the abeam point to the fix. In nil wind conditions, the straight inbound and outbound legs will each take 1 min, as will the rate 1, 180° turn inbound. The complete hold, i.e., from outbound the fix back to the fix, will take 4 min, which is important to remember when planning an approach time.

A complete holding pattern (above 14,000 ft) will, if flown perfectly, be timed at 4 min from the abeam point to the fix. In nil wind conditions, the straight inbound and outbound legs will each take $1^1/_2$ min, and the rate 1, 180° turn inbound will still take 1 min.

With DME available the outbound leg may be expressed in terms of distance. Where this is done, care should be taken that at least 30 s is available on the inbound leg after completion of the inbound turn, and that the slant range is taken into account.

NB: When a clearance is received specifying a departure time from the holding fix, then the pilot should adjust his or her pattern within the limits of the established holding procedure in order to leave the holding fix at the time specified.

Hold-Base Turn Alternative Procedure   The outbound leg of a holding pattern has an alternative procedure when the inbound track of a hold is the same as the NDB or ILS approach, and the procedure is published as such.

- The outbound leg of the hold can be extended to a predetermined distance or time.

NB: Maintain the track by using twice the drift for 1 min 30 s and then by single drift if required. At the end of the outbound leg the QDM should be approximately ±15° from the inbound axis.

- Descend on the outbound leg per the procedure.
- Commence a rate 1 turn onto the final approach track.

Horizontal Situation Indicator (HSI) Instrument   *See* HSI (Horizontal Situation Indicator) Instrument.

Horn Balance   *See* Hinge/Horn Balance.

Hot Start   A *hot* start is one in which the engine ignites and reaches self-sustaining rotational speed, but the combustion is unstable and the EGT rapidly rises past its maximum limit. The causes of a hot start are

- Overfueling (throttle open)
- Air intake/exhaust blocked
- Tailwind causing the compressor to run backward
- Seized engine (e.g., ice blockage)

The indication of a hot start is that the EGT rises rapidly toward its maximum limit. The actions required for hot starts are as follows:

1. Close fuel lever/stop fuel delivery before the EGT limit has been reached.
2. When the engine rpm has slowed to its reengagement speed, motor over the engine to blow out the fuel (for approximately 60 s).

Howgozit   *See* Fuel Howgozit.

HSI (Horizontal Situation Indicator) Instrument   The horizontal situation indicator (HSI) is a sophisticated primary navigation instrument. It comprises a remote indicating compass that displays the aircraft's directional magnetic heading and an ILS/VOR display that gives an easy-to-understand display of the aircraft's situation to the selected VOR radial or ILS localizer and glide slope, relative to the aircraft's magnetic heading. Because the basis of the HSI is the remote indicating compass, the instrument is a head-up display; i.e., the aircraft's magnetic heading accommodates the position at the top of the dial, regardless of the aircraft's selected course.

The mode, either ILS or VOR, in which the HSI works is determined automatically by the VHF/NAV signal selected and received by the aircraft; i.e., selecting a VOR frequency, the HSI displays VOR indications, etc.

The HSI instrument is found in most modern aircraft and consists of the following:

1. *Remote indicating compass.*   This compass is continuously and automatically aligned with magnetic north. Therefore the HSI shows the magnetic heading at the top of the display at all times (i.e., the head-up display) and usually has a heading bug.

2. *Combined course track and deviation bar.*   The vertical course track bar is superimposed on the compass dial display and is set by a course selector, either on an MCP or by the HSI. It is free to rotate about the compass dial, to enable it to be aligned with the compass bearing of the selected track. The deviation bar comprises the center section of the course track bar and indicates the horizontal angular deviation of the aircraft from the selected track, in the same manner as the OBI and the ILS indicators.

NB: When the deviation bar is aligned with the course bar, the aircraft is on its selected track.

3. *Localizer dot scale.*   This is on a rotating card that moves with the course track bar to maintain its 90° abeam angle. The measurement of the localizer dot scale is as follows:

VOR mode: One dot represents 2° localizer deviation (10° or more full scale).

ILS mode: One dot represents $\frac{1}{2}°$ localizer deviation ($2\frac{1}{2}°$ or more full scale).

NB: EHSI (electronic HSI), on EFIS displays may show two different localizer scales, depending upon the amount of localizer deviation experienced. These scales may vary between different systems.

4. *Glide slope/dot scale.* The glide slope information on the HSI is shown on a scale to the right of the instrument. The scale and the indications are the same as displayed on other ILS display instruments, namely, one dot represents a glide slope deviation of 0.15° and full dot scale deviation represents a glide slope deviation of 0.7° or more.

The glide slope needle shows the position of the actual glide slope, and the center of the dot scale represents the aircraft's relative position to the glide slope. However, the drawback to this glide slope indication is that it does not produce a crossbar with the localizer deviation bar. However, aircraft with HSI instruments also usually have flight director bars on the (E)ADI which provide an easy-to-read crossbar indication for the pilot to follow.

NB: When the HSI is tuned to a VOR, the glide slope bar is removed from sight and a glide slope warning flag is present.

The HSI always acts as a command instrument, because the head-up display of the RIC ensures the correct orientation at all times, whenever the HSI is tuned to a VHF signal, either VOR or ILS. By flying toward the deflected deviation bars, the aircraft can center the course track and glide slope bars and thereby regain track.

In addition, the HSI, especially the EHSI, can display additional useful information from the INS/IRS, such as wind direction and ground speed, and NDB/ADF navigation information.

Humidity (Relative Humidity) Humidity is the water vapor in the air, and relative humidity is a measure of the amount of water vapor present in a parcel of air compared to the maximum amount the parcel can support (i.e., when the air is saturated) at the same temperature. Relative humidity is usually expressed as a percentage; i.e., relative humidity is 100 percent when the air is saturated.

The amount of water vapor that a parcel of air can hold depends upon its temperature, in that warm air is able to hold more water (in a vapor or liquid state) than cold air. In other words, cooler air supports less water vapor.

Hung Start A *hung start* occurs when the engine ignites but does not reach its self-sustaining rotational speed. (Self-sustaining speed is an rpm engine speed at and above which the engine can accelerate on its own, without the aid of the starter motor.)

The cause of a hung start is insufficient airflow to support combustion, due to the compressor not supplying enough air because of one or a combination of the following:

- High-altitude, low-density air
- Hot conditions, low-density air
- Inefficient compression
- Low starter rotational speed

The indications of a hung start are

- High EGT, above normal
- Engine rpm below normal self-sustaining speed

The actions required for hung starts are to

1. Close fuel lever/stop fuel delivery.
2. Motor over the engine to blow out the fuel (for approximately 60 s).

NB: To gain a successful start in hot and high conditions, you have to introduce more air into the engine. Adjusting the fuel supply does not help.

$$\text{Increasing the fuel} = \text{rpm decreases and EGT increases}$$

$$\text{Decreasing the fuel} = \text{rpm increases and EGT decreases}$$

Hydraulic(s)   Hydraulics is the science relating to the behavior of liquids under various conditions. Hydraulic pressure is created when you attempt to compress fluids against a solid obstruction, such as a piston head. (Liquid is incompressible except at very high pressures of 4000 psi and above, whereas air is highly compressible.)

This pressure acts in all directions as a force per unit area.

Barmah's hydraulic press uncovered two facts:

1. The smaller the area under load (force), the greater the pressure generated.

$$\text{Pressure (force per unit area)} = \frac{\text{force}}{\text{area}}$$

For example, if a force of 100 lb is applied to a piston area of 2 in^2, it will produce a pressure of 50 psi.

2. The larger the area under pressure, the greater the load (force) available.

$$\text{Force (total load available)} = \text{pressure} \times \text{area}$$

For example, if a piston has an area of 4 in^2 and it can support a pressure of 50 psi, it can balance a force (load) of 200 lb (4 $\times$ 50).

Further, hydraulic piston work = force $\times$ distance; thus for a given fluid pressure, the force produced can be varied by adjusting the piston size, and the resultant linear motion will vary inversely in proportion to the size of the piston head.

A basic hydraulic system consists of the following:

1. A reservoir
2. A pump
3. A filter
4. A selector or control valve
5. Hydraulic jack or actuators
6. Accumulator

A hydraulic pump delivers a flow of fluid from a reservoir into the system via a filter to protect the system from foreign particles. A valve system provides pressure/thermal release; nonreturn valves and directional control valves are connected to a mechanical service lever (i.e., landing gear) that directs fluid flow via either a supply line or a return line to an associated side of a movable piston (hydraulic jacks/actuators). The actuator converts the fluid flow into a linear or rotary motion because the fluid pressure causes the piston head to travel backward and forward in its cylinder when the pressure is greater on one side than the other, until an equilibrium of pressure forces is reached and the piston becomes stationary (known as *locked*) or the piston has reached its dead stop. (Intermediate stops are used for a service with several different settings, i.e., flaps.) At any point the service selector lever/valve can be placed in the off position to maintain the pressure forces on both sides of the piston head and thereby lock in that particular position.

An accumulator that stores hydraulic fluid under pressure is usually incorporated into the hydraulic system for the following reasons:

1. To provide an emergency supply of fluid into the system in the event of a pump failure
2. To dampen pressure fluctuations
3. To allow thermal expansion
4. To prolong the period between cutting in and out of the automatic cutout valve and so reduce its wear

As the power required for operating different services—flaps, spoilers, undercarriage, nosewheel steering, powered flying controls, etc.—varies according to their size and loading, a gearing effect must be provided. This is easily achieved by varying the size of the jack pistons, with the hydraulic system pressure remaining constant.

The common aircraft systems that use a hydraulic force are

1. Landing gear
2. Brakes/antiskid system
3. Steering
4. Flying controls
   a. Ailerons
   b. Elevator
   c. Rudder and yaw damper
   d. Trailing-edge flaps
   e. Leading-edge flaps and slats
   f. Spoilers—flight and speed brakes
5. Stairs
6. Doors

The advantages of a hydraulic system over a pneumatic system are as follows:

1. Hydraulic fluid is incompressible, this makes the system respond instantly and be more efficient.
2. Hydraulic fluid that leaks from the system is easier to detect visually than is air from a pneumatic system.

Hydraulically Operated Control Surfaces   *See* Control Surfaces—Hydraulically Operated.

Hydraulic Fluid   Hydraulic fluid must

1. Be incompressible up to 4000 psi.

2. Have good lubricating properties for metal and rubber.

3. Have good viscosity (with too great a viscosity, the thicker the fluid, the greater its friction, especially under cold temperatures; with too low a viscosity, the thinner the fluid, and the easier it is for the fluid to escape around seals, etc.).

4. Have high boiling and low freezing temperatures, e.g., a temperature range of +80 to −70°C.

5. Be compatible with materials used for the seals, rings, etc. (the hydraulic system manufacturer would recommend a certain type of hydraulic fluid). This type of fluid should always be used and never mixed with any other type of fluid.

6. Have a flash point above 100°C.

7. Be nonflammable.

8. Be chemically inert.

9. Be resistant to evaporation.

10. Not sludge or foam.

11. Have good storage properties.

12. Be noncorrosive.

13. Be well priced and easily available.

The three main types of hydraulic fluids are

1. DTD 585, which is a refined mineral-based oil (petroleum). It is not fireproof and is used with synthetic rubber seals.

2. Lockheed 22, which is a castor-based oil (vegetable). It is not fireproof and has poor high-temperature characteristics. It is used with seals made of natural material.

3. Skydrol, which is a phosphate ester-based oil. It is fire-resistant, and it is used with synthetic rubber seals (butyl).

Using the wrong oil, even as a mixture, will cause damage to the seals, will result in leaks and a rise in temperature due to low fluid levels, and may cause corrosion.

Hydroplaning   *See* Aquaplaning/Hydroplaning.

Hypermetropia   Hypermetropia is the condition of long-sightedness. It is associated with a shorter-than-normal eye that results in the image being formed behind the retina, making images of close objects blurred.

Hypertension   Hypertension is high blood pressure.

Hyperventilation   Hyperventilation is the condition of *overbreathing,* i.e., in excess of the ventilation level required to remove carbon dioxide, causing changes in the acid-base balance of the body.

Hyperventilation means an excess of oxygen. Hyperventilation is caused by the following:

1. Hypoxia

2. Anxiety

3. Motion sickness

4. Vibration

5. Heat

6. High *g* force

7. Many other causes to a lesser extent

The indications/symptoms of hyperventilation are

1. Dizziness

2. Tingling sensation, especially in hands and lips.

3. Visual disturbances, i.e., tunneling or clouding.

4. Feelings of hot or cold.

5. Anxiety, which creates a circle of cause and effect.

6. Impaired performance—often dramatically reduced performance.

7. Loss of consciousness. However, afterward respiration returns to normal, and the individual recovers.

These indications/symptoms are very similar to the indications and symptoms of hypoxia. Below an altitude of 10,000 ft, symptoms for hyperventilation should be diagnosed as hyperventilation and treated accordingly, i.e., by regulating the rate and depth of respiration to reduce the individual's overbreathing. Normally this is accomplished with deep breaths into a paper bag. However, some medical bodies now suggest to simply allow the patient to hyperventilate to the point of passing out, during which time the body will regain a normal breathing pattern for when the patient comes round.

Above an altitude of 10,000 ft, the same symptoms should be diagnosed as hypoxia and treated accordingly, by giving oxygen to the individual. This approach is taken because (1) it is very difficult to distinguish between the effects of hypoxia and those of hyperventilation, and (2) since hypoxia is possible above 10,000 ft, the worst scenario—hypoxia—should be assumed.

Hypoxia Hypoxia is a human condition which occurs when the oxygen supply available to the human tissues is insufficient to meet their needs, especially respiratory requirements. Hypoxia means a lack of oxygen.

Respiration brings oxygen into the body and removes carbon dioxide. However, hypoxia is prevalent in aviation (i.e., respiration brings less oxygen into the body) as a result of high-altitude conditions with the associated fall in ambient air pressure/density, which has a corresponding fall in the partial pressure of the oxygen in the atmosphere. That is,

$$\text{Partial pressure of oxygen} = \begin{cases} 3.08 \text{ psi} & \text{sea level} \\ 1.70 \text{ psi} & \text{at 15,000 ft} \\ 1.10 \text{ psi} & \text{at 25,000 ft} \end{cases}$$

Hypoxia is caused by the following:

1. *Altitude.* The greater the altitude, the greater the hypoxia and the more rapid its progression. This is so because respiration brings less oxygen into the body at high altitudes, because of the decrease in the ambient air pressure/density,

which has a corresponding fall in the partial pressure of oxygen in the atmosphere. Susceptibility to hypoxia includes the following effects.

2. *Time.*   The longer the exposure time to a lack of oxygen, the greater the effect of hypoxia.

3. *Exercise.*   This increases the demand for oxygen and hence increases the degree of hypoxia.

4. *Illness.*   This increases the energy demand of the body and therefore of oxygen, and hence this causes an increase in the degree of hypoxia experienced.

5. *Fatigue.*   It lowers the threshold for hypoxia symptoms.

6. *Drugs and alcohol.*   They reduce the body's tolerance of altitude and therefore increase an individual's susceptibility to hypoxia.

7. *Smoking.*   Carbon monoxide, produced by smoking, binds to hemoglobin with a far greater affinity than oxygen and therefore has the effect of reducing the hemoglobin for oxygen transport, exacerbating any degree of hypoxia.

An individual's time of useful consciousness with hypoxia is approximately

30 min at	18,000 ft
2 to 3 min at	25,000 ft
45 to 74 s at	30,000 ft
20 s at	40,000 ft
12 s at	45,000 ft

The indications/symptoms of hypoxia are as follows:

1. *Cyanosis.*   Blue color develops in the parts of the body where the blood supply is close to the surface, e.g., lips and fingernails.

2. *Apparent personality change.*   There is a change in outlook and behavior, i.e., euphoria, aggression, or loss of inhibitions; it is like a drunken state.

3. *Impaired judgment.*   There is loss of self-criticism; one is unaware of reduced personal performance.

4. *Muscular impairment.*   One has poor muscular control.

5. *Memory impairment.*   Especially short-term memory is impaired, which makes drills difficult to complete.

6. *Sensory loss.*   Vision, especially color, is lost early because the eyes require oxygen to work. This is followed by impairment of touch, orientation, and hearing.

7. *Hyperventilation.*   The individual tends to overbreathe in an effort to get more oxygen.

8. *Impairment of consciousness.*   Extreme hypoxia can cause the individual to become semiconscious or even unconscious.

The degree of these indications/symptoms varies with altitude, because of the different levels of oxygen partial pressure. Therefore the pilot should be aware that these indications/symptoms develop slowly at low altitudes and very quickly at high altitudes.

Hypoxia is treated by increasing the individual's oxygen supply. Therefore pilots should be familiar with the appropriate oxygen drills, whenever they are flying at an altitude where hypoxia can occur, i.e., above 10,000 ft.

Icing   The formation of ice (icing) is the change of state of water from liquid to solid form when the temperature is less than the freezing point of water (0°).

NB: There are three states of water: liquid (water drops), gas (water vapor), and solid (ice).

Ice can form from either of the other two states of water:

1. Water vapor, i.e., by sublimation, causing hoar frost.

2. Water droplets, i.e., by freezing rain or by freezing supercooled water drops (SWDs).

NB: Airframe icing occurs in freezing clouds when SWDs are present, causing rime and/or clear ice.

Icing from water drops, which are present in clouds and as rain, poses a risk to aircraft in freezing conditions, i.e., when the TAT (in flight) or OAT (on the ground) is between $+10$ and -40 to $-45°C$. Therefore the following conditions are an indication of the icing risk present.

1. *Clouds.*   The severity of the icing present depends upon the amount of water drops within a particular cloud type. Cumulus and cumulonimbus clouds have moderate to severe icing conditions. This is so because these clouds have large quantities of large SWDs above the freezing level.

NB: Clear icing with an OAT of 0 to $-23°C$, rime icing with an OAT of -23 to $-45°C$.

Tropical cumuliform clouds have more severe icing conditions, because tropical clouds have more moisture present than temperate cumulus and cumulonimbus.

Stratocumulus clouds have moderate icing conditions which can even be experienced above the freezing level, due to a lower moisture level and vertical extent of the cloud. Nimbostratus clouds have moderate icing conditions, experienced from 0 to 210°C OAT, band, due to the abundance of suspended water drops. Fog and low stratus (AC & As) clouds have slight icing conditions, which are experienced from 0 to $-10°C$ OAT, due to the small suspended water droplets. Cirrus clouds have no airframe icing conditions because they contain ice particles, which do not stick to an airframe except in the most exceptional circumstances.

2. *Freezing rain.*   This also presents an icing risk, the severity of which depends upon the amount of rainfall.

NB: Water vapor icing by sublimation produces only slight hoar frost conditions, which are generally not considered an icing risk, except in extreme circumstances.

The different types of airframe icing are as follows:

1. *Hoar frost* is a rough, pitted, frostlike ice distributed evenly over a surface. Hoar frost can form directly, by sublimation, on the airframe when an aircraft is parked in subzero temperatures or when a cold soaked aircraft descends from a high altitude into moist warmer air. However, this form of icing tends to evaporate as the airframe warms up during the descent. In some cases, however, hoar frost buildup on the inner wing tanks can persist for some time, especially if the tank is

full, due to the slow heating properties of the fuel keeping the fuel tank surface cold. This has caused incidents on quick-change rounds, even at very warm airports, where hoar frost over the inner wing tank comes off during the takeoff run, causing sudden flameouts of rear-mounted engines. Although hoar frost is not as dangerous as clear ice, it can obscure vision through a cockpit window and possibly affect the lifting characteristics of the wings.

2. *Clear ice* is a sheet of solid, clear/glazed ice with very little air enclosed. Clear ice is formed when only part of a SWD, with a temperature between 0 and $-23°$C, freezes when it strikes a cold aircraft surface. This initial freezing process releases latent heat, which in turn reduces the rate of freezing of the remaining liquid water, which then spreads backward by the airflow over the airframe and coalesces with water from other partially frozen water drops, before fully freezing on the cold airframe or propeller surfaces. The surface of clear ice is smooth, usually with undulations and lumps. It can alter the aerodynamic shape of aerofoils quite dramatically and reduce or destroy their effectiveness.

3. *Rime ice* is a mixture of tiny ice particles and trapped air, giving a rough, opaque crystalline deposit that is fairly brittle. Rime ice occurs when small supercooled liquid water droplets between -23 and $-45°$C freeze on contact with a surface whose temperature is below zero. However, because the drops are small, the amount of water remaining after the initial contact freezing is insufficient to coalesce into a continuous clear ice sheet. Therefore rime ice often forms on the leading edges and can affect the aerodynamic qualities of an aerofoil or the airflow into an engine intake. It does cause a significant increase in weight.

4. *Rain ice* is freezing rain or drizzle that falls from any sort of cloud and freezes if it meets a surface (airframe) that is below $0°$C. Freezing rain in a front falling from warm air into cold air beneath presents a serious hazard, as the subsequent icing can be heavy and clear. This rain ice is particularly bad in front of a warm front, and it will occur below the freezing level in the warm sector and above the freezing level in the cold sector. If the $0°$C isotherm is at or below ground level in the cold air, you can get freezing rain on the ground, and this icing presents a serious hazard to aircraft taking off.

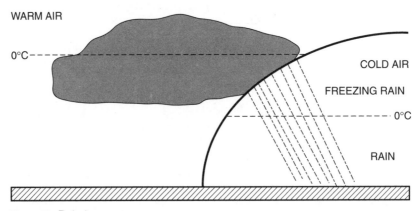

Figure 98   Rain ice.

Ice particles in the atmosphere (i.e., crystals, snow, and hail) do not stick to the airframe, except sometimes at temperatures near zero.

Ice buildup on an aircraft or within its engine induction system can be a serious hazard to flight safety, particularly at slow phases of flight, say takeoffs and landings, because of the following effects:

1. *Adverse aerodynamic performance effects.* As ice builds up on an aircraft's aerofoils (i.e., wings), it seriously disrupts the airflow pattern, causing it to separate at a significantly lower angle of attack than it would from a clean wing. This results in
   a. Reduced stalling angle
   b. Reduced maximum lift capability
   c. Increased stall speed
   d. Increase in drag
   e. Reduced amount of lift (loss of lift) at a given angle of attack, especially at higher angles of attack
2. *Control surface effects.* Ice formation may freeze on control surfaces, leading to restricted movement and even loss of control in extreme circumstances.
3. *Increase in aircraft weight effects.* The buildup of ice on the aircraft will increase the overall weight of the aircraft, with all the associated effects of higher weights, such as higher stall speed and extra lift required. In addition, this uncontrolled increase in weight may
   a. Change the position of the aircraft's cg
   b. Unbalance the various control surfaces and propellers, causing severe vibration and/or control difficulties
4. *Reduced engine power.* Ice buildup in the engine intake, or piston engine carburetor, can restrict the airflow into the engine, causing a power loss and even engine failure in extreme circumstances.
5. *Vent blockage effects.* Ice buildup on pitot and static probes will produce errors in the aircraft's pressure-driven flight instruments, such as the ASI, altimeter, and VSI.
6. *Degraded navigation and radio communication.* If ice builds up on navigation and radio aerials, then these systems will be degraded and unreliable.

An ice-laden aircraft may even be incapable of flight. If the ice or frost buildup on the leading edges and upper surfaces of the wings is too great, then the lift produced is reduced and insufficient to balance the increased aircraft weight. Or if the ice builds up on the engine intake, it will degrade the power output to the extent that it is insufficient to balance the increased drag from the ice on the airframe.

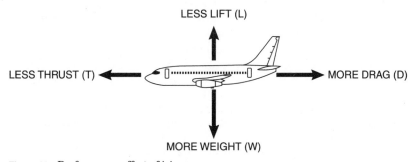

Figure 99   Performance effect of icing.

The purpose of preflight preparation for icing conditions is to ensure that the whole aircraft is free from deposits of frost, ice, and snow. When necessary, use a deicing fluid to achieve effective removal of any frost or ice and to provide a measure of protection against any further formation.

First, fit engine blankets and pitot/static covers as required before deicing with engines off. Ensure that all orifices and guards are cleared of snow or slush, e.g., APU inlets, pressurization inlets and outlets, and exposed operating mechanisms, in particular leading edges, control surfaces, flaps, slats, etc., and their associated hinges and gaps.

NB: An ingress of moisture, snow, and rain to door mechanisms and seals is more likely to occur when doors are open. The time they are left open should be kept to a minimum, and a check for contamination should be made prior to departure.

Second, deice the aircraft, ensuring that both wings receive similar treatment and that no deposits are left on control surfaces which may put them out of balance, with the consequent risk of flutter. Also make sure that no deposits are left in operating mechanisms, hinges, and gaps that may refreeze during flight and jam the controls.

NB: The efficiency of the deicing fluid under varying atmospheric conditions is dependent upon the correct mixture strength and methods of application. For example, a fluid diluted with water will effectively remove the ice; however, its ability to prevent further formation will be significantly reduced. In fact, under certain conditions the fact that the aircraft's surfaces are wetted may actually enhance the accumulation of wet snow, particularly if there is any significant delay between deicing and takeoff.

After the aircraft has been deiced, consider the application of a stronger mix to the wing leading edges to provide further anti-icing protection.

NB: When used as an anti-icing agent, the FDP fluid should be sprayed onto the aircraft cold and undiluted, either before the onset of icing conditions or after hot deicing has been carried out. This will leave a film of fluid approximately 0.5 mm thick on the surfaces and will give protection overnight in all but the most severe weather conditions.

Third, check all your deicing and anti-icing systems as well as systems that can be affected by icing conditions.

- Check the airframe condition for ice accumulation.
- Check the indications, and by visual means if possible, that the anti-icing and deicing systems (e.g., inflatable boots) are operating satisfactorily.
- Check that pitot and static sources are not iced up. If practical, check the operation of pitot heaters before flight, and be aware of the function of the alternate static source and its corrections.

NB: Iced-up pitot or static sources will cause lost or intermittent indications.

In addition, you should apply any further procedures that your flight manual may recommend. When taking off from a runway contaminated with snow or slush, in icing conditions you should consider the conditions at your preflight preparations, through the start-up, taxi, the actual takeoff, and the after-takeoff phase of flight.

Operation from a snow- or slush-covered runway involves a significant element of risk. Therefore operations from contaminated runways should be avoided wherever possible. However, if you are committed to a departure in these conditions, then consider the following:

1. Deice the aircraft, either on stand prior to the engine start or when taxiing through a deicing station.

NB: Aircraft "taxi through" deicing stations are presently being used which deice an aircraft with the engines running. Winter environmental conditions and the manner of application create potentially unsafe conditions if an incorrect deicer solution is inadvertently sprayed into the engines/APU inlets or contacts the exhaust when the engines or APU is running. APU and engine bleeds should be closed during such operations to minimize the risk of contamination of the cabin environment.

2. Ensure that engine anti-ice is switched on during ground operations in icing conditions.

NB: In the absence of guidance from your aircraft's flight manual, engine icing is considered possible if the OAT is $+10°C$ or less with visible moisture.

Use carburetor heat and propeller deicing if appropriate.

3. During taxiing in icing conditions, the use of reverse thrust on podded engines should be avoided, as this can result in ice contamination of the wing leading edges, slats, etc. For the same reason, stay a good distance behind the aircraft ahead. In no circumstances should an attempt be made to deice an aircraft by positioning it in the wake of the engine exhaust of another aircraft.

4. Before takeoff, ensure that the wings are not contaminated by ice or snow. Operate all the appropriate anti-icing and deicing systems (e.g., engine, wing, propeller, carburetor, probes) before and after takeoff, in accordance with the aircraft's flight manual. Takeoff power should be monitored on more than one instrument, say, the EPR and N1 gauges. This is done in case there is icing on any engine probes, which causes an overreading on the main engine power instrument. Monitoring a second instrument can cross-check the reliability of the main instrument.

Icing during flight can occur at any time of the year. At temperatures just below freezing where large water droplets are present, glaze ice can accrete quickly on aircraft surfaces. It is difficult to detect; up to 2 in of ice has remained undetected by crews. Equally, melting ice can run back and refreeze on surfaces other than the leading edges. Both of these types of ice buildup in flight can rapidly degrade the aircraft's performance and controllability. Therefore a pilot should take the following precautions when icing conditions are possible:

1. Pilots should avoid icing conditions for which the aircraft is not approved.
2. Keep probe heating on when airborne.
3. Crews should visually check for the buildup of ice on the airframe at regular intervals, and they should select wing anti-icing when required.
4. Crews should monitor the air temperature and signs of visible moisture regularly and select engine anti-icing when required.

NB: Switch on and off the engine anti-icing systems separately to guard against simultaneous engine flameouts from ice ingestion.

5. Be aware of the influence of ice buildup on performance and controllability.
   a. The performance effect of wing contamination includes
      (1) Reduced stalling angle of attack
      (2) Increased stalling speed
      (3) Reduced maximum lift capability

(4) Reduced amount of lift at a given angle of attack, especially at higher angles of attack

b. Controllability difficulties of the aircraft, due to ice buildup, can be manifested in many different situations. For example, the deployment of flaps may cause a reduction in longitudinal controllability when ice is present on the tailplane. When this happens, the tailplane can stall with subsequent loss of control from which recovery may be impossible in the time/height available. Therefore if tailplane icing is suspected, then it is wise not to deploy flaps. Also be aware of a lateral imbalance, if you were to only deice one wing; on some types this can occur with an engine failure and the cross-bleed closed.

In the following conditions, you should expect engine icing and therefore turn on the engine anti-icing systems.

On the ground whenever the outside air temperature (OAT) is $+10°C$ or lower with visible moisture present.

NB: Static air temperature (SAT) measurements are usually considered to represent ground OAT.

In flight during the climb and cruise, whenever the total air temperature (TAT), which is SAT plus the kinetic effect of the aircraft, is colder than $+10°C$ but warmer than $-40°C$, with visible moisture present.

NB: You should not turn on the engine anti-icing systems in the climb or cruise when the SAT is colder than $-40°C$. This is because only ice crystals are present in the atmosphere at these extreme temperatures. These crystals do not pose an icing threat, because they have no liquid state that can "stick" to the aircraft.

In flight descent, when the TAT is colder than $+10°C$, with visible moisture present.

NB: This is true even with an SAT colder than $-40°C$. This is so because in a descent the temperature will increase, and the SAT will become warmer than $-40°C$, creating a real risk of SWD icing. Therefore the engine anti-icing should be switched on at the top of descent, whenever the TAT is colder than $+10°C$ and visible moisture is present.

Icing—Carburetor   *See* Carburetor Icing.

Icing Fluids   The two main types of anti-icing fluids used for deicing on the ground are

1. *Type I fluids* (unthickened). These fluids have a high glycol content and a low viscosity. The deicing performance is good; however, they provide only limited protection against refreezing.
2. *Type II fluids* (thickened). These fluids have a minimum glycol content of approximately 50 percent and due to the thickening agent have special properties which enable the fluid to remain on the aircraft surfaces, known as the holdover time. The deicing performance is good; in addition, protection is provided against refreezing and/or buildup of further accretions, when exposed to freezing conditions. Therefore it is also an anti-icing fluid. Its anti-ice holdover time depends on the following factors:

   a. Type of snow
   b. Wet or dry snow

NB: Wet snow will result in a shorter holdover time.

c. Airframe temperature

d. OAT

e. Amount of precipitation

The deicing fluid applied would be described as follows: Type of fluid (I or II), percent of fluid-to-water mixture, and holdover time expected.

IFR Flight Levels *See* Flight Levels (IFR).

ILS—Instrument Landing System The ILS is a precision-approach radio aid that gives slope and track guidance to enable low-minima approaches for suitably equipped aircraft.

An ILS has two separate ground transmitters, the localizer and the glide slope.

The localizer, which produces an (azimuth) beam, is to provide tracking guidance along the extended runway centerline (i.e., azimuth guidance left and right of the extended runway centerline). The localizer consists of two overlapping beams, one modulated to 90 Hz (yellow area) left of the centerline and one at 150 Hz (blue area) right of the centerline. If there is more 90-Hz than 150-Hz modulation, the needle on the ILS cockpit display instrument is deflected to the right to give a "Fly Right" indication; and if there is more than 150-Hz modulation, it is deflected to the left. When the 90- and 150-Hz modulated signals are equal, the indicator shows a "Fly Straight Ahead" signal. The localizer transmitter is usually in line with the centerline at the end of the runway, and the signal is protected 10° either side of the centerline up to a height of approximately 6000 ft out to 25 nm, and 35° either side of the centerline out to 17 nm.

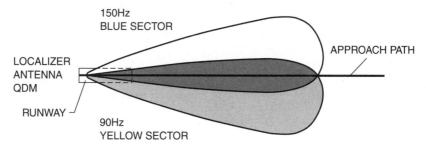

Figure 100 ILS localizer beams.

The glide slope produces an (elevation) beam to provide vertical guidance toward the runway touchdown point, i.e., vertical guidance above and below the glide slope. The glide slope is normally set at an angle of 3° to give a reasonable rate of descent glide path. It consists of two overlapping beams, one modulated to 90 Hz above the glide slope and the other modulated to 150 Hz below the glide slope.

See Fig. 101 at the top of the next page.

The glide slope works in a similar way to the localizer. If the signal received is dominated by 90-Hz modulation, it gives a "Fly Down" indication; and if the signal received is dominated by 150-Hz modulation, it gives a "Fly Up" indication. When the 90- and 150-Hz modulated signals are equal, the indicator shows an on glide slope signal. The glide slope transmitting aerial is usually situated about 300 m in from the runway threshold, to ensure adequate wheel clearance over the airfield

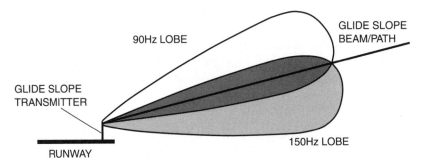

Figure 101 ILS glide path beams.

fence. Therefore when flying a glide slope, the aim is to touch down not on the runway's "piano keys" but on the touchdown zone near the point where the glide slope intersects the runway. The coverage of the glide slope signals extends to 8° either side of the localizer centerline out to 10 nm.

The aircraft needs its own ILS receiver to decipher the localizer and glide slope information and then to display it on either a purpose-built ILS indicator or an HSI instrument. Both are command instruments, and by making adjustments to maintain the localizer centerline and the glide slope path, the aircraft will arrive at minimas exactly in line with the runway for landing. That is, fly toward the deflected needles to center both the localizer and the glide slope as a center cross-display.

The ILS localizer works in the 108- to 112-MHz VHF band, which it shares with the VOR. To avoid confusion with the VOR signals, the ILS uses frequencies at odd 100- and 150-kHz spacing. The ILS glide path uses the 329.3- to 335-MHz ultra-high frequencies at 150-kHz spacing. The glide path frequency is automatically selected when its paired VHF localizer channel is selected. The ILS normally has associated short-range marker beacons, and/or colocated DME information, which provides additional distance-to-touchdown information. Up to three marker beacons are used on an ILS that sends out fan-shaped beams that point directly upward. The outer marker is roughly positioned at the point at which the glide slope is intercepted, that is, 4 to 5 nm from touchdown. The middle marker is usually positioned just before the CAT I decision height, that is, $\frac{1}{2}$ to $\frac{3}{4}$ nm from touchdown, and the inner marker is positioned just before the threshold is crossed. Marker beacons are largely being replaced with DME equipment that gives a continuous range from touchdown. The DME is usually paired with the ILS frequency so that it is automatically selected with the ILS.

Normally ground installation and aircraft are fitted with CAT I ILS equipment. The errors that an ILS experiences are as follows:

1. *False glide slope.* The ground transmitter creates a mirror 150-Hz sidelobe that overlaps the top of the 90-Hz main lobe to produce a false glide slope at approximately twice the angle of incidence above the real glide slope, that is, 3° real glide slope, 6° false glide slope. By attacking the real glide slope from underneath, you ensure that you never encounter the false glide slope. However, if you were to follow the false glide slope, it should be easily recognized by the aircraft's high rate of descent, typically 1500 ft/min for a 6° glide slope.

2. *Phantom signals.*   Outside of a localizer's protected range, its signal cannot be relied upon because the signal received is likely to be from another station using a similar frequency. This occurrence is rather common because of the small frequency range set aside for all the ILS stations.

3. *Backcourse approaches.*   An aircraft attempting an approach using a backcourse beam would display exactly the opposite of the correct indications, especially on simple ILS or VOR indicators. However, the HSI display can be corrected by either (1) rotating the display 180° to make the backcourse indications act in the correct sense or (2) switching to backcourse on aircraft with backcourse selectors. This system reverses the signal internally to show the correct indications.

There are many different methods that achieve the aim of flying an ILS. The following description is one method that is offered as an example only. There are many variations of flying an ILS, and the reader should be aware of these when contemplating this procedure. Flying an accurate ILS approach requires a fast instrument scan and good coordination.

Positioning   An ILS approach may be arrived at by means of a published approach procedure, visual self-maneuvering, or radar vectors. Whichever method is used, the aircraft should be in a position at which interception of the localizer and glide path is possible. Ideally a final intercept heading of 30°, with a speed that is low enough to allow a tight turn (approximately 160 to 180 kn for a typical jet aircraft) straight onto the localizer, should be the aim. In addition, sufficient time should be allowed to establish on the localizer track before intercepting the glide path. This is accomplished by intercepting the localizer at a point well before the glide slope. For a procedural ILS, if you are at the assigned altitude during the base turn, you should have enough time to capture the localizer before the glide slope. However, for visual self-positioning and also for backing up radar vectoring, you should always consider the following points:

1. Always know the distance at which you expect to intercept the glide slope, and plan to intercept the localizer before this point.

2. Back this up further by monitoring the position of the glide slope indication. On a final intercept heading, and even prior to starting any procedure turn inbound to the localizer, ensure you have *at least* one-half "Up" deviation on the glide slope indicator.

Localizer   When you are established on a good intercept heading, concentrate on capturing the localizer. Display the ILS on your main navigation instrument. Be aware of the wind conditions, and calculate the approximate *wind-corrected* heading required to maintain the localizer track. When the localizer bar starts to move, call "Localizer Alive" (on a 30° intercept heading the lubber line is on the fully extended localizer bar). Begin turning as the localizer bar reaches the tip of the lubber line, and keep the lubber line aligned with the localizer bar during the turn.

NB: Some anticipation of the LOC may be necessary during strong tailwind conditions. Remember to continue the turn onto the wind-corrected heading to use as an initial datum, and a smooth localizer interception should be achieved. Make the R/T call "Call Sign...Localizer Established."

NB: The R/T call "Localizer Established" can be made when the localizer bar is within one dot deviation on the ILS navigation instrument.

Once you are established on the localizer, make small correction immediately if any movement of the localizer is apparent, and maintain good localizer tracking down to the decision height. Two further points should be mentioned:

- The wind is likely to *back* and decrease during the descent. Therefore you need to vary the drift to maintain the localizer track.
- The localizer beam width gets progressively narrower and therefore more sensitive as the aircraft gets closer to the touchdown point. Therefore any corrections should become smaller for the same localizer deviation as you track down the localizer.

Glide path The aim should be to establish on the localizer track before intercepting the glide path; i.e., make sure that the glide path is above the aircraft. When the glide path needle begins to move, call "Glide Slope Alive."

NB: The rate of movement of the glide path needle will vary according to the effect of the headwind or tailwind. As the glide path needle approaches the aircraft's horizon (namely, the center of the ILS navigation instrument), check the glide slope takeoff distance, from the plate or self-calculated distance, smoothly reduce power for the descent, and pitch down to maintain the glide path.

NB: The missed-approach altitude is usually set once the glide path has been captured.

The basic technique for flying the glide path is to make pitch changes with reference to the artificial horizon and vertical speed indication to maintain the glide path. Maintain good glide path tracking down to the decision height, making small correction immediately if any movement from the glide path is apparent.

NB: Cross-check the vertical speed to confirm that you have not captured a false (6°) glide slope, which would be represented by a vertical speed descent that is double that expected.

When necessary, make power changes to maintain the approach speed. Confirm that the height over the outer marker is correct and within limits on both altimeters, and start the stopwatch. Make the necessary RT position report, "Call Sign...Outer Marker Inbound."

NB: Outer markers are being phased out, and this check is being replaced by a 4-nm final-approach check at most airports.

Two further points should be appreciated:

1. The wind is likely to decrease during the descent, so be prepared to adjust the power to maintain approach speed.
2. The glide path beam width gets progressively narrower and therefore more sensitive as you get closer to the touchdown point. Therefore any corrections should become smaller for the same glide path deviation as you descend on the glide slope.

You can descend on the glide path only when you have been cleared for the ILS procedure and you have captured the localizer, within ±5°C.

Missed approach At the decision height the missed-approach procedure must be commenced if not visual with the runway or approach lead in lights.

You can calculate the distance from the threshold at which you would intercept the ILS glide slope by dividing the height by the glide slope angle times 100. You will get the distance from the threshold, in nanometers, at which you should intercept the glide slope. For example,

$$3000 \text{ ft} / (3 \times 100) = 10 \text{ nm approximately}$$

Likewise, if you wish to calculate the height at which you would intercept the glide slope at a particular distance from the threshold, multiply the distance, in nanometers, by the glide slope angle times 100. For example,

$$(10 \text{ nm}) (3 \times 100) = 3000 \text{ ft approximately}$$

You can calculate the glide slope angle, given a gradient, by using the following formula:

$$\text{Glide slope angle} = \text{gradient } \% \times 0.57$$

You can calculate the approximate rate of descent (ROD) for a 3° glide slope by using the formula:

$$5 \times \text{ground speed} = \text{ROD (ft/min) required for a 3° glide slope}$$

$$5 \times 140 = 700 \text{ ft/min ROD}$$

ILS—Backcourse Approach   Some ILS installations provide a mirror image of the localizer beam on the reciprocal runway, known as a *backcourse beam*. As with any mirror image, right is left and left is right, and an aircraft attempting an approach using a backcourse beam would display exactly the opposite of the correct indications unless the aircraft were fitted with a backcourse *selector* that reverses the signal internally to show the correct display. Backcourse procedures can be found in the United States and Canada, but are prohibited in the United Kingdom and most of Europe.

ILS CAT I, II, and III Limits   *See* CAT I, II, and III ILS Limits.

ILS Indicator   A purpose-built ILS display instrument is a further development of the omni bearing instrument (OBI). It still acts as a VOR (VHS omni range) navigation display when a VOR frequency is selected, but can also act as an ILS display instrument when an ILS frequency is selected, to guide a landing aircraft along both a localizer track and a glide slope descent path.

The ILS indicator is a "track up" display, like the OBI. The selected track (localizer) accommodates the position at the top of the dial, regardless of the aircraft's heading. The ILS indicator differs from the OBI as follows:

1. *Localizer deviation bar and dot scale.*   The track deviation bar (vertical needle) and the horizontal dot scale represent the aircraft's horizontal angular deviation from the ILS localizer beam.

One dot represents an ILS localizer deviation of ½° and a full five-dot scale represents a total ILS localizer deviation of 2½° or more.

2. *Glide slope deviation bar and dot scale.*   The ILS also displays the aircraft's angular deviation from the ILS glide slope beam using a horizontal needle/bar superimposed on a vertical dot scale in the middle of the display. The glide slope needle/bar shows the position of the actual glide slope, and the center of the dot scale represents the aircraft's relative position to the glide slope. Thus the pilot's aim is to maintain the glide slope bar/needle on the center dot of the glide slope scale.

The measurement of the glide slope dot scale is as follows: One dot represents a glide slope deviation of 0.15°; full dot scale deviation represents a glide slope deviation of 0.7° or more. The glide slope display is automatically activated once the ILS has been tuned and selected. Both the localizer and the glide slope displays should be used as command instruments. That is, fly toward the deflected needles to center both the localizer and glide slope tracks as a center cross display.

It is common in modern aircraft for the ILS display to also double as the VOR display. When the instrument is being used for the VOR, i.e., a VOR frequency is selected, the glide slope horizontal bar is usually removed out of view and a glide slope warning flag is present.

ILS—Marker Beacons *See* Marker Beacons.

Increased V2 Increased V2 is a technique of improving an aircraft's climb gradient performance, in the second segment, by increasing the V2 climb speed. Increasing the V2 base speed increases lift, because lift is a function of speed, and for a given weight an increased V2 speed will provide an increase in lift, therefore producing a greater net climb gradient. Thus an increased V2 climb allows higher obstacles to be cleared in the second-segment climb-out profile.

An increased V2 speed requires a greater than normal takeoff distance, thereby reducing the horizontal distance from reference zero to the obstacle and thus increasing the gradient required to clear it. The closer the obstacle is to reference zero, the greater is this detrimental effect, which offsets the improved climb gradient achieved by the increased V2 speed. Therefore, the increased V2 improvement in obstacle-limited weight conditions is much less for close-in than for distant obstacles.

An increased V2 climb profile technique can only be used if the takeoff weight is not restricted by field length limits, so that some of or all the excess field length can be used to increase takeoff speed V_R and thereby V2 speed.

The increased V2 technique is used for the following reasons:

1. To achieve a greater obstacle clearance performance, with an improved takeoff climb gradient, without reducing takeoff weight. This is achieved because lift is a function of speed, and for a given weight an increased V2 speed will provide an increase in lift, therefore producing a greater net climb gradient. Thus an increased V2 climb allows higher obstacles to be cleared in the second-segment climb-out profile, without reducing takeoff weight.

2. To allow a higher aircraft takeoff weight that achieves the standard (minimum) takeoff climb gradient which corresponds to the required obstacle clearance gradient. The increased V2 speed will increase the lift generated to achieve the minimum required climb gradient to clear any obstacles for the heavier aircraft weight. Thus an increased V2 climb allows higher takeoff weights that still meet the obstacle clearance requirements.

The use of increased V2 techniques is usually prohibited on wet runways.

Indicated Airspeed (IAS) The IAS is a measure of dynamic pressure translated to a speed and displayed to the pilot on the airspeed indicator (ASI) usually in knots per hour. IAS = dynamic pressure ($\frac{1}{2}\rho V^2$). However, because the design of the ASI is calibrated to ISA mean sea-level (MSL) conditions, it does not compensate for the reduction of density with altitude and/or warmer temperatures. In fact, IAS is only equal to the aircraft's true airspeed (TAS) at ISA MSL conditions. Therefore at a constant IAS climb the aircraft's TAS increases with altitude; and at a constant TAS climb, IAS will decrease with altitude.

IAS has greatest importance with regard to the performance of the aircraft, i.e., stall speed, ROC, lift/drag ratio, etc., which are all a function of IAS, which itself is a function of dynamic pressure.

INS/IRS An INS is an onboard, self-contained inertia navigation system that can provide continuous information on the aircraft's position without any external assistance.

An inertia reference system (IRS) is a modern INS that usually has a greater integration into the FMS, and it provides the aircraft's actual magnetic position and heading information with reference to the FMS required position and heading.

The directional acceleration information provided from the INS accelerometers and gyros is calculated by the position computer that determines the aircraft's present latitude and longitude position, provided a correct initial position has been inputted.

The general principle of all INSs is that they measure from an initial inputted position the aircraft's inertia movement, as a great circle track direction and distance, to continuously determine its up-to-date position.

The components of an INS are

1. Accelerometers

2. Gyroscopes

3. Position computer

The aircraft moves in three dimensions, but the navigation equipment is only interested in acceleration in the horizontal plane. Therefore the key to the whole INS arrangement is the *accelerometers.*

1. *Accelerometers.* Either an INS will have three accelerometers in all three axes, which are literally strapped down to the aircraft's structure (strapped down INS), or it will have two accelerometers, of north/south and east/west axes, mounted on a stable horizontal platform (stable platform INS). The most common type of accelerometer is the E & I bar pendulum linked to a feedback system.

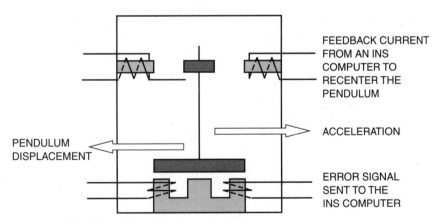

Figure 102 E & I bar accelerometer.

The I bar pendulum is free to move in one plane only, i.e., north/south; and as the aircraft accelerates in this plane, the inertia of the pendulum causes it to be displaced in the opposite direction. This results in a reduced air gap between the E & I bar which is transmitted as an error signal to the INS computer. In response the computer signals a current to the feedback coils to return the pendulum to its central position. The amount of current required to do this is directly proportional to the acceleration experienced along the plane of direction.

Thus the two or three accelerometers can sense the aircraft's acceleration, which the position computer can integrate as a velocity, which can be further integrated to find a distance traveled. Furthermore via interpretation of which axes accelerometer measures a positive distance, the position computer can determine an approximate direction of travel.

NB: The strapped down INS accelerometer provides an additional vertical speed reading.

2. *Gyroscopes.* An INS will have three gyroscopes (north, east, and azimuth) arranged at right angles to each other. The gyroscopes used will be either ring laser gyros, for the strapped down INS, or the conventional gimbals-type gyro used for the stable platform type of INS.

The gyroscope's primary role is to maintain the navigation system's orientation to true north, which thereby enables the inertia navigation system to determine the true compass direction of flight.

NB: The gyros can also feed information to the attitude indicators and the radar stabilization system.

3. *Position computer.*   The INS has a central position computer that is fed with information from the accelerometers, gyros, and altimeter. The position computer integrates the acceleration information from the accelerometers into a velocity and then into a distance as 1 nm for 1 min of north/south arc change of latitude distances, and for east/west distances (departure) into a change of longitude by the following formula: change of longitude = E/W distance × 1/cos (latitude). The INS position computer will then take the change of latitude and longitude as determined by the accelerometer, the direction of flight as determined by the gyros, and the altitude reading from the altimeter. It adds this information to the original inputted position and also allows continually for earth rotation and transport wander, using correction rates equal to the error rates that you would expect from a space gyro, to calculate a continuously updated present position.

The INS/IRS setup procedure is typically carried out using the *mode selector unit* (MSU) and the *control display unit* (CDU). First the aircraft must be aligned to true north. This is initiated by the MSU mode selector switch being placed to the align position, during which time the aircraft must not be moved. The aircraft's present position (ramp position) must then be inserted via the CDU before the alignment sequence is completed, followed by en route way points, which can be inserted from either CDU. Once the alignment is complete (approximately a 10-min cycle), the "Nav Ready" light on the MSU will illuminate, and before pushback, the mode selector should be moved to the Nav position, thus completing the INS setup procedure. The mode selector can be selected straight to Nav, which is not an approved checklist method, but the inertia navigation will align normally, and the system will only automatically transfer to the Nav mode when the aircraft's present position is inserted.

NB: If the aircraft is moved before alignment is complete and/or navigation is selected, then the aircraft's alignment has to be restarted from the beginning of the cycle, resulting in a time delay. Therefore, pilots should always make sure that the alignment is complete and Nav is selected before they move.

An inertia navigation system is aligned to true north by its gyroscopes. However, the orientation to true north is different for the two different main types of inertia navigation systems. The strapped down inertia navigation system has three ring laser gyroscopes arranged at right angles to each other, which sense the earth's rotation on start-up to define true north. Once the aircraft's position has been inserted, the true north reference is maintained as the aircraft moves by using the interference patterns of the laser beams to detect movement.

The stable platform inertia navigation system has three gimbal gyros—north, east, and azimuth. Each senses the earth rotation in one axis only and initially orients the platform to a coarse true north alignment. Then the INS computer uses the gyro's sensed information, combined with information on the aircraft's velocity and latitude, to run an *earth rotation program*. This program continually corrects for the earth's rotation and transport wander to accurately define true north and thereby to maintain the platform's alignment to true north at all times. This type of inertia navigation system is known as the *north-aligned system*.

NB: The stable platform is initially leveled by axes motors to the point where the north/south and east/west accelerometers sense no gravity, which means that the platform is level. The gyro and INS computer will then keep the platform level. This allows the platform's orientation to true north to be consistent and thereby the aircraft's directions of flight to be consistently correct.

INS/IRS systems find magnetic north by applying a stored magnetic variation to the calculated true north.

The advantages of an inertia navigation system are as follows:

1. It is a totally global system. It enables an aircraft to fly great circle tracks and to navigate accurately across vast expanses of open sky where no ground base navigation aids are available, such as in the north Atlantic or the Pacific Ocean.
2. It is a completely self-contained system and therefore is free from external navigation aids and atmospheric errors.
3. It is a very accurate system.

NB: Modern aircraft are fitted with two or more independent inertia navigation systems, which has the advantage that their positions can be compared for possible system errors.

4. Inertia navigation systems that employ ring laser gyros have the following advantages:
   a. Warm-up time is shorter.
   b. They suffer no real wander.
   c. They do not precess.
   d. They are extremely accurate.

The disadvantage of an inertia navigation system is its errors, which can fall into one of three categories.

1. *Bounded* errors are errors that do not keep increasing with time, or increase and decrease in a cycle.
   a. Schuler loop is only common to a stable platform INS that has been programmed to remain horizontal as the aircraft moves around the surface of the earth. If disturbed, it will oscillate about a true level for a period of 84.4 min. This error causes the ground speed reading to vary over the 84.4-min period although the mean ground speed is the real ground speed. The error exists at the first accelerometer level, i.e., acceleration, which can be passed up through the integration to affect velocity and distance, resulting in a distance error during the Schuler loop cycle; but the error returns to zero at the end of the cycle.
   b. North alignment error will produce bounded velocity errors.

2. *Unbounded* errors are errors that continue to increase as time goes on.
    a. Initial-position errors inputted to the system will create unbounded errors in velocity and position.
    b. North alignment error will produce serious unbounded errors in position.
3. Inherent system errors
    a. The INS position computer makes no allowance for a distance between two points being greater at height than on the surface, because of the curvature of the earth. Fortunately these differences are not large.
    b. The INS position computer also makes no allowance for the fact that the earth is not a true sphere.

As a result, accumulated errors will cause the IN position at the end of the flight to be different from the ramp position. The magnitude of this accumulated error is known as the *radial error rate,* and it should be checked at the end of each flight to determine whether inertia navigation is operating within its defined limits of accuracy.

An INS/IRS is a better system than GPS for providing navigation information mainly because of the following:

1. Downgrading is imposed on the normal GPS. This is especially true for en route navigation over large expanses of land and sea because the absence of ground installations precludes the use of differential GPS. Therefore only normal GPS is available, and its errors and downgrading make the GPS less accurate than an INS/IRS.
2. An INS/IRS is the only truly onboard, self-contained system; therefore it is not prone to external influences, i.e., natural effects or transmitter errors such as those which affect the GPS.

Therefore the INS/IRS is a better navigation system because it is more self-contained and suffers fewer errors than the GPS.

Instantaneous Vertical Speed Indicator (IVSI)   The IVSI was designed to counter the time lag error experienced by simple VSI. It uses two spring-loaded dashpots in the static line prior to the capsule that causes an immediate differential pressure to be sensed due to their inertia at the start of a climb or descent. Once the aircraft is established in a climb or descent, the dashpots are centered by their springs; and when the aircraft starts to level out, the opposite inertia of the dashpots produces an immediate change in the reading on the IVSI display.

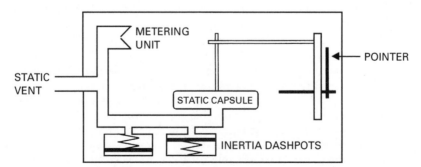

Figure 103   Instantaneous vertical speed indicator.

The advantage of the IVSI is the immediate display of any change in the aircraft's ROC or ROD.

The disadvantage of an IVSI is that the dashpots which sense the vertical acceleration of the aircraft are also affected by the acceleration in a turn. Therefore the IVSI has an error that it initially shows as a ROC when applying large angles of bank, i.e., over 40° angle of bank. However, if the turn is maintained, the IVSI will stabilize to zero, but then indicates a ROD as the aircraft rolls out of the turn.

Insulator   *See* Electrical Insulator.

Inter   Inter relates to the *intermittent* variations to the general forecasted weather, which are more frequent than "tempos," i.e., conditions fluctuate almost constantly.

NB: An inter can relate to improvements as well as deterioration in wind, visibility, weather, or clouds.

Once the inter weather events are finished, the original prevailing weather reasserts itself.

Intertropical Convergence Zone (ITCZ)   *See* ITCZ—Intertropical Convergence Zone.

Inverted Dive   A dive is recognized by the following characteristics:

1. A low pitch attitude on the artificial horizon

2. Descent indications on the altimeter and vertical speed indicator

3. An increasing airspeed on the airspeed indicator

An inverted dive is recognized by the following additional characteristic: Loose equipment will fly around the flight deck, and the pilots will be hanging onto the shoulder harness straps.

The recovery from an inverted dive is achieved as follows:

- Roll the aircraft with the ailerons for approximately 4 to 8 s.

NB: Newer jet aircraft types will tend to require only 4 s, because their ailerons are much more effective than those on older aircraft types.

This will roll the aircraft through approximately 180° and will leave the aircraft near enough the right way up in a spiral dive.

- Then perform a normal spiral dive recovery.

IRS   *See* INS/IRS.

ISA (International Standard Atmosphere)   The International Civil Aviation Organization (ICAO) ISA conditions at sea level are

Temperature   +15°C (ELR −2°C per 1000 ft)

Pressure   29.92 in mercury (inHg) or 1013 hPa (mbar) (decreases with altitude)

Density   1225 g/m^3 (density is influenced by temperature and pressure)

ISA conditions are a theoretical measuring stick so that we can compare the real and ever-changing atmospheric conditions against an expected standard.

NB: The actual atmosphere can differ from the ISA in many ways. Sea-level pressure varies from day to day, indeed from hour to hour, and the temperature also fluctuates at all levels. The variation of ambient pressure throughout the atmosphere, both horizontally and vertically, is of great significance to pilots as it affects the operation of the altimeter. Usually the temperature, pressure, and density all decrease with an increase in altitude.

Island Holding Fuel   Island holding fuel is a quantity of fuel uplifted, usually in place of diversion fuel, which allows an aircraft to hold over a destination aerodrome for an extended time. It is often associated with sorties to remote islands, e.g., Easter Island, where there is no diversion option and there is a possibility of a delayed landing due to adverse weather patterns.

Isobar   An isobar is a line on a meteorological chart that joins places of equal sealevel pressure. On meteorological charts, isobars are normally spaced at 2-mbar intervals that form patterns surrounding areas of high- and low-pressure systems.

Isoclinic Lines   Isoclinic lines, or isoclines, are lines joining points of equal dip.

Isogonals   Isogonals are lines joining points of equal magnetic variation on the surface of the earth. Magnetic variation can vary between 18° and 0°.

Isotherm   An isotherm is a line joining places of the same mean temperature.

Isothermal Layer   An isothermal layer is one where the air remains at the same temperature through a vertical section of the atmosphere, say at a constant 10°C between 5000 and 10,000 ft.
   NB: Remember that temperature usually decreases with altitude.

ITCZ—Intertropical Convergence Zone   The intertropical convergence zone (ITCZ) is where converging air masses meet near the thermal equator. And like the thermal equator, the ITCZ movement is a function of seasonal heating, which is much greater over the land than over the sea. That is, over South America and southern Africa, the ITCZ movement is large, especially in the summer season, whereas over the Atlantic Ocean its movement is small, or in other words it is stable. The effect of the ITCZ determines the weather pattern over a significant portion of the globe.
   The ITCZ differs from weather fronts associated with traveling depressions in that its converging trade winds start off dry, having subsided in the subtropical highs, but quickly pick up moisture when they cross any sea areas and hence become moist and unstable. The considerable uplift at the convergence zone produces predominantly cumuliform clouds with cumulonimbus clouds (CBs) reaching 50,000 ft on occasions. When active, the ITCZ may have nimbostratus (NS) and altostratus (AS) clouds with embedded convective cells giving heavy rain, violent turbulence, and severe icing even at high levels. On the plus side, these conditions are patchy, and it is usually possible to find a way though the ITCZ at a medium or low level. The ITCZ can vary from 30 to 300 mi wide, and its associated weather is also variable, with the worst weather usually found on the trailing side.

Jet/Gas Turbine Engine   Frank Whittle described the theory behind the jet engine as the *balloon theory*. "When you let air out of a balloon, a reaction propels the balloon in the opposite direction." This of course is a practical application of Newton's third law of motion. A jet/gas turbine produces thrust in a similar way to the piston engine propeller combination, by propelling the aircraft forward as a result of thrusting a large weight of air rearward.

$$\text{Thrust} = \text{air mass} \times \text{velocity}$$

Early jet engines worked on the principle of taking a small mass of air and expelling it at an extremely high velocity. Later gas turbine engines have evolved into taking and producing a large volume mass of air and expelling it at a relatively slow velocity (e.g., high bypass engine).

The jet engine, or aero thermodynamic duct (to give it its real name), has no major rotating parts and consists of a duct with a divergent entry and convergent or convergent-divergent exit. When forward motion is imparted to it from an external source, air is forced into the engine intake, where it loses velocity or kinetic energy, and therefore its pressure energy increases as it passes through the divergent duct. The total energy is then increased by the combustion of fuel, and the expanding gases accelerate to the atmosphere through the outlet converging duct, thereby producing a propulsive jet.

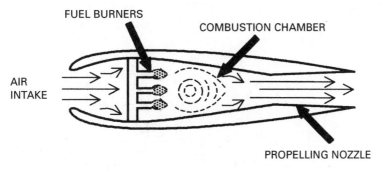

Figure 104   The jet engine.

A jet engine is unsuitable as an aircraft power plant because it is incapable of producing thrust at low speeds, as it requires a forward motion itself before it produces any thrust.

The gas turbine engine has avoided the jet engine's inherent weakness by introducing a turbine-driven compressor that produces thrust at low speeds. Therefore the aircraft power plant is in fact a gas turbine engine and will be subsequently referred to as such.

The gas turbine is essentially a heat engine using air as a working fluid to provide thrust by accelerating air through the engine, increasing its kinetic energy. To obtain this increase, the pressure energy is increased first by a *compressor,* followed by the addition of heat energy in the *combustion chamber,* before its final conversion back to kinetic energy in the form of a high-velocity jet efflux across the *turbine.* (This provides extra shaft power to drive a conventional frontal propeller or fan, to compress extra air to provide more jet flow, as in a ducted fan and bypass engines.) The airflow is then finally exhausted through the *exhaust* nozzle duct.

The mechanical arrangement of the gas turbine, as detailed below, is in series so that the combustion cycle occurs continuously at a constant pressure.

1. *Compressor.* The compressor is a rotating part that is driven by the engine turbine via a connecting shaft and acts as a supercharger for the gas turbine engine. The axial-flow compressor replaced the early-evolution centrifugal compressor, because of its ability to accommodate a number of stages in series, which created higher compression ratios, thereby increasing its pressure energy and improving its all-around efficiency.

2. *Combustion chambers.* They produce heat energy in the form of hot accelerating gases by igniting the fuel and the extensive volume of air supplied by the compressors. The combustion chamber is not an enclosed space, and therefore the pressure of the air does not rise, like that of the piston engine during combustion, but its volume does increase. This process is known as heating at a constant pressure.

3. *Turbine.* The turbine has the task of providing the power to drive the compressor and accessories. The flow of the hot combustion gases across the turbine causes it to rotate and, via a connecting shaft, causes the compressors or propeller (for turboprop aircraft) to rotate. Therefore the engine is self-sustaining.

4. *Exhaust system.* The exhaust system is a convergent duct, which converts the gas back into kinetic energy by accelerating the remaining gases to atmosphere as a propulsive jet.

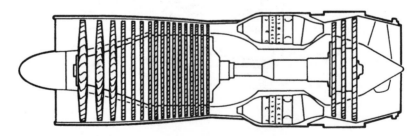

SINGLE-SPOOL AXIAL FLOW TURBOJET

Figure 105   Gas turbine engine—single-spool axial-flow compressor. (*Reproduced with kind permission of Rolls-Royce Plc.*)

The development of the jet/gas turbine engine has seen its evolution from the early pure jet design, to the centrifugal compressor gas turbine design, to the single-spool axial-flow gas turbine design, to the twin-spool axial-flow gas turbine design, to the bypass gas turbine design, to the fronted fan gas turbine design, and then to the triple-spool axial-flow turbofan gas turbine design.

There are four types of jet/gas turbine engines, whose individual efficiency, excluding the helicopter's shaft gas turbine engine, is measured in terms of the propulsive efficiency against airspeed.

1. *Turboprop.*   The propeller's efficiency is greatest at slow speeds, say, below 350 mph, due to its ability to displace a large air mass to generate forward thrust. However, its efficiency rapidly decreases above a medium speed, say, above 350 mph, because of the disturbance of the airflow caused by the high supersonic blade tip speeds of the propeller.

2. *Bypass turbojet.*   The bypass ducted fan engine deals with a larger relative airflow mass and a lower jet velocity than the pure jet engine. Thus it provides a propulsive efficiency which is comparable (albeit slightly less efficient) to that of the turboprop at slow speeds and which exceeds that of the pure jet engine especially at high speeds, where it is designed to be most efficient. The high bypass ratio is simply a further development of the bypass principle and is generally more efficient than the low-bypass-ratio engine.

3. *Pure turbojet.*   The propulsive efficiency of the pure jet depends on its forward airspeed; therefore, it is extremely inefficient at slow speeds and efficient only at extremely high speeds. However, these efficient high speeds are greater than the normal range of most modern airline transport aircraft.

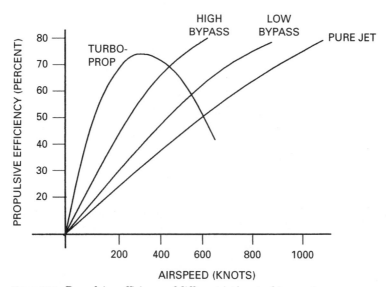

Figure 106   Propulsive efficiency of different jet/gas turbine engines.

4. *Shaft.*   The gas turbine shaft is a helicopter rotor engine.

Frank Whittle invented the jet aero engine as a means of increasing an aircraft's attainable altitude, airspeed, reliability, and to a lesser extent maneuverability, for the military. He then developed his design into the jet/gas turbine engine for two main reasons:

1. To achieve higher altitudes and thereby airspeed, because the propeller aircraft had limited altitude and speed capabilities. Whittle realized that you had to fly higher where the drag was less, to achieve higher speeds.

NB: A propeller-driven aircraft was unable to achieve these heights, because the thin air hindered the efficiency of the propeller, as it was unable to grip or catch the air and throw it backward, which prevented it from achieving high altitudes. Also, low air density meant low resistance on the propeller. Therefore the propeller rpm would become supersonic, causing blade compressibility losses which resulted in a loss of forward motion and thereby loss of attainable altitude.

2. As a more simplistic and therefore reliable engine, because the piston engine was a very complicated engine with many moving parts and so was unreliable. Whittle designed a simple engine that is self-sustaining with controllable thrust. A forward compressor that works as a supercharger squashes the air and then delivers it to a combustion chamber, where it is ignited and then accelerated rearward over a turbine, which turns the forward compressor via a common shaft, before being expelled rearward. This employs Newton's third law of motion, which states that every action has an equal and opposite reaction.

While the motivation for the development of the jet/gas turbine engine for airliners was simply that only the jet engine could produce the qualities required to satisfy the demands made by the modern airliner, namely, a higher total power output, which could obtain high-altitude, high-speed conditions for heavier aircraft, resulting in a greater range and less travel time. The piston engine, however, produced only a fraction of the thrust horsepower of the jet engine, mainly because of propeller compressibility losses and the restrictive cylinder power output, which restricted it to low attainable altitudes, speeds, and aircraft weights.

Jet/Gas Turbine Engine—Bleed Valves *See* Bleed Valves—Jet/Gas Turbine Engine.

Jet/Gas Turbine Engine—Fuels *See* Fuels.

Jet/Gas Turbine Engine—High Altitudes *See* Cruise—High-Altitude (Jet).

Jet/Gas Turbine Engine—Most Efficient It is most efficient at high altitudes and high rotational speeds.

Jet/Gas Turbine Engine—Noise *See* Noise—Jet/Gas Turbine Engine.

Jet/Gas Turbine Engine—Thrust to Thrust Lever Position The thrust lever produces greater engine thrust from its movement near the top of its range than at the bottom. On the jet/gas turbine engine, thrust is proportional to

- Compressor rpm (mass airflow)
- Temperature (fuel/air ratio)
- The efficiency of the compressor at varying rotational speeds

An engine's operating cycle and gas flow are designed to be at their most efficient at a high rotational speed, where it is designed to spend most of its life. Therefore as rotational speed rises, mass flow, temperature, and compressor efficiency all increase, and as a result more thrust is produced, say, per 100 rpm, near the top of the thrust lever range than near the bottom.

In practical terms, this translates to differing thrust output per inch of thrust lever movement. At low rpm speed (near the bottom of its range), a 1-in movement of the thrust lever could produce only 600 lb of thrust; but at a high rpm (near the top of its range) a 1-in movement of the thrust levers could typically produce 6000 lb of thrust. For this reason if more power is required at a low thrust lever setting, then a relatively large movement/opening of the thrust lever is required, i.e., when initiating a go-around/overshoot. But if one is operating at the top of the thrust lever setting, a large reduction or increase in thrust would require only a relatively small movement of the thrust lever. Obviously an appreciation of the jet/gas turbine engine's response characteristics helps the pilot's understanding and operation of the engines.

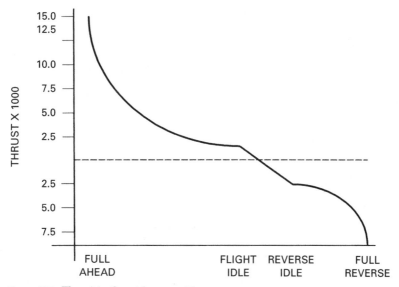

Figure 107 Thrust to thrust lever position.

Jet/Gas Turbine Engine—Upset Disturbed or turbulent airflow will cause a jet/gas turbine engine to be upset and to stall. This occurs because a jet/gas turbine is designed to operate using a clean uniform airflow pattern which it obtains within the aircraft's normal operating attitude. However, beyond the aircraft's normal angles of incidence and slip, and/or in extremely severe weather turbulence, the engines can experience a variation in the ingested air's pressure/density volume, angle of attack, and velocity properties. This changes the incidence of the air onto the compressor blades, causing the airflow over the blades to break down and/or induce aerodynamic vibration. This upsets the operation of the engine, causing it to stall. The stall can be identified by increases in total gas temperature (TGT), engine vibration, and rpm fluctuation.

There are two types of stall:

1. *Low rpm speed, intake/forward compressor stall.* The compressor blades will stall, choking the engine, because the angle of incidence of the air relative to the blade is too high (positive incidence stall), which is a typical front compressor stage problem at low rotational speed. Therefore to prevent the airflow stalling in the forward compressor section at low rpm speeds, variable inlet guide (stator) vanes are progressively closed in order to maintain an acceptable angle of incidence of the airflow onto the following rotor blades. Conversely as the

engine rpm speed increases, the guide vanes are opened. When the engine is at 100 percent rpm speed, the guide vanes are fully opened.

2. *High accelerating rpm speed, rear compressor stall.* The compressor blades will stall because the angle of incidence of the airflow relative to the blade is too low (negative incidence stall), which is typical of a rear compressor stage stall problem at high rpm speeds. Therefore to prevent the airflow stalling and causing the engine to choke in the rear compressor, interstage bleed valves provide additional cover by draining away excessive volumes of air, when the engine is at a high or accelerating rpm speed.

If the stall is only mild, then it is not immediately dangerous; but in a bad case the engine will be completely upset, resulting in substantially all the thrust being lost and extreme EGT, and causing the engine to surge.

Aircraft with rear-mounted fuselage engines are more likely to suffer intake stalling than other aircraft. This is so because at angles of incidence approaching the stall, the airflow pattern over the wings breaks down, generating disturbed air which is ingested by the engines, which are ideally placed to receive it. This leads to intake stalling which is corrected for by the automatic switching of the auto ignition system that is triggered by the stall warning system. This at least stops the engine from running down while the aircraft is being recovered to a safe attitude.

Therefore the design features of the engine that safeguards against a stall are

1. Inlet guide vanes (especially for forward compressor stalls)

2. Bleed valves (especially for rear compressor stalls)

3. Auto ignition system (especially for rear-mounted engines)

However, if an aircraft were still unfortunate enough to suffer a stall, then a recovery technique would normally involve adjustment of the aircraft's attitude to within its safe limits and closing the throttles smoothly and slowly (for severe stalls or engine surges).

In addition to disturbed air, other conditions can upset an engine:

1. *Contaminated air,* e.g., bird strikes, can cause structural damage to the engine. Therefore caution should be exercised in its further use after ingestion of any foreign objects. Lower-thrust operation or even an engine shutdown may be required. A ground inspection is a must.

2. *Volcanic ash* can upset the engine and cause it to flame out due to the lack of air intake. Therefore volcanic ash should be avoided at all costs, even if this means doing a 180° turn. If, however, you do stumble into a volcanic ash cloud, then you should leave immediately by the quickest route. If you need to restart your engines, then a common drill would be to turn on all engine air bleed demand systems, e.g., anti-ice systems, to remove the ash from the engine when relighting.

3. *Heavy ice ingestion* can upset the engine and cause it to flame out, which often occurs when blocks of ice are broken off the engine cowling by the anti-ice system. Therefore to prevent this from happening to all the engines at once, de-ice only one engine at a time and ensure that its high-energy ignition system is activated first.

Jetstream Jetstreams are simply narrow bands of high-speed upper thermal winds at very high altitudes. The official definition of a *jetstream* is a strong narrow current on a quasi-horizontal axis, in the upper tropopause or stratosphere, characterized by strong vertical and/or lateral wind shear (CAT). The wind speed must be greater than 60 kn for a wind to be classified as a jetstream. Jetstreams are typically 1500 nm long, 200 nm wide, and 12,000 ft deep; and their speed is directly proportional to the thermal gradient. That is, the greater the thermal gradient, the greater the speed of the jetstream.

Jetstreams are driven by thermal gradients and therefore are found wherever the thermal gradient is high enough. There are two bands of rapid temperature changes, i.e., high/maximum thermal gradient, in each hemisphere, that are marked enough to produce a jetstream.

1. At the polar front, around the 60° latitude, where the polar air meets the subtropical air. This is a polar front jetstream, and it is the most marked thermal gradient to be found, especially when it is over land in the winter.

2. At the intertropical front, where the subtropical air meets the tropical air. This is known as the intertropical front jetstream.

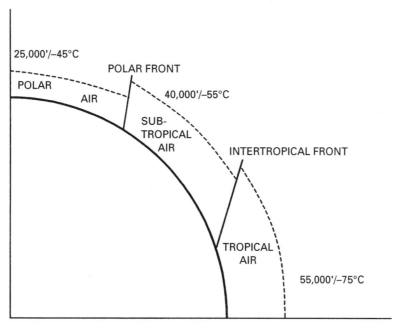

Figure 108 Global temperature gradient locations for jetstreams.

The jetstream exists just below the tropopause in the warm air of a pressure system (i.e., in the subtropical air at the polar front and in the tropical air at the intertropical front) but appears on the surface chart to be in the cold sector. This is because of the slope of the front with height.

See Fig. 109 at the top of the next page.

The jet moves with the front, i.e., south in the winter. And its direction is not always westerly, as the jet follows the pressure system; in fact the polar front jetstream can blow from 190° to 350° around the polar front. However, its overall direction is generally regarded as being from west to east.

The maximum wind shear/CAT associated with a jetstream can be found level with or just above the jet core, in the warm air, but on the cold polar air side of the jet.

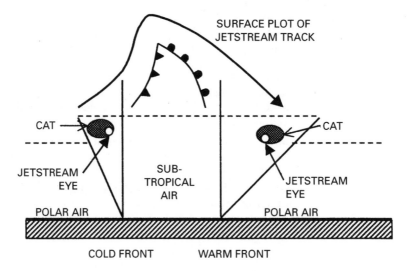

Figure 109   Plan and section view of a frontal system and jetstream.

Katabatic Wind A katabatic wind is a local valley wind that flows down the side of a hill. It is formed when the side of a hill cools rapidly, when the sun sets, which in turn cools the layer of air next to the ground, making it more dense, resulting in an airflow down (subside) the side of the hill. This is a katabatic wind.

Krueger Flaps These are leading-edge wing flaps used to increase the wing chamber and therefore increase the maximum coefficient of lift.

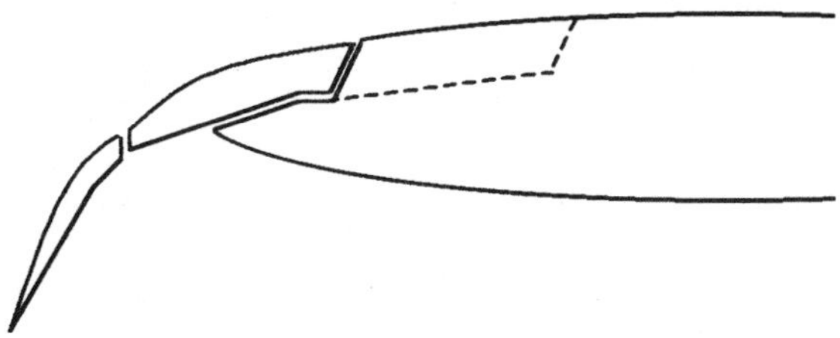

Figure 110 Krueger flaps.

Lambert's Projection Chart   A Lambert projection is a form of conic projection, with the difference that the cone cuts into the earth model, and therefore the chart is not a projection in the true sense of the word, but is mathematically produced. The Lambert projection has a maximum 24° spread of latitude.

STANDARD PARALLEL

PARALLEL OF ORIGIN

STANDARD PARALLEL

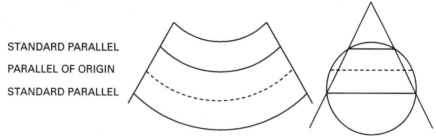

Figure 111   Lambert's conical projection chart.

Parallel of origin   On Lambert's conical chart, convergency is constant at all latitudes, but is only correct at the *parallel of origin* where the cone is a tangent to the earth model.

Convergency = change of longitude × sin (parallel-of-origin standard parallels)

The earth globe touches the conical projection at two points, which denote the two standard parallels on the chart. Scale is correct at the standard parallels. However, because of the chart's modest spread of latitudes, the differences in scale and convergency across the chart are small enough to be considered insignificant. Therefore it is assumed that scale and convergency on a Lambert chart are correct at all latitudes; and because the chart convergency is close to earth convergency, other chart properties are the same as earth properties. That is, rhumb lines are curved, and great circle tracks are straight lines.

Lambert's projection charts are therefore orthomorphic because the process has the same scale in all directions and a right-angle graticule. The main advantage of using a Lambert chart is that you can plot radio bearings as a straight line without applying a conversion angle.

Landing—on a Contaminated (Wet) Runway   For landing on a contaminated runway, there are four distinct phases that you must consider, in the following order:

1. *Planning.*   Be equipped with certain essential information and make an assessment of the criticality of the landing.
   a. Ensure that the landing distance available is greater than the landing distance required on a wet runway for your landing weight.

NB: Flight performance manuals will contain data on the landing distance required for your weight and the condition of the runway.

Also consider the width of the runway. A narrow, wet runway will be less forgiving in these conditions; namely, skidding off the side of the runway is a distinct possibility.

b. Know your approach and aquaplaning speed. Therefore you will know the speed band after landing at which aquaplaning can commence. With this information, plan accordingly the full use of your antiskid devices, thrust reversers, spoilers, and brakes, to minimize the probability of aquaplaning.

NB: From your preflight walk-around you will know the condition of your tires; i.e., the more worn the tire, the more likely it will be to aquaplane. If the conditions for landing are marginal, i.e., landing distance available (LDA) is only slightly more than landing distance required (LDR), and you have a high approach speed with a low aquaplane speed (large aquaplane speed band) with worn tires, then it would be good airmanship to decide not to attempt a landing in wet conditions.

c. Consider the effect of any crosswind and transverse slope of a runway. This is necessary because drainage can be seriously affected by winds above 10 to 15 kn. Thus a crosswind could have a significant effect on water depth on one side of the runway, and therefore aquaplaning may be more likely on that side and hence asymmetric braking action may be required. Most runways drain adequately under normal rates of precipitation, and heavy precipitation would rarely last long, therefore a delay of 15 to 20 min should be sufficient to allow really critical water depths on the runway to clear. If, after having assessed the criticality of the landing, you decide that the landing can be made without undue risk, start the approach.

2. *Approach.* There is no acceptable operational variation of the approach technique that will give you a shorter landing distance than from the standard approach technique. Any aiming short, ducking under the glide slope, or reducing the speed below the target approach speed will just tend to create more problems. Therefore the only object of the approach is to fly a normal approach, which gets you to the threshold at the correct height, correct speed, correct rate of descent, correct power and attitude, etc.

3. *Threshold.* At the threshold, if you are not totally satisfied with your position, then don't hesitate and go around—especially if the crosswind for these conditions is above its limit. However, if you are happy with your position and the weather conditions at the threshold, proceed with the landing.

4. *Landing.* Having committed to the landing, be positive and don't try to be fancy.
- Do not indulge in a prolonged flare, as it wastes distance, which is valuable in wet conditions.
- A firm touchdown is required to break through the water.

NB: A really smooth touchdown might induce an aquaplane if the aquaplane conditions are critical.

- Once the main wheels are on the ground, push the control column forward to get the lift off the wings and the weight onto the wheels.
- Pull thrust reversers and spoilers immediately, to get their maximum worth, and hold them as long as required.
- Monitor the auto brakes and the antiskid system. If they appear to be malfunctioning, then employ manual braking. However, be aware that on a wet runway you may initially get an impression of brake failure on many aircraft types.

- For manual braking, start as soon as all the wheels are on the ground. Increase brake pressure progressively, to just below the onset of any skid. If a skid is felt, release the brakes to allow the wheel to spin; then lightly reapply the brakes.

Aquaplaning accidents on contaminated runways are avoidable accidents. This is so because you can assess all the information and make an informed, common-sense decision before proceeding. If you are ever in doubt, then you should not attempt a landing, but should hold and wait for an improvement, or even divert.

Landing—Crosswind As with all flight procedures, planning is the first and probably the most important aspect of the task. The most important part of planning for the crosswind landing is to ensure that the crosswind component on the selected runway does not exceed the most limiting of the aircraft, company, or your own personal limits.

It is particularly true in the crosswind case that a good landing requires a good approach. This extends back around the circuit, but especially on the base leg, where the circuit truly becomes the approach to land. The effect of the wind starts to take on its own individual importance during a crosswind landing because on the base leg you will experience either a headwind or a tailwind, which requires specific consideration, as follows.

Base leg A tailwind on the base leg will increase your speed over the ground and tend to carry you past the runway extended centerline. For this reason you should show some anticipation and

- Commence your descent early.
- Begin your turn onto finals early.
- Continue the turn onto finals beyond the runway heading to allow for drift.

A headwind on the base leg will decrease your speed over the ground. For this reason you should

- Delay descent.
- Delay turn onto finals.
- Stop the turn short of the runway heading to anticipate the expected drift.

There are three accepted crosswind landing techniques:

1. The crab method
2. The wing-down method
3. The combination method (a crab approach and a wing-down landing)

The crab method The crab method of crosswind landings employs laying off drift all the way down the final approach and through the flare. On final approach, track down (crab) the extended centerline by heading into wind; i.e., lay off drift.
NB: Because of the crab angle needed to maintain the extended centerline, the runway will appear to one side of the aircraft's nose, but will still look symmetric.
The wings should be kept level except when you are adjusting the crab angle.
NB: Wind strength often decreases near the ground, so continual adjustments to heading will have to be made to maintain your track-down final approach. This is especially the case in strong and gusty conditions.
During the flare, reduce the power and raise the nose normally, and continue to maintain the track down the centerline by crabbing into the wind. If any sideways

drift appears to develop before touchdown, it can be counteracted by a small amount of wing down into wind and keeping straight with rudder.

Perform the flare normally in a crosswind landing, but do not prolong the hold-off; otherwise sideways drift could develop.

Just prior to touchdown, the aircraft should be placed on the ground, with the wheels aligned with the runway. Therefore,

- Align the aircraft (wheels) with the centerline direction with smooth but positive rudder pressure just prior to touchdown.
- Hold the wings level with aileron to avoid any sideways drift across the runway before the wheels touch.

On the landing run, keep straight with rudder, lower the nosewheel to the ground, and keep the wings level with progressive into-wind aileron. Hold the nosewheel on the ground to obtain positive steering to allow for greater directional control.

Judgment and timing are important when using the crab method. Failing to remove the crab angle prior to landing will result in the wheels touching down sideways, placing stress on the landing gear. Removing it too early will allow a sideways drift to develop, resulting in landing downwind of the centerline. In both cases the landing will feel heavy. A reasonable touchdown can only be achieved with fine judgment in removing drift and contacting the ground smoothly.

The wing-down method The wing-down method sees the aircraft track down the extended centerline by slipping all the way down the final approach, flaring, and touching down. Because the aircraft tracks forward toward the runway in a slip, this maneuver is called a forward slip.

On final approach, track down the extended centerline by forward slipping.

- Lower the into-wind wing a few degrees.
- Apply opposite rudder pressure to stop the aircraft turning, thereby aligning the aircraft's longitudinal axis with the runway centerline.

  NB: If the aircraft starts to drift downwind across the runway, you have to

- Lower the into-wind wing a few degrees further.
- Keep the aircraft straight along the runway centerline with slightly greater opposite-rudder pressure.

If the aircraft starts to slip into wind across the runway, you have to

- Raise the into-wind wing a few degrees.
- Keep the aircraft straight along the runway centerline with slightly less opposite-rudder pressure.

During the flare, the wing-down and opposite rudder is held on through the flare. Reduce the power and raise the nose normally, and continue to maintain the track down the centerline by slipping into the wind. If any sideways drift seems to be developing before touchdown, it can be counteracted in the same manner as on the approach. Perform the flare normally, but do not prolong the holdoff; otherwise sideways drift could develop.

During the touchdown, the wing-down and opposite rudder is held on through the touchdown.

- Touchdown will be made on the into-wind wheel first, because of the wing being down into the wind.

- Maintain directional control with the rudder, as there may be a tendency for the aircraft to yaw into the wind when the wing-down wheel touches the runway first.

- The other main wheel will follow onto the ground naturally.

- Throughout the maneuver the aircraft will be tracking straight down the runway with its longitudinal axis aligned with the centerline. No sideways drift across the runway should be allowed to develop.

On the landing run, keep straight with the rudder and lower the nosewheel to the ground, and keep the wings level with progressive into-wind aileron. Hold the nosewheel on the ground to obtain positive steering to allow for greater directional control.

The advantage of the wing-down method is that less judgment and timing is required in the actual touchdown, since the aircraft is aligned with the runway centerline throughout the flare and the touchdown. There is no crab angle to remove and no sideways drift. It is of lesser importance if the aircraft touches down a little earlier or a little later than expected.

Combined method The combined method uses the crab approach followed by a wing-down landing. The advantages of this method are a more comfortable and balanced (efficient) approach and an easier crosswind landing if prior to touchdown the wing-down method is employed by lowering the into-wind wing while simultaneously applying opposite rudder to align the aircraft's wheels with the runway centerline. The transition to the wing-down method can vary from about 100 to 20 ft depending upon the individual pilot's experience. However, in strong and gusty crosswinds, it is better to introduce the wing-down method earlier than in calm conditions.

On final approach, crab down the extended centerline by heading into the wind; i.e., lay off drift. The wings should be kept level except when adjusting the crab angle.

NB: This is preferable to the wing-down technique, which employs crossed controls, with the aircraft out of balance, which is both inefficient and uncomfortable.

Approaching flare height, transition should be made to the wing-down method, at approximately 100 to 20 ft above the runway, by simultaneously

- Lowering the into-wind wing to prevent sideways drift

- Using smooth rudder pressure to align the aircraft wheels with the runway centerline

During the flare, the wing-down and opposite rudder is held on through the flare. If any sideways drift seems to be developing before touchdown, it can be counteracted by slipping into wind with aileron and opposite rudder to maintain the runway centerline. Reduce the power and raise the nose normally. Perform the flare normally, but do not prolong the holdoff; otherwise sideways drift could develop.

At touchdown, the wing-down and opposite rudder is held on through the touchdown.

- Touchdown will be made on the into-wind wheel first, because of the wing being down into the wind.

- Maintain directional control with the rudder, as there may be a tendency for the aircraft to yaw into the wind when the wing-down wheel touches the runway first.

- The other main wheel will follow onto the ground naturally.
- Throughout the maneuver the aircraft will be tracking straight down the runway with its longitudinal axis aligned with the centerline. No sideways drift across the runway should be allowed to develop.

On the landing run, keep straight with the rudder, lower the nosewheel to the ground, and keep the wings level with progressive into-wind aileron. Hold the nosewheel on the ground to obtain positive steering to allow for greater directional control.

A crosswind from the left of the aircraft's nose, in the northern hemisphere, is more difficult for the following reason. The wind backs in direction during a descent in the northern hemisphere. Therefore a crosswind from the left of the aircraft's nose becomes greater nearer the ground, because it backs in direction from A to B, resulting in a lower headwind component and a greater crosswind component.

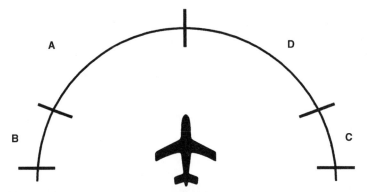

Figure 112   Variation of crosswind direction in a descent.

In contrast, a right-side crosswind component will become less nearer to the ground, as it backs, from C to D, resulting in a greater headwind component.

In the southern hemisphere, a crosswind from the right creates the more difficult crosswind landing because the wind veers in direction during a descent.

Landing Distance Available (LDA)   LDA is the distance available for landing, taking into account any obstacles in the flight path, from 50 ft above the surface of the runway threshold height (fence) to the end of the landing runway.

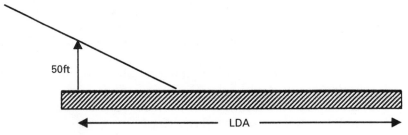

Figure 113   Landing distance available.

Landing Distance Required (LDR)   Landing distance required (LDR) is the distance required from the point where the aircraft is 50 ft over the runway (i.e., threshold *fence* height) at a maximum VAT threshold speed to the point where the aircraft reaches a full stop. LDR is normally factored by a safety margin, usually 8 percent. If reverse thrust is inoperative, LDR is increased by an additional safety margin, usually 10 percent, for jet aircraft, although reverse-thrust performance is not factored into the original LDR. If the antiskid system is inoperative, then the LDR is increased by a large safety margin, normally about 50 percent.

Landing Gear   Most aircraft use the *tricycle layout,* where the two main undercarriage units are positioned just aft of the center of gravity and support up to 90 percent of the aircraft's weight and the entire initial landing shocks. The nosewheel unit keeps the aircraft level and in most cases provides a means of steering.

The main purpose of retractable landing gear is to improve aircraft performance by reducing the drag created by extended gear in flight.

The landing gear is operated by the crew via a gear lever in the cockpit. A system of green and red lights for each wheel indicates whether the associated wheel is down and locked (green) or up and locked (red). Gear retraction is normally effected by a hydraulic system, although pneumatic or electrical systems can also be used. Often a hydraulic system is the primary source of gear retraction power, and an electrical system or an emergency accumulator hydraulic power source of gear retraction is used as a backup system. In all cases the retracted gear is secured in its stowage position by mechanical locks. The gear is usually extended by using the power source that retracts the gear, i.e., hydraulics, although sometimes the gear is extended by simply releasing the mechanical locks and the extension is effected by gravity. Once the gear is down, mechanical locks hold the gear in position. However, in the main, gravity extension is only used as a backup procedure.

In addition, a means is normally provided to protect from retracting the gear on the ground, via a ground/air sensor that locks the gear lever in the down position when the aircraft is on the ground. However, an override switch (normally on the underside of the gear lever) is provided for emergency use, when belly flopping is a better option in a nonnormal situation.

Landing Performance   An aircraft's landing performance is subject to many variable conditions including these:

1. *Aircraft weight.*   An increased aircraft weight results in a greater landing distance required (LDR). An increased aircraft weight has the following variable effects that increase the landing distance required:
   - The stalling speed is increased for higher aircraft weight, so the minimum approach speed V_{REF}/VAT ($1.3V_S$) must be higher. A higher approach speed requires a greater landing distance to come to a full stop.
   - The greater aircraft weight and higher approach speeds result in a greater momentum of the landing aircraft and therefore requires greater distance to stop.
   - The increased weight means the kinetic energy ($\frac{1}{2}mV^2$; m = mass) is higher and the brakes have to absorb this greater energy, which increases the landing distance required.

Overall an increase in weight increases the LDR. As a guideline, a 10 percent increase in weight requires a 10 percent increase in landing distance, i.e., a factor of 1:1. Fortunately most performance tables or graphs allow the pilot to extract the performance figure applicable to the aircraft's actual gross weight. When an aircraft's weight is found to be limiting, i.e., too heavy to meet a landing performance requirement, the only answer is to lower the aircraft weight. This can be accomplished either before the flight or by dumping fuel prior to landing, if allowed (usually only allowed for emergency landings), until the revised aircraft weight can achieve the landing performance required.

NB: This is so, assuming no other, longer runway is available.

2. *Aircraft flap setting.* Increased flap settings decrease the landing distance required. The use of increased flap settings has the following variable effects that decrease the landing distance required.

- Flap deployment increases lift and reduces the stalling speed V_S. Therefore the approach speed VAT/V_{REF} ($1.3V_S$) is less, which results in a shorter LDR.
- The higher the flap deployment, the greater the aerodynamic drag that helps to slow the aircraft down, which results in a shorter LDR.
- The higher the flap setting, the steeper the approach path, the lower the forward velocity and momentum on landing, which results in a shorter LDR.

3. *Aerodrome pressure altitude.* A high-pressure altitude aerodrome (high elevation) decreases an aircraft's performance which results in an increased LDR. The higher the pressure altitude of the aerodrome, the lower the air pressure and the lower the air density. A low air density reduces an aircraft's engine and aerodynamic (C_L) performance which results in an increase in the landing distance required for a given aircraft weight.

Ergo, *high* means a decrease in performance that results in either a greater LDR or a lower landing weight (LW).

4. *Air density/density altitude* (temperature and pressure altitude). An increase in density altitude (decrease in air density) decreases an aircraft's performance which results in an increased LDR. Density altitude is a function of pressure altitude and/or temperature; and the higher the density altitude, the lower the air density. High-pressure altitude and/or temperatures decrease the air density. The low air density associated with a high-density altitude decreases the aircraft's engine and aerodynamic (C_L) performance, which results in an increased LDR for a given aircraft weight due to the following effects:

- With low air density, the true airspeed will be greater to provide the same lift force than for lower-density altitudes. This is so because to produce the required aerodynamic lift force [remembering our lift formula $L = C_L \times \frac{1}{2}\rho V^2$ (TAS) $\times$ span], the increase in TAS compensates for the decrease in air density. Therefore the touchdown speed will be higher as the density altitude becomes higher, and so the amount of kinetic energy to be dissipated to stop the aircraft is greater. Hence a longer landing distance is required.
- Also the lower the air density, the lower the drag generated by the drag devices deployed on landing. Therefore the stopping action is less effective as the density altitude becomes higher, and hence a longer landing distance is required.

Ergo, hot and high mean a decrease in performance that results in either a greater LDR or a lower LW.

An increase in pressure altitude of 1000 ft or an increase of 10°C will increase the LDR by approximately 5 percent, or a factor of 1.05.

In addition, the lower air density at high-density altitudes reduces the WAT (weight, altitude, temperature) go-around performance of the aircraft. This is so because the engine (including the propeller) will not produce as much power/thrust owing to the reduced ambient air density, and therefore the mass of air entering the engine is lower, causing a decrease in engine thrust produced. This results in a decrease in the aircraft's go-around and climb performance.

If the WAT go-around performance is limiting, then the aircraft's landing weight has to be restricted to a lower weight that meets the required go-around climb performance. Landing performance tables and graphs allow for variations in the pressure altitude and the ambient temperature of the aerodrome, i.e., the density altitude.

5. *Humidity.* High humidity decreases air density which decreases an aircraft's aerodynamic (C_L) and engine performance and results in an increased LD required for a given aircraft weight. This is so because when humidity is high, lighter water molecules replace some of the air's heavier molecules, which lowers the overall air density. And the lower the air density, the lower the aircraft's aerodynamic performance, which requires a higher TAS on the approach and generates less drag on landing, which increases the landing distance required for a given aircraft weight. In addition, the low air density reduces the engine thrust, which affects the aircraft's go-around performance and can restrict the aircraft's landing weight if the go-around performance is limiting.

Ergo, hot, high, and humid mean a decrease in performance that results in either a greater LDR or a lower LW.

6. *Wind.* Wind has a profound effect on the landing performance of an aircraft. An aircraft may experience a headwind (HW), tailwind, or crosswind.

A headwind reduces the landing distance required for a given aircraft weight, or it permits a higher LW for the LDA. This is so because the ground speed (GS) is reduced by the headwind for the same TAS, e.g.,

$$TAS = 120 \text{ kn}$$

$$HW = 20 \text{ kn}$$

$$GS = 100 \text{ kn}$$

Therefore the aircraft has a lower touchdown speed from which it has to stop. Additionally a headwind provides a resistance to the motion of the aircraft that is used as an efficient braking drag. These effects result in a shorter landing distance required for a given aircraft weight.

Ergo, the greater the headwind, the greater the aircraft's landing performance.

Furthermore, the wind shear effect on the approach to landing is important. An approach into wind experiences a decrease in the headwind value as you get closer to the ground.

NB: This is so because the wind speed drops off near the ground due to the effect of ground friction.

Therefore the aircraft experiences a dropoff in performance and a tendency to sink and possibly undershoot the aiming point as you get closer to the ground.

Ergo, you have to be vigilant on the final approach with a headwind and prepared to compensate for a loss of performance by increasing thrust and/or pitch.

A tailwind increases the landing distance required for a given aircraft weight, or it requires a lower LW for the LDA. This is true because the ground speed is increased by the tailwind for the same TAS, e.g.,

$$TAS = 120 \text{ kn}$$

$$HW = 20 \text{ kn}$$

$$GS = 140 \text{ kn}$$

Therefore the aircraft has a higher touchdown speed from which it has to stop. Additionally the braking drag is less effective; in fact, the tailwind has a pushing effect on the aircraft which is completely the opposite effect required to assist a landing aircraft. Therefore a longer landing distance is required.

Ergo, the greater the tailwind, the worse the aircraft's landing performance.

Furthermore, the wind shear effect on the approach to landing is important. An approach with a tailwind experiences a decrease in the tailwind value as you get closer to the ground.

NB: This is so because the wind speed drops off near the ground due to the effect of ground friction.

Therefore the aircraft experiences an increase in performance (the same effect as an increasing headwind) that produces a tendency to float farther down the runway.

As a guideline, a tailwind component of 10 percent of the landing speed will increase the landing distance required by 20 percent, or a factor of 1:2. However, downwind landings are *not* recommended even when the tailwind component has been factored into the LDR, because it is still possible for the aircraft to overrun the runway because of the cumulative tailwind effects on the aircraft, such as tendency to land long, higher touchdown speeds, and poor drag/braking action.

Not more than 50 percent of the reported headwind or not less than 150 percent of the reported tailwind should be used to calculate the landing performance.

These adjustments provide a safety margin to the reported wind, which covers acceptable fluctuations of the actual wind experienced (i.e., the actual headwind not being as strong as reported or the actual tailwind being stronger than reported) to ensure that the aircraft needs its takeoff run, distance, and climb performance or landing distance requirements.

NB: Most performance charts and tables already factor in these adjustments.

It is extremely bad airmanship to attempt a landing with a tailwind.

The aircraft must not land in a crosswind that exceeds its certified maximum crosswind limitation for the aircraft type, to safeguard directly the directional and lateral control and indirectly the takeoff run performance of the aircraft. If the certified crosswind limitation of the aircraft is exceeded, it has the following effects:

- Directional control is reduced. The effect of the crosswind on the aircraft's keel surface causes the aircraft to weathercock into the wind. This can be balanced by the aerodynamic lift force of the rudder. However, above the certified maximum crosswind limit, the rudder has an insufficient force to balance the weathercock effect and maintain an adequate directional control ability to an unacceptable level (VMCG/A).
- Lateral control is reduced. A further effect of the crosswind is to lift the into-wind wing, which has to be held down with aileron. However, above the certified maximum crosswind limit, the aileron will have insufficient force to balance the lift force on the into-wind wing; therefore the aircraft's lateral control is reduced.

- Drag is increased. Above the maximum certified crosswind limit, some aircraft types may be able to maintain directional and/or lateral control with maximum deflected rudder and/or aileron control surfaces, but suffer from an increase in drag on these surfaces. This results in an unacceptable decrease in acceleration during the takeoff run that is detrimental to the aircraft's landing performance.

7. *Runway length, slope, and surface (including wet or icy conditions).*   The length of the available runway is one of the performance limitations that restricts the maximum weight of the aircraft. The longer the runway, the greater the aircraft's stopping action and the higher its VAT/V_{REF} approach speed can be. And because the VAT/V_{REF} speed is related to the aircraft's weight, it can be seen that the longer the runway available, the greater the aircraft's landing weight can be.

NB: Other factors, such as the maximum permissible structural certificate of airworthiness (C of A) weight of the aircraft, may be more limiting than the runway length.

A low-friction runway surface increases LDR for a given aircraft weight, because the surface does not permit effective braking.

NB: A low-friction surface can be a contaminated hard surface, such as with snow and standing water, or nonhard surfaces, such as grass and mud.

A particular runway surface phenomenon that significantly increases LD is aquaplaning. Because of the possibility of aquaplaning on a wet runway, landing into a headwind is highly recommended.

NB: Landing performance graphs and tables (i.e., of field length) normally allow for such variations in surface conditions.

A downward runway slope requires a longer LD for a given aircraft weight, or a lower LW for a given LDA. This is so for the following reasons:
- It takes longer for the aircraft to touch down from the 50-ft threshold (fence) height, because the runway is falling away beneath the aircraft.
- The downslope maintains the aircraft's momentum, and therefore the aircraft's braking is less effective.

An upward slope requires a shorter LD for a given aircraft weight, or allows a higher LW for a given LDA. This is so for the following reasons:
- It takes a shorter distance for the aircraft to touch down from the 50-ft threshold (fence) height, because the runway is rising beneath the aircraft.
- The upslope dissipates the aircraft's momentum, and therefore the aircraft's braking is more effective.

NB: Landing performance graphs and tables (i.e., of field length) normally allow for such variations in runway slope.

The various landing performance charts and tables are presented logically and allow you to enter known conditions, such as temperature and pressure altitude, and to proceed through the chart or table to end up with an answer for takeoff distance required, given a set aircraft weight. Or you can find the highest allowable aircraft weight, given a set takeoff distance available for the ambient conditions.

Landing Performance—Limitations   The normal performance limitations for an aircraft's landing profile relate to the aircraft's weight. That is, the aircraft must weigh less than the most restrictive weight to meet all the various performance criteria during the landing profile. This weight is known as its *maximum landing* *weight* (MLW). The normal landing performance is based upon the determination of the most restrictive MLW from the following criteria:

1. Certified maximum structural landing weight (C of A limit).
2. Weight, altitude, and temperature (WAT) limit that allows the aircraft to meet the necessary climb gradient performance to clear any obstacles in the go-around path with one engine inoperative.
3. Field length available, i.e., the most restrictive runway length available, given both still-air and wind conditions on all available runways. Normally a landing safety factor is mandatory for public transport flights.
4. Determination of the performance speeds, given the calculated most limiting aircraft weight.
   a. VAT/V_{REF}.

NB: Performance *V* speeds are a function of the stall speed V_S, which itself is a function of the aircraft weight. Ergo, *V* speeds are weight-determined.

Landing performance calculations are divided into two basic groups.

- Those planned on scheduled landing conditions, i.e., calculated before takeoff. Normally variable factors such as temperature, wind, and slope are restricted from being used for scheduled landing conditions, because of the large time span between the planned or scheduled landing and the execution of that landing.

- Those planned on in-flight landing calculations, such as when you are in radio contact with the destination aerodrome and the variable factors, such as temperature, wind, and slope, can be more accurately computed.

Landing—Visual Clues The main visual clue during a landing is the pilot's judgment of a 3° glide slope; i.e., a flat-looking runway means you are low, and a steep-looking runway means you are high. This judgment only comes with experience. Initial judgment of an appropriate glide slope may be facilitated by aids such as VASIs and PAPIs or by positioning the aircraft at predetermined heights above known ground features or distance from touchdown points.

During the final-approach stages, the extended sides of the runway intersect at the horizon, and texture gradients in the surrounding terrain also indicate horizon location. However, these cues are accurate only when the terrain and runway are level. Sloping runways and terrain may produce incorrect estimates of horizon location to the pilot and result in an inaccurate judgment of the approach slope. To prevent the final-approach angle from varying, the pilot should aim at the projected impact point, because this is the point on the ground away from which visual texture flows. As long as the visual texture flows away from this point, and as long as the visual angle between this point and the horizon remains constant, the approach is progressing normally.

Land/Sea Breezes Land/sea breezes are a product of surface heating and atmospheric convection currents that produce small local airflow circulation cells.

Land/sea breezes are particularly important to aviators, in connection with coastal airports, because they can determine the wind direction at a local level, which may be considerably different from the general wind direction. In addition, they may cause some wind shear and/or turbulence to be experienced as an airplane passes from one body of air to another.

Sea breezes   These occur during the daytime, usually between midday and late afternoon on hot, sunny days, when the land heats up more quickly than the sea. The air above the land becomes warmer and rises, causing the pressure in the warmer air column over the land to be greater than that at a similar height over the sea. This is so because a warmer temperature makes the air less dense and lighter in weight, which allows the column of air to rise with the effect of a greater actual distance (height) for a 1-mbar change of pressure. This results in a height (e.g., 2000 ft) of the air over the land having a higher pressure level (e.g., 1000 mbar) than that of the colder air over the sea (e.g., at 2000 ft/998 mbar). Therefore a pressure difference is born at height (i.e., 2000 ft) that produces an airflow at height from the land to sea. This induces an opposite pressure difference at the surface. Namely, the outflow of air at height cools, gets denser and heavier, and therefore descends over the sea, creating a high surface pressure. This induces a flow of air at the surface from the sea to the land (which has a low surface pressure, due to the rising convective current air), which is the sea breeze.

These small thermal systems associated with sea breezes can operate 10 to 15 nm either side of the coast, with surface winds of around 10 kn. The vertical extent of the sea breeze is usually only 1000 to 2000 ft. Hence it is quite possible that the wind direction and speed at circuit height, which are not influenced by the sea breeze, are quite different to the surface wind. Therefore a considerable change in wind direction and/or speed and possibly some wind shear and/or turbulence may be experienced in the circuit when you descend into the sea breeze system. In addition, the cooler sea breeze moving over the land can bring fog or mist with it, causing visibility problems at coastal airports.

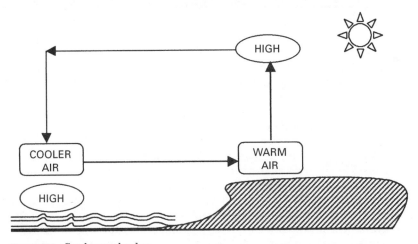

Figure 114   Sea breeze by day.

Land breezes   These occur at night, when the land cools more quickly than the sea, causing the air above it to cool and subside. The air over the sea, however, retains its heat better and becomes the warmer of the two columns of air, and therefore its air rises. The pressure in the warmer air column above the sea is greater than that over the land, at height, so the air flows at height from the sea to the land. This induces an opposite pressure difference at the surface between the land and sea. Namely, the outflow of air at height cools, gets denser and heavier, and therefore descends over the land, creating a high surface pressure. This induces a flow of air at the surface from the land to the sea (which has a low surface pressure, due to the rising warmer air), which is the offshore land breeze.

Because the turbulent layer is thinner at night, the first movement of air, in this case from the sea to the land, will be at a lower height than during the day. In addition, a land breeze could hold a sea fog offshore early in the day; but as the land warmed up, the land breeze would die out and a sea breeze could develop that brings in the sea fog, creating visibility problems at coastal airports.

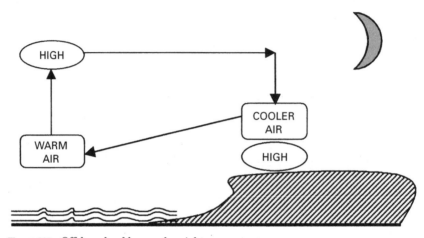

Figure 115   Offshore land breeze, by night.

Latent Heat   Latent heat is the heat energy, measured in calories, absorbed or released when water changes from one state to another.

NB: There are three states of water: (1) water vapor (gas), (2) water liquid (cloud, mist, fog, rain, etc.), and (3) water solids (ice).

When water changes to a higher-energy state, i.e., from ice to liquid to vapor, it absorbs/uses latent heat energy (from the surrounding atmosphere/properties). And when it moves to a lower-energy state, i.e., from vapor to liquid to ice, it releases latent heat energy (into the surrounding atmosphere/properties). This can be important in certain forms of airframe ice formation.

For example, 80 cal of latent heat is released with the change of state of 1 g of water to ice. If this water were supercooled, at say $-20°C$, then it would already have 20 cal less latent heat energy than it ought to have at $0°C$. When it then reverts to ice from water, this is made up by the latent heat being released by the change of state of water to ice. Therefore the 20 cal needed will change 20/80, or $\frac{1}{4}$, of the water drops to ice. The remaining three-fourths of the water runs back along the airframe, rapidly cooling and freezing as a clear sheet of ice.

Lateral Stability   Lateral stability is the tendency for the aircraft to return to a laterally level position, around the longitudinal axis, on release of the ailerons in a sideslip.

Two principal features make an aircraft naturally laterally stable:

1. *Wing dihedral.*   The airflow due to a sideslip causes an increase in the angle of attack (lift) on the lower (leading) wing and a decrease in angle of attack on the raised wing, because of the dihedral angle.

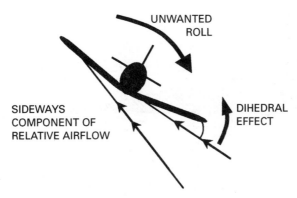

Figure 116   Dihedral—sideslip angle of attack.

The lower wing thus produces an increase in lift because of the increased angle of attack, and the raised wing produces less lift. The difference in lift causes a rolling moment, which tends to restore the wing to its laterally level position.

2. *Side loads produced on the keel surface.*   When the aircraft is sideslipping, a side load will be produced on the keel surface, particularly the fin. This side load will produce a moment to roll the aircraft laterally level, which in general terms is stabilizing. The magnitude of this effect is dependent on the size of the fin, but regardless its effect is small compared to other laterally stabilizing effects.

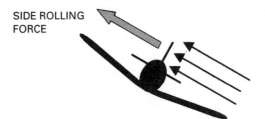

Figure 117   Keel surface side loads—sideslip.

LDA (Landing Distance Available)   *See* Landing Distance Available (LDA).

LDR (Landing Distance Required)   *See* Landing Distance Required (LDR).

Leading-Edge Slats   These increase the wing's chamber area and MAC, thereby increasing its maximum C_L, which reduces the aircraft's stall speed.

Lenticular (Mountain) Clouds   *See* Mountain (Lenticular) Clouds.

Lift Lift is the phenomenon generated by an aerofoil due to pressure differences above and below the aerofoil.

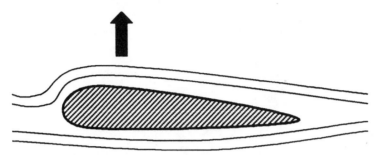

Figure 118   Aerofoil-generated lift.

NB: An aerofoil is cambered on its top side and flat on its bottom side. Therefore the airflow over the top of the aerofoil has to travel farther and thereby faster than the airflow below the aerofoil. This causes the pressure below the aerofoil to be greater than that above, creating a pressure difference which results in an upward *lift* force.

The formula for lift is

$$\text{Lift} = \tfrac{1}{2}\rho \, V^2 \, S \, C_L$$

where $\tfrac{1}{2}\,\rho$ = one-half the value of air density
V^2 = airflow velocity squared
S = wing span area
C_L = coefficient of lift

The combined values of these properties determine the amount of lift produced.

The greatest lift force is generally produced during the takeoff.

NB: Lift is caused by a pressure difference above and below the wing, and the size of the difference determines the amount of lift produced.

The difference in pressure experienced is affected by the functions of lift, which are

1. Configuration (flap setting)

2. Speed of airflow over the wing

3. Angle of attack

which are optimized during the takeoff stage of flight, plus

4. Air density

During the takeoff phase of flight, the configuration, speed of airflow over the wing (high *V*2 speed), and angle of attack are maximized, by employing a maximum angle (*V*x) climb profile or gradient (i.e., vertical distance/horizontal distance) after rotation (within its operating envelope). This provides its greatest lift component, especially in terms of lift against weight (as the aircraft will be heaviest at takeoff). For example, the Airbus 320 has a normal takeoff gradient of 2.4 percent and a go-around gradient of only 2.1 percent, making its takeoff gradient the greater.

However, for some aircraft types, a higher rate of climb may be achieved during a go-around under certain conditions, such as lower weight, higher speed, intermediate approach flap, and a higher power setting.

See also High-Lift Devices.

Lift Coefficient   *See* Coefficient of Lift (C_L).

Lighting Designations—on Obstacles   *See* Obstacle—Lighting.

Lightning   Lightning is most likely to occur when the outside air temperature (OAT) is $+10$ to $-10°C$ and within or close to a thunderstorm associated with a mature-stage cumulonimbus (CB) cloud. Lightning is a discharge of potential static electricity that has built up in a cloud (normally a CB cloud) to its critical level. The air along the path of the lightning experiences intense heating. This causes it to expand violently, and it is this expansion which produces the familiar clap of thunder. Either lightning can be contained within the cloud, or it can travel from the cloud to the ground. Obviously it can be a hazard to aviation.

Lights—Airfield Beacon   *See* Airfields—Beacon.

Lights—ILS Approach Lights   *See* Approach Lights—for an ILS Runway.

Lights—Navigation and Anticollision Lights (on an Aircraft)   *See* Navigation and Anticollision Lights—on an Aircraft.

Lights—Runway   *See* Runway Lights.

Lights—Taxiway   *See* Taxiway—Lights.

LNAV and VNAV   The LNAV and VNAV are FMC functions, which are predetermined or custom-made route profiles, based on the FMC's managed performance and navigation data. They are stored in the FMC databases so that they can be selected via the CDU to be flown, either as an autopilot and auto throttle modes of operation, or as a guide for manual flying.

The LNAV (lateral navigation) function guides the aircraft's lateral movement and is available from takeoff to localizer capture: and the VNAV (vertical navigation) function guides the vertical path of the aircraft, including climb and descent profiles.

A stored company route will contain a complete lateral and vertical flight profile, which can be amended by the crew if required.

Loading—Maximum Range   When loading an aircraft to obtain maximum range, you would load it with an aft center-of-gravity position/hold loading for aircraft (especially jet/swept-wing) with a nose-up en route attitude. This will allow it to achieve its maximum possible range. An aft cg position is normally accomplished by utilizing the aft cargo hold, which gives the aircraft its nose-up en route attitude naturally; therefore the stabilizer can remain streamlined to the airflow and produce no relevant drag (i.e., aft cg = 6° nose-up attitude = 0° elevator/stabilizer deflection).

Thus the aircraft can be operated at its optimum thrust setting to obtain its maximum range, without having to use excessive engine thrust to compensate for drag.

NB: It is beneficial, even when the cg is aft of its optimum position, because the stabilizer would produce a greater lift force (to produce a downward pitching moment of the nose, to gain its en route attitude). This is beneficial to the aircraft's overall performance because it increases the aircraft's lift capabilities, whereas a forward cg has a detrimental effect on the aircraft's performance.

However, a few aircraft with a nose-down en route attitude would require a forward cg position.

Loading (Weight and Balance)   *See* Weight and Balance (Loading).

Local Speed of Sound (LSS)   The speed of sound, Mach 1.0, does not remain a constant value of knots per hour because it is influenced by temperature; therefore the speed of sound decreases with a decrease in temperature. The speed of sound for a particular temperature is known as the *local speed of sound*. The local speed of sound can be found from

$$\text{LSS} = 38.94 \sqrt{\text{absolute temperature}}$$

At ISA MSL conditions, LSS is equal to 660 kn/h, and we know that temperature decreases with an increase in altitude. Therefore the LSS will also decrease with an increase in altitude (as temperature decreases, the LSS decreases).

Longitudinal Stability   Longitudinal stability is an aircraft's natural ability to return to a stable pitch position around its lateral axis after a disturbance.

When an aircraft is in equilibrium, the tailplane in general will be producing an up or down load to balance the moments about the cg. (It is assumed that the elevator remains in its original position throughout during any disturbance in pitch.) If the aircraft is disturbed in pitch (say, nose up), there will be a temporary increase in the angle of attack. The increase in tailplane angle of attack produces an increase in tailplane lift, which will cause a nose-down pitching moment. (The tailplane is thus able to produce a stabilizing moment due to a displacement in pitch, as long as the cg remains within its limits.) The wings also experience this increase in angle of attack, resulting in the wings producing an increase in lift. The moment and the direction of the moment produced by this lift will depend on the relative positions of the center of pressure (cp) and the cg. If the cp is in front of the cg, the increase in wing lift will produce a destabilizing (nose-up) moment. As such the longitudinal stability will decrease because of the natural tendency to increase lift after a disturbance. However, the tailplane moment should restore the aircraft's longitudinal stability if the cg is within its range limits. If the opposite is true and the cp was behind the cg, then any increase in lift will produce a natural stabilizing (nose-down) moment. As such the longitudinal stability is positive even without any help from the tailplane. The aircraft will remain longitudinally stable if the cp remains behind the cg and/or the cg is within its limits. However, the relative positions of the cp and the cg will affect the degree of the aircraft's natural longitudinal stability.

See Fig. 119 at the top of the next page.

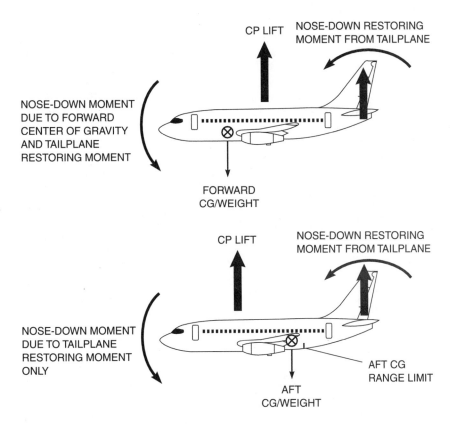

Figure 119   The cg and cp positions, longitudinal stability.

NB: The longitudinal stability of an aircraft is its natural ability to return to its stable pitch position. This should not be confused with the aircraft's maneuverability in pitch provided by the elevator, which is governed by cg range limits.

Long-Range Cruise   This is a speed significantly higher than the maximum range speed, that is, 10 kn (Mach 0.01), which results in a 1 percent mileage loss, at a constant altitude. The long-range cruise schedule requires a gradual reduction in cruise speed as gross weight decreases with fuel burnoff.

Loran C   Loran C is a long-range navigation system that uses ground-based beacons operating in the low-frequency (LF) band and using surface waves to achieve ranges of more than 100 nm. It is available in North America, the north Atlantic, Europe, and the Mediterranean.

Loran is similar to decca. Loran operates a series of transmitting stations arranged in chains to determine hyperbolic position lines. Each chain includes a master station with two to four slave stations arranged around the master station at ranges of 600 to 1000 nm. Loran calculates the differential range by pulse technique by measuring the time difference between the pulses sent from master and slave stations to automatically calculate the aircraft's position. Output from the Loran system can be integrated with other onboard navigation systems containing way points of a predetermined route. This allows it to automatically navigate the aircraft to its flight planned route and to even automatically select and deselect required Loran chains as a navigation source. Alternative information such as range and bearing to way points, track, and track error can also be displayed in flight.

The accuracy of Loran C can be as good as 0.1 to 0.2 nm in areas of good coverage, but this degrades as you move away from the stations. The ICAO maximum allowed error is ±0.5 nm at 900 to 1000 nm over sea for 95 percent of its use.

Low-Pressure System   A low-pressure system, which is also known as a *depression,* has the following characteristics.

1. *Pressure gradient.*   In a low-pressure system, the surface barometric air pressure rises as you move away from its center; in other words, its pressure drops as you move toward its center.

NB: A low is depicted on a weather chart by a series of concentric isobars joining places of equal sea-level pressure, with the lowest pressure value at its center.

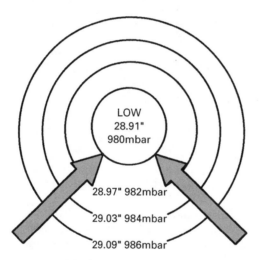

Figure 120   A low's pressure gradient.

The pressure gradient (which is depicted by the closeness of the isobars) relates to the wind speed; i.e., a strong pressure gradient = close isobars = a strong wind.

2. *Airflow pattern.*   The surface pressure at the center of a depression is lower than that at the surrounding horizontal areas; i.e., it is a low-pressure system. However, the pressure gradient at its upper levels is reversed, creating a high-pressure system, where the pressure at the center is higher than that in the surrounding horizontal areas.

NB: It is important to realize and remember that a surface low-pressure system is associated with a high-pressure system higher up.

However, the center of a depression has a greater pressure at the surface than above it. Because the airflow/wind is essentially pressure-driven, these phenomena cause an inflow of air toward the center of the low-pressure system at its surface, known as *convergence.* This causes the surface air pressure at the center of a low-pressure system to become greater than that of the air above, and so the air rises/ascends (unstable) at the center of a depression. At the top upper level it flows outward, as a result of the air pressure at the center of the depression being greater than at the surrounding areas, i.e., a high-pressure system.

The airflow pattern of a depression can be categorized as follows:

- Convergence (inflow) at the lower layers
- Rising air at the center
- Divergence (outflow) in the upper layers

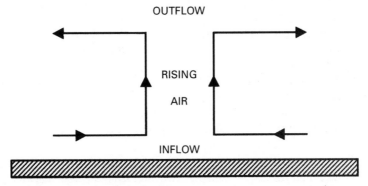

Figure 121   Airflow pattern in a depression.

3. *Wind direction.*   The wind circulates anticlockwise around a low-pressure system in the northern hemisphere and clockwise in the southern hemisphere.

NB: Flying toward a low in the northern hemisphere, an aircraft will experience right (starboard) drift.

However, it should be remembered that above a surface low there is an upper high system. Therefore the wind direction will reverse at height in a low-pressure system.

4. *Air mass.*   A low-pressure system will usually be made up of at least two different air masses; a cold air mass and a warm air mass, with cold and warm fronts.

5. *Movement.*   Low-pressure systems are generally more intense than highs, because they are more concentrated, in terms of area, and have a stronger pressure gradient (pressure change with distance). The more intense the depression, the *deeper* it is said to be, i.e., the lower the pressure at its center. Therefore lows move faster across the surface of the earth and tend to have a shorter life span than highs.

6. *Weather.*   In general the weather associated with a low-pressure system is characterized as follows:

- Cloud formation and related weather, i.e., precipitation, etc. This is due to the adiabatic cooling experienced by the ascending air in a depression. Any instability in the rising air may lead to the vertical development of cumuliform clouds accompanied by rain showers.
- Possible good visibility (except in rain showers). This is so because the vertical motion will tend to carry away most of the particles suspended in the air.
- Moderate to strong winds. A low-pressure system typically has a marked pressure gradient, which represents the strength of the wind.

- Frontal weather. Because fronts are associated with low-pressure systems, frontal weather is normally the most prominent weather associated with a depression.

The different types of depressions (low-pressure systems) are developed and classified according to their trigger phenomenon. Namely:

1. *Frontal depression,* i.e., frontal interaction between two different air masses.

2. *Thermal depression.* A surface such as concrete (city area) reradiates heat energy efficiently, causing thermal currents. These thermal currents produce an area of low pressure at the surface, which thereby attracts a surface airflow from a higher-pressure area around it, which is in turn heated and becomes thermal currents. Isolated rainstorms often characterize a thermal depression in the afternoon.

3. *Tropical storm depressions.* Surface heating over large areas in the ITCZ produces tropical storm depressions.

4. *Orographic depression.* Downwind of a mountain range, an area can exist that is shielded from the wind blowing over the mountains. This phenomenon creates a local low-pressure area.

In the northern hemisphere you would experience a drift to the right (starboard) when flying from a high- to low-pressure system. This is true because the wind circulates anticlockwise around a low in the northern hemisphere; therefore, flying toward the center of a low-pressure system, you have a wind from the left (port) causing you to drift to the right (starboard).

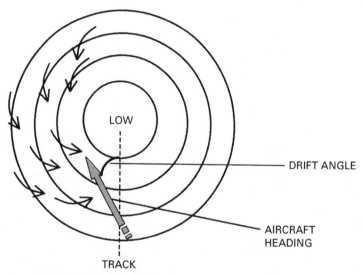

Figure 122   Right/starboard drift—low-pressure area ahead.

In the southern hemisphere when you are flying from a high- to low-pressure system, the opposite is true. You would experience a drift to the left (port). This is true because in the southern hemisphere the wind circulates clockwise around a low, and anticlockwise around a high. Therefore the airflow/wind experience will be coming from the right, causing you to drift to the left (port).

When flying from a high- to a low-pressure system, you will experience a change in atmospheric pressure, causing a "Barometric Pressure Error" in the altimeter. Therefore when you fly from high to low (pressure), *beware below,* because your actual flight level is lower than your altimeter-indicated level. In other words, the altimeter overreads. Conversely, the opposite is true when you fly from low- to high-pressure areas.

This high-to-low mnemonic applies equally to temperature values.

Low-Pressure System—Frontal   *See* Frontal Depressions.

Low-Visibility Takeoff   With high-intensity runway center lights available, the RVR minimum for takeoff is 150 m RVR at the touchdown point (or 125 m under some authorities), and the midpoint RVR must be at least equal to the touchdown RVR. The reported stop end RVR can be less than the takeoff minima.

With *no* high-intensity runway center lights available, the RVR minimum for takeoff is 200 m RVR at the touchdown point, and the midpoint RVR must be at least equal to the touchdown RVR. The reported stop end RVR can be less than the takeoff minima.

MABH (Minimum Approach Breakoff Height) Minimum approach breakoff height is the lowest height of the wheels above the ground such that if a go-around were initiated without external references in normal operation, the aircraft would not touch the ground during the procedure. Or, with a critical engine failure, it can be demonstrated that an accident is improbable during a go-around.

Mach Crit (M_{Crit}) The aircraft's Mach speed at which the airflow over a wing becomes sonic is called the *critical Mach number*. At or near the critical Mach number, the drag begins to increase because of the air's compressibility effect. This increase in drag is known as *wave drag* because it is caused by the formation of shock waves on the aircraft. Subsonic aircraft experience a rapid rise in drag above the critical Mach number, and because the aircraft's engines do not have the available power to maintain its speed and lift values under these conditions, the aircraft suffers a loss of lift.

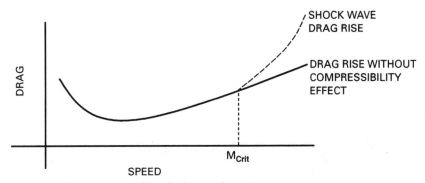

Figure 123 Drag rise caused by shock wave formation.

The aircraft's Mach speed is lower than the airflow speed over a wing. A typical M_{Crit} speed of $0.72M$ experiences sonic Mach 1 airflow speed over the upper surface of the wing.

As a swept-wing aircraft passes its M_{Crit} speed, the center of pressure will move rearward for two reasons.

1. The shock waves on the wing's upper surface occurs toward the leading edge, because of the greater chamber which creates the greatest airflow velocity to be experienced at this point. This upsets the lift distribution chordwise and causes a rearward shift in the center of the lift (cp).

2. The swept wing tends to experience the shock wave effect at the thick root part of the wing first, causing a loss of lift inboard; therefore the lift force now comes from predominantly the outboard part of the swept wing, which is farther aft because of the wing sweep.

The characteristics of M_{Crit} are as follows:

1. Initial Mach buffet, caused by the shock waves on the upper surface of the wing as the aircraft approaches M_{Crit}, is usually experienced.

2. An increase in drag due to the breakdown of airflow causes the stick force to change from a required forward push to a neutral force and then a required pull force as the aircraft approaches and passes $M_{\text{Crit.}}$

3. A nose-down change in attitude (Mach tuck) occurs at or past $M_{\text{Crit.}}$

4. There is a possible loss of control.

Mach Meter   The Mach meter measures the airspeed relative to the speed of sound. In essence the Mach meter is a combined ASI and altimeter, which comprises the following:

1. A capsule feed with pitot pressure inside an ambient static pressure feed case acts as an ASI and measures dynamic pressure as the airspeed.

2. A sealed capsule contains ISA mean sea-level (MSL) conditions inside the ambient static pressure feed case, which acts as an altimeter, by measuring the static pressure, which it relates to an altitude.

$$\text{Mach meter reads ratio} = \frac{\text{pitot} - \text{static}}{\text{static}}$$

These two functions inside the Mach meter are linked via a ratio arm which itself acts upon a ranging arm to ultimately move the pointer/digital display on the instrument face.

Although the local speed of sound is dependent only on temperature, no temperature sensors are found in the Mach meter, as it utilizes the mathematically proven fact that

$$\text{Mach number} = \frac{\text{dynamic pressure}}{\text{static pressure}}$$

Mach number (MN) is usually displayed in digital form in a window on the ASI face; therefore, the MN and the corresponding IAS are easily assessed.

Mach Meter—Errors   The Mach meter suffers from the following errors:

1. *Instrument error* is caused by inaccuracies in the Mach meter's construction, especially the friction of its moving parts.

2. *Pressure error* is caused by variations of the static and pitot probe relative position with respect to the airflow over the aircraft's surface. This can lead to sensed pressure readings that are not representative of the free atmosphere when the airflow pattern is disturbed at different airspeeds, angles of attack, and configurations. (This is also known as position or configuration error.)

However, these errors are extremely small, and therefore the indicated Mach number speed can be read as the true Mach number speed.

3. *Blocked pitot-static system.*   A static line blockage means that the static pressure in the MN instrument case remains a constant value. Therefore at a constant altitude the MN will read correctly, but in a descent the MN will *overread,* due to an increase in the pitot pressure in the capsule, against the trapped low static pressure of the higher altitude. In a climb the MN will *underread,* due to a decrease in the pitot pressure in the capsule, against the trapped high static pressure of the low altitude.

A pitot line blockage means that the total pressure in the MN instrument capsule remains a constant value. Therefore at a constant altitude the MN reading will not change, even if the airspeed does, due to the trapped pitot pressure in the capsule, against a constant altitude static pressure. In a descent the MN will *underread*, due to an increase in the static pressure in the case, against a constant pitot pressure. In a climb the MN will *overread*, due to a decrease in the static pressure in the case, against a constant pitot pressure.

The blockage is normally caused in flight by ice formation or small insects covering the pitot or static port. The actions for a blocked pitot/static system causing an unreliable MN reading would be to

- Ensure that the pitot-static probe anti-ice heating (pitot heat) is *on,* if applicable.
- Use an alternative, such as a static source or an air data computer, if applicable.
- Use a limited flight panel, i.e., standby ASI.
- Fly at a correct attitude and power setting.

NB: If the blockage has been caused by ice formation and you have an unserviceable or no pitot heat system, then you should avoid icing conditions.

NB: The Mach meter does not suffer from density or temperature errors because its design of a built-in altitude capsule and its ratio to the dynamic pressure measured compensate for density and temperature variations.

Mach Number (MN)   The Mach number is a true airspeed indication, given as a percentage relative to the local speed of sound, e.g., one-half the speed of sound, or 0.5 Mach. In other words, MN is a calculation of the true airspeed (TAS) relative to the local speed of sound (LSS).

$$\text{MN} = \frac{\text{TAS}}{\text{LSS}}$$

Mach number is a function of the LSS, which is influenced by temperature; therefore the main influence on MN is temperature. As temperature decreases, the LSS decreases. LSS = 38.94/absolute temperature. If the LSS decreases for a constant true airspeed, then the Mach number must rise, because MN = TAS/LSS. Therefore MN is a function of the behavior of TAS and the LSS. With an increase in altitude at a constant IAS, the MN increases because TAS increases with altitude, due to a decrease in air density, and the LSS decreases, due to the drop in air temperature.

Mach number is used as a speed reference at high altitudes, usually above 26,000 ft, as the MN becomes the aircraft's limiting speed, in preference to IAS. (The MN limiting speed is the aircraft's certified maximum speed that retains its required level of handling.) That is, up to approximately 26,000 ft an aircraft would climb at a constant IAS against an increasing MN. At approximately 26,000 ft the MN speed reaches the aircraft's MN limiting speed. Therefore above 26,000 ft the aircraft is flown at a constant MN with a decreasing IAS for an increase in altitude. Mach number decreases in a long-range cruise as weight decreases at the same flight level.

With a constant MN, the TAS will reduce with altitude, but the RAS (IAS) will reduce more.

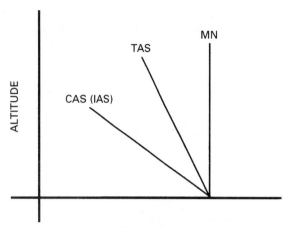

Figure 124   Constant MN against TAS and RAS (IAS).

If you climb at a constant CAS (IAS), your MN will increase. This is so because TAS increases because air density decreases with altitude, and LSS decreases because of a reduction in air temperature with an increase in altitude.

Therefore, given that MN is a function of TAS divided by LSS, the MN will increase during a constant-CAS (constant-IAS) climb.

Given two aircraft flying at different flight levels at the same MN, the aircraft at the lower altitude has a higher TAS/CAS (IAS), because LSS decreases as temperature decreases with an increase in altitude.

Therefore, to maintain the same MN at a higher altitude, the TAS must be reduced; and if the TAS is reduced, the CAS (IAS) will be reduced. Therefore, the aircraft at the lower altitude has the higher TAS/CAS (IAS).

When climbing through an isothermal layer at a constant TAS, the MN will remain constant. This is so because an isothermal layer has a constant air temperature, and therefore the LSS is constant. (LSS and TAS are constant; therefore MN must also remain the same value.)

If an aircraft climbs at a constant TAS through an inversion layer, then its MN will decrease more than normal. This is true because the LSS increases due to the increase in temperature of the inversion layer, and TAS is constant, therefore MN will decrease.

Mach Trimmer   The purpose of a Mach trimmer is to automatically compensate for Mach tuck (longitudinal instability) above M_{Crit} speed.

A Mach trimmer is a system that artificially corrects for Mach tuck above the aircraft's M_{Crit} speed by sensing the aircraft's speed and signaling a proportional upward movement of the elevator or variable incidence stabilizer to maintain the aircraft's pitch attitude throughout its speed range up to MDF.

NB: Mach trimmers allow for an aircraft's normal operating speed range to be above its M_{Crit} speed.

A Mach trimmer may become active in any one of three speed bands.

1. Above MMO, it will only operate when the aircraft is forced into an overspeed condition.

2. Below MMO but above its normal maximum cruise Mach number, it will operate more frequently especially if a high Mach cruise speed is chosen.

3. Within the normal Mach cruise range, it will be active most of the time.

In the event of a Mach trimmer failure, there is usually an imposed reduced
MMO value so that a margin is retained below the Mach speed at which the onset
of instability occurs.

Mach Tuck   This is the nose-down pitching moment that an aircraft experiences as
it passes its critical Mach number $M_{Crit.}$ Mach tuck is a form of longitudinal instability that occurs because of the center of pressure's rearward movement behind
the center of gravity, which induces the aircraft to pitch down (or the aircraft's
nose to "tuck"). Also, because the M_{Crit} shock wave causes a reduction in the downwash over the tail, the balancing force available from the tailplane is considerably
reduced, and this further aggravates the aircraft's nose-down (tuck) attitude. The
aircraft's Mach tuck becomes more marked as its airspeed increases.

Magnetic Compass Instrument   The magnetic compass, also known as the direct
reading compass, is the primary source of directional information in all types of
aircraft, and it displays the *compass heading*. It comprises a freely suspended horizontal magnet attached to a compass card, which is enclosed in a liquid-filled
case. The magnet will swing so that its axis points roughly north/south, and the
aircraft moves around the magnet so that the compass heading of the aircraft is
read off the compass card against a lubber line on the instrument case.

The direct reading compass is affected by two magnetic fields:

1. The earth's magnetic field, which is generated by the magnetic poles and which
causes a "variation" from a true north (heading) to a magnetic north (heading)

2. Any other magnetic field generated around the compass, which causes the compass magnet to deviate from precisely indicating magnetic north to show a compass north (heading)

Therefore, to calculate the true heading, the compass heading requires the application of a compass deviation and a magnetic variation error.

Magnetic Compass Instrument—Errors   The magnetic compass suffers from the following errors:

1. *Acceleration errors,* which are a function of dip. A change of airspeed, either
an acceleration or a deceleration, will cause errors in the displayed direction of the
magnetic compass, namely, *apparent turns,* especially on easterly and westerly
headings. This occurs because the magnet's center of gravity (cg) point is offset
from the pivot point, to counteract the dip of the magnet on the pole-seeking side.
Thus the effect of inertia through the magnet's cg causes the magnet to swing, and
because the compass card is attached to the magnet, it indicates a turn/new direction even though there has been no real change in direction.

See Fig. 125 at the top of the next page.

That is, it indicates northern hemisphere on a easterly/westerly direction, accelerating indicates an apparent turn to the north, and decelerating indicates an
apparent turn to the south.

NB: In the southern hemisphere, the residual dip is to the south pole so the cg
of the magnet is to the north of its pivot point, thereby reversing the effect.

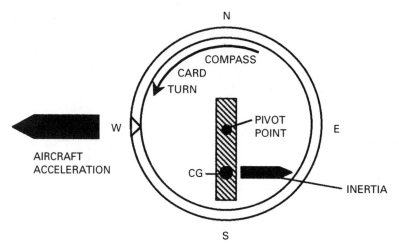

Figure 125   Compass magnet, acceleration/deceleration inertia.

Thus, east/west accelerations produce an apparent turn to the south, and east/west deceleration produces an apparent turn to the north.

Once a steady speed has been established, the magnet will settle down and the compass will read correctly. There are no acceleration/deceleration errors on north/south headings, because the magnet pivot and cg point lie on the same line as the acceleration/deceleration, and so the magnet is not exposed to any inertia forces.

2. *Turning errors.*   There are two types.

The *turn acceleration error* is also a function of *dip*. In a turn the aircraft has a centripetal force acting on it, which acts toward the center of the turn through the pivot point of the compass magnet. The compass magnet's cg point, which is offset from the pivot point to counteract the dip of the magnet, is left behind due to the inertia, which makes the compass swing. Because the compass card is attached to the magnet, it induces a compass orientation error and thereby a compass heading error during turns through northerly and southerly directions.

See Fig. 126 at the top of the next page.

That is, in the northern hemisphere, when turning through north the compass heading indication tends to lag or underread, therefore undershooting the compass heading. When turning through south, the compass heading indication tends to lead or overread, therefore overshooting the compass heading.

NB: In the southern hemisphere, the opposite is true because the cg is on the north side of the pivot point of the compass magnet.

That is, when turning through north, the compass heading indication tends to lead or overread. When turning through south, the compass heading indication tends to lag or underread.

There are no turning acceleration errors when turning through east/west directions as the cg is in line with the pivot point along the radius of turn. Therefore the magnet is not exposed to any inertia forces as a result of the centripetal force experienced by the aircraft.

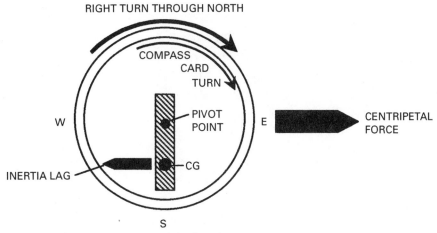

Figure 126   Compass magnetic turning inertia.

The second turning error is the *liquid swirl error*, an effect that relates to the liquid in the compass that opposes, due to its viscosity, the movement of the compass in any direction. This makes the compass lag or underread in all turning directions. This is, however, a much smaller turning error, but it acts in series with the turning acceleration error, so it either increases lag or decreases the lead of the turning errors.

Magnetic Dip   Magnetic dip is the natural phenomenon of the vertical component of the magnetic field over the earth's surface and its effect on the magnetic compass.

The earth's magnetic field has two components/forces. A horizontal force parallel to the earth's surface is utilized to align the compass with magnetic north and therefore to determine direction, and a vertical force causes the needle to dip down. At the magnetic equator the horizontal force is dominant; therefore there is no dip, and the compass is accurate. However, as you move closer to either of the magnetic poles, the vertical component increases and this induces the magnetic bar in the compass to dip down to align itself vertically with the magnetic field.

See Fig. 127 at the top of the next page.

An actual pole is not at the end of a magnet, but some distance inside; and the shorter, fatter magnets, such as the earth, have their poles nearer the center, which is why there is a greater degree of dip for the degree of the latitude. That is,

- At the magnetic equator there is no dip.
- At approximately 60° latitude there is about 65° to 70° of dip.
- At the magnetic poles there is 90° of dip.

Therefore, when the vertical force of the magnetic field is dominant and there is little horizontal force, i.e., at high altitudes (above 60° latitude), extreme dip is present and the magnetic compass is very unreliable.

This is obviously unsatisfactory. Therefore, to counteract the dip and to keep the magnetic compass at least approximately horizontal, a high pivot point is designed

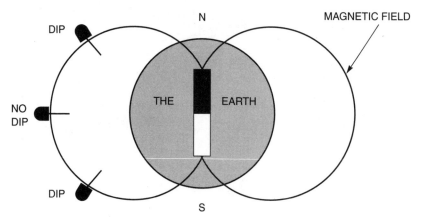

Figure 127   Magnetic dip is stronger near the poles.

with the magnet suspended below, like a pendulum. When the magnet tries to align itself vertically with the field and it tries to dip down, the magnet's cg moves backward, into a position on the opposite side of the suspension point; i.e., in the northern hemisphere it's south of the pivot point, and in the southern hemisphere, north of the pivot point. This sets up a balancing couple between the supporting force of the suspension point and the weight force acting through the cg.

This couple acts against the magnetic dip force to reasonably balance the compass magnet in the horizontal plane, with only a small amount of residual dip remaining, that is usually less than 3°.

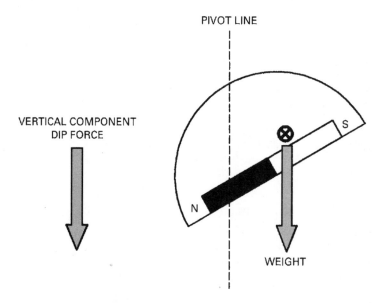

Figure 128   Magnet pendulous suspension.

However, the offset cg creates other acceleration and turning errors.

Magnetic Direction   Magnetic direction is measured with reference to magnetic north.
NB: The magnetic north pole is near to the geographic true north pole but is not in exactly the same place.

Magnetic Field—Earth   The earth has a magnetic iron core, which makes the earth act as a giant magnet with north and south magnetic poles. The magnetic poles are slightly offset from the geographic poles, and the earth's surface is covered with a resultant weak magnetic field that radiates from its magnetic poles. Because the true pole and magnetic pole are not coincident, the true and magnetic meridians that radiate from their respective poles are also not coincident. The angular difference between a corresponding true and magnetic meridian is called *variation.* If the magnet points slightly to the east of true north, then the variation is said to be east (plus); and if the compass points to the west of true north, then the variation is said to be west (minus). That is,

Magnetic heading + easterly variation = true heading

with variation east magnetic least, variation west magnetic best.

Magnetic Variation   This is the difference between the direction of magnetic north and true north. It is therefore the difference between all true headings and their corresponding magnetic headings at any point. Variation is not a constant across the earth but varies in magnitude from place to place.
    Lines of equal variation are called *isogonals,* and variation is described as either east (plus) or west (minus).
    NB: The magnetic pole experiences a gradual change in position with the passage of time.

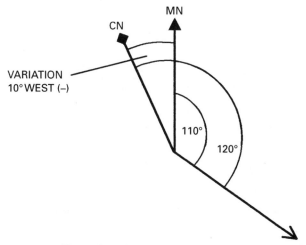

Figure 129   Magnetic variation.

Magnetism   Magnetism occurs in iron and steel metals. The molecules in the metal have north and south magnetic poles that attract the opposite pole of other molecules and that combine to produce the metal bar as a single magnet, itself with opposite north and south poles and a magnetic field. The magnetic field produced by the magnet is represented by curved lines that emanate from one pole and terminate at the opposite pole.

The properties of magnetism are that (1) opposite poles attract, and like poles repel; and (2) magnetic fields represent lines of magnetic force.

Magnetos   Magnetos are used on aero internal combustion engines to supply the high-tension voltage necessary to cause an electric spark at the spark plug. Magnetos are self-contained electrical machines, combining the principles of a generator and a transformer. Normally two magnetos are fitted (in a dual ignition system, each cylinder receives an electric spark from each magneto) and both are driven by the engine to convert mechanical energy to electric energy.

Maneuverability Margin/Envelope   The maneuverability margin falls between an aircraft's upper and lower speed limits. It is

- Between the aircraft's stall speed *VS* at the bottom end of its speed range and its VDF/MDF speed at the top end of its speed range   *or*

- Between 1.2 or 1.3*VS* (representing a safe operating limit above the stall) at the bottom end of its speed range and VMO/MMO at the top end of its speed range

Because an aircraft's normal operating range is between 1.2 or 1.3*VS* and VMO/MMO, it is fair to say that an aircraft's maneuverability margin/envelope lies between these speed limits.

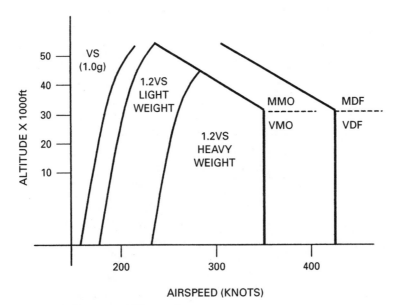

Figure 130   Maneuverability graph.

Notice that the stall speed increases with aircraft weight which therefore reduces its maneuverability envelope. This is important because an aircraft's weight changes (usually decreases) during a flight, and therefore the higher stall speed margin of the envelope results in a reduced maneuverability margin for the aircraft in the lower speed range at the beginning of its flight when the aircraft is heavy. However, the maneuverability margins are further affected by altitude because

- The stall speed increases slightly with altitude.
- MMO becomes more limiting at approximately FL260 and above, which reduces the Mach airspeed.

These maneuverability speed margins are further reduced depending on the amount of g pulled. The VMO/MMO speed is not affected, but the 1.2 or 1.3VS is affected as stall speed increases with weight and pulling excess g in effect increases the aircraft's weight. Therefore the maneuverability margin/envelope is decreased as

1. The maneuverability margin is significantly reduced as the stall speed is increased.
2. Absolute ceiling is reduced in altitude as a consequence of item 1.

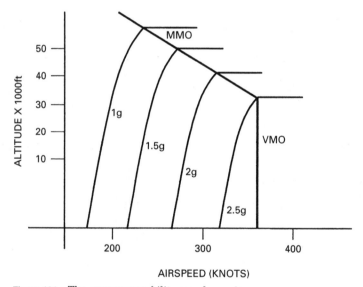

Figure 131 The g maneuverability speed margins.

The absolute ceiling also has a performance margin in the form of a lower maneuver ceiling. At the absolute 1g ceiling, the aircraft's maneuverability becomes increasingly restrictive; therefore, a lower maneuver ceiling allows enough margin for approximately a 1.5g maneuver performance.

Finally the maneuverability speed limit margins consist of

1. Low-speed stall buffet (1.2 or 1.3VS) margin against the actual weight with an approximate 1.5g maneuver performance
2. VMO/MMO high-speed margin
3. Maneuver ceiling

Maneuvering Speed   *See* V_a Speed.

MAP (Missed-Approach Point)   The MAP for a nonprecision approach is defined by a fix—a facility, a distance from a fix or a time from a fix, or the start of the inbound descent (MAPt).

You cannot descend below the calculated MDA(H) unless you become visual; however, you can level off and maintain the minimum decision altitude (MDA) and track into the MAP in the hope of becoming visual. If not visual at the MAP, a missed approach should be initiated.

NB: Most commercial operators would carry out a descent that mirrored a constant glide slope that reached the MDA at the MAP. So the missed approach is initiated at the MDA.

However if a turn is specified in the missed-approach procedure, then it should not be initiated until the aircraft has passed the MAP and is established in the climb.

Marker Beacons   There are three:

1. Outer marker, blue light, low-pitch dashes/the letter O, which is - - - in Morse code
2. Middle marker, amber light, medium-pitch dots and dashes/the letter C, which is -·-· in Morse code

NB: The Morse letter M, or - -, is not used because except for its pitch it would be indistinguishable from O when transmitted repeatedly.

3. Inner marker, white light, high-pitch dots/the letter I, which is ·· in Morse code

Marshaling Signals from a Marshaler to an Aircraft

Description of signal	Pictorial of signal	Meaning of signal
1. Right or left arm is down, with the other arm moved across the body and extended to indicate the position of another marshaler.		Proceed under guidance of another marshaler.
2. Arms are repeatedly moved upward and downward, beckoning onward.	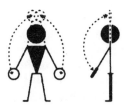	Move ahead.

3. With right arm down, left arm is repeatedly moved upward and backward. to port.The speed of arm movement indicates the rate of turn.

Open up starboard engine or turn

4. With left arm down, right arm is repeatedly moved upward and backward. The speed of arm movement indicates the rate of turn.

Open up port engine turn to starboard.

5. Arms are repeatedly crossed above the head. The speed of arm movement indicates the urgency of the stop.

Stop.

6. A circular motion is made with the right hand with either (a) the left arm pointing to the appropriate engine or

Start engines.

(b) the left hand overhead with the number of fingers extended indicating the number of the engine to be started.

7. Arms are extended, with the palms facing inward, then swung from the extended position inward.

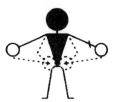

Chocks are inserted.

8. Arms are down, with the palms facing outward, then swung outward.

Chocks are removed.

9. One arm and hand are placed level with the chest, then moved laterally with the palm downward.

Cut engines.

10. Arms are placed down, with the palms toward the ground, then moved up and down several times.

Slow down.

11. Arms are placed down, with the palms toward the ground. Then move either the left or the right arm up and down several times, indicating that the motors on the left or right side should be slowed down.

Slow down engine(s) on indicated side.

12. Arms are placed above the head in a vertical position.

This bay.

13. The right arm is raised at the elbow, with the arm facing forward.

All clear: marshaling is finished.

14. Raise arm, with fist clenched horizontally in front of body, then extend fingers.

Release brakes.

15. Raise arm, with fingers extended horizontally in front of body, then clench fist.

Engage brakes.

Marshaling Signals from the Pilot to a Marshaler

Description of signal	Meaning of signal
1. Raise the arm and hand with fingers extended horizontally in front of face, then clench fist.	Brakes engaged.
2. Raise the arm with fist clenched horizontally in front of face, then extend fingers.	Brakes released.
3. With arms extended with palms facing outward, move hands inward to cross in front of face.	Insert chocks.
4. With hands crossed in front of face, palms facing outward, move the arms outward.	Remove chocks.
5. Raise the fingers on one hand, with the number of fingers indicating the number of the engine to be started. ■ Outer port engine is the number 1 engine. ■ Inner port engine is the number 2 engine. ■ Inner starboard engine is the number 3 engine. ■ Outer starboard engine is the number 4 engine.	Ready to start engine number (as indicated).

Mass Balance   A mass balance is another form of aerodynamic balance control on a control surface. The hinge moment/air load force experienced by a deflected control surface tends to rotate the control surface back to its neutral position, but it is balanced by a mass weight that keeps it in its deflected position. So the stick *control force* which reflects the overall air load force/hinge moment experienced on the control surface is reduced to a manageable level.

The main purpose of a mass balance is to prevent control surface flutter. Flutter is an oscillation of the control surface, which can occur due to the bending and twisting of the structure under loads as a consequence of the cg point of the control surface being behind the hinge. Its inertia causes it to oscillate about the hinge line when the structure distorts. Therefore flutter may be prevented by adding a weight mass to the control surface in front of the hinge line, which brings the cg point of the control surface closer to the hinge line.

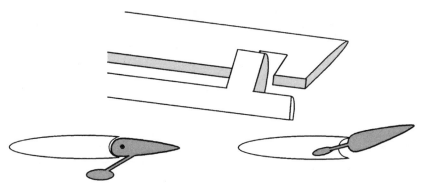

Figure 132   Mass balance forward of the hinge line.

Maximum Angle (*Vx*) Climb Profile   The maximum climb angle (*Vx*) is the steepest angle or highest gradient of climb used to clear close-in obstacles, over the shortest horizontal distance.

Maximum Brake Energy Speed   *See* VMBE Speed.

Maximum Continuous Thrust   Maximum continuous thrust is simply that—the maximum permissible engine thrust setting for continuous use, expressed as either N1 or EPR.

Maximum Endurance   Maximum endurance is achieved by flying at the maximum endurance airspeed, which is the indicated airspeed that relates to the thrust required to balance the minimum drag (VIMD) experienced by the aircraft. Minimum drag (VIMD) is the lowest point/airspeed on the drag curve. Also thrust is a product of engine power, and fuel consumption is a function of the engine power used. The aircraft flying at its maximum endurance speed thereby has the lowest fuel consumption in terms of pound weight of fuel per hour, and hence produces the longest flight time for a given quantity of fuel. Broadly speaking, this speed remains constant with altitude and is used for any delaying action, for instance, holding at a destination airfield.

See Fig. 133 at the top of the next page.

The total drag generated by an aircraft is high at both high and low airspeeds. At high airspeeds the total drag is high because it experiences a lot of profile drag, and at low airspeeds the total drag is high because it experiences a lot of induced drag. Minimum drag occurs at an intermediate speed (VIMD). This is presented by an aircraft's drag curve.

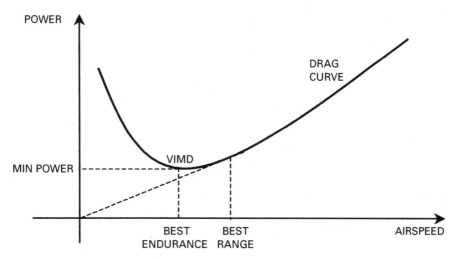

Figure 133 Best endurance and range speeds.

The drag curve varies according to aircraft type, especially between jet and propeller-driven aircraft, because each aircraft type has different drag characteristics. Therefore the airspeeds which relate to the maximum endurance and range vary with aircraft type. Also a particular drag curve changes for different operating weights, and therefore the maximum range and endurance will also be different.

Maximum Range Maximum range is achieved by flying at the maximum range cruise airspeed, which is the indicated airspeed that balances a value of drag slightly higher than the minimum drag point (best endurance speed). This speed achieves a greater range for a given quantity of fuel because the benefits of the increased airspeed outweighs the associated increase in drag and fuel consumption. At the maximum range airspeed, the ratio of power required (to produce thrust to balance drag) to airspeed is minimum.

NB: Airspeed is the rate of covering distance, and power required is the rate of burning fuel to produce thrust to balance drag.

The ratio occurs on the total drag curve where the line from the origin is a tangent to the curve. It is more common to use the maximum range speed en route. See Fig. 133 for best endurance and range speeds.

The total drag generated by an aircraft is high at both high and low airspeeds. At high airspeeds the total drag is high because it experiences a lot of profile drag, and at low airspeeds the total drag is high because it experiences a lot of induced drag. Minimum drag occurs at an intermediate speed (VIMD). This is presented by an aircraft's drag curve.

The drag curve varies according to aircraft type, especially between jet and propeller-driven aircraft, because each aircraft type has different drag characteristics. Therefore the airspeeds that relate to the maximum endurance and range vary with aircraft type. Also a particular drag curve changes for different operating weights, and therefore maximum range and endurance will also be different.

Maximum Range Cruise   This is the speed at which, for a given weight and altitude, the maximum fuel mileage is obtained. It is difficult to establish and maintain stable cruise conditions at maximum range speeds.

Maximum Range—Loading   *See* Loading—Maximum Range.

Maximum Rate (*Vy*) Climb Profile   The maximum or best rate of climb Vy is the highest vertical speed that gains height in the shortest time.

Maximum Service Ceiling   The maximum service ceiling is an aircraft's imposed en route maximum operating altitude/flight level, which provides a safety margin below its absolute ceiling.

An imposed service ceiling is important because it ensures the safe operation of a flight by preventing an aircraft from experiencing the coincident characteristics of the Mach number and prestall buffet at the higher absolute ceiling (coffin corner). These two qualities are, in practice, difficult to distinguish between and therefore a danger to the flight. The imposition of a service ceiling below the absolute ceiling protects against the aircraft encountering such a situation during normal operations.

Maximum Takeoff Thrust   Maximum takeoff thrust is simply that—the maximum permissible engine thrust setting for takeoff, expressed as either an N1 or EPR figure. Maximum takeoff thrust is the highest thrust setting of the aircraft's engine when the highest operating loads are placed on the engine. However, as a protection to the engine, maximum takeoff thrust settings have a time limit on their usage, namely, 5 min for all engines working and 10 min with an engine failure.

NB: Some authorities allow a 10-min time limit with all engines operating.

Mayday   The radio-telephone spoken term *Mayday* indicates that an aircraft is threatened by grave and immediate danger and requests immediate assistance. A typical Mayday call from a pilot would be

1. Mayday, Mayday, Mayday
2. ATC name, e.g., Atlanta Control
3. Aircraft call sign
4. Message, including relevant points such as failures, actions being taken, and requirements

MDA (Minimum Decision Altitude)   *See* Minimum Decision Altitude (MDA).

MDF   *See* VDF/MDF Speed.

MEA (Minimum En route Altitude)   The minimum en route altitude is the safe altitude within the airway, that is, 5 nm either side of the airway centerline, and a minimum altitude at which radio reception is guaranteed.

Medication   The following medications are not compatible with flying duties and will therefore temporarily ground a pilot:

- Antibiotics, e.g., penicillin

- Tranquilizers, antidepressants, and sedatives
- Stimulants, which are used to maintain wakefulness or to suppress appetite
- Antihistamines, which are used to combat colds, hay fever, etc.
- Drugs for the relief of high blood pressure
- Analgesics to relieve pain
- Anesthetics, which are used for local, general, or dental purposes, usually requiring approximately 24 h before a pilot may return to flying duties

NB: Some authorities and individual companies may have further medications that prevent pilots from flying.

It is also recommended that active pilots not donate blood, because it reduces, at least temporarily, the ability to move energy-giving oxygen around the body.

Any medication in your system, whether prescribed or self-purchased, is a reason for not flying. A condition whose symptoms are bad enough to warrant medication could affect an individual's performance, and no drug exists that is completely free of side effects. However, most companies have a list of allowable prescribed medications and limits that can be used during duty periods.

Illegal drugs are obviously not allowed. In fact, spot checks are now carried out on pilots which can lead to criminal prosecution and licenses being revoked, if illegal drugs are found in a pilot's system. So be warned.

Mercator Projection Chart   A Mercator projection is one of the earliest reliable charts, created by a light projection of an earth globe model onto a cylinder of paper wrapped around the globe model that touches only at the equator. Therefore only the equator has the correct scale and convergency.

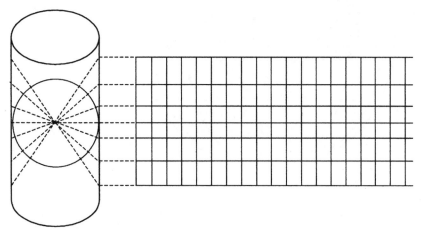

Figure 134   Mercator projection chart.

This type of projection produces a chart with the following properties:

1. The meridians are distorted as parallel straight lines.
2. East/west distances (or departure), which decrease with latitude on the earth, are constant on the Mercator chart.
3. Scale expands with an increase in latitude.

4. Rhumb lines are straight lines, and great circle tracks are curved lines.

The Mercator projection is therefore *not* orthomorphic because its scale is not constant. However, over small distances it is considered to be orthomorphic because changes in its scale are so small that it is considered to be constant.

The disadvantages of a Mercator projection are as follows:

1. Great circle radio bearings can only be plotted as straight lines after a conversion angle has been applied.

2. Long distances cannot be measured accurately because of scale expansion.

3. It is unusable above 70° latitude.

Meridians of Longitude   These run from pole to pole and converge at both poles. Meridians of longitude are described by their angle away from the prime meridian, namely, the *Greenwich meridian*. The units of measurement used to describe a meridian's position are degrees, minutes, and seconds, and named east or west depending on where they lie in relation to the prime meridian. The maximum longitude of 180° is known as the *Greenwich antimeridian*. Additionally meridians at 15° intervals in an easterly direction are classified as *alpha* to *Zulu*. The Greenwich meridian is known as Zulu.

Figure 135   Meridians of longitude.

A change of latitude, north/south distances, along a meridian of longitude is measured as 1 nm equal to 1 min (60 minutes = 1 degree).

NB: North/south distances can be measured across a pole where points lie on opposite meridians of longitude.

Metar   A metar is a written coded routine aviation weather report for an aerodrome. It is an observation of the actual weather, given by a met observer at the aerodrome.

NB: Cloud base in a metar is given above ground level (AGL).

Weather observations are taken at many aerodromes and issued as a metar on a routine basis. Normally a metar is issued on the hour and half-hour.

A code nearly identical to that used for terminal aerodrome forecasts (TAFs) is used, with two noticeable differences:

1. The time of the observation is specified in a metar as a four-figure group; for example, 0855 = 08.55 h UTC (Coordinated Universal Time) (as opposed to a TAF where a four-figure group represents the period of commencement to the end of the valid period of the forecast).

2. An additional two temperatures are given that represent the actual temperature and the dew point temperature. For example, 09/08 means 9°C is the actual temperature and 8°C is the dew point temperature.

NB: Metars may be supplemented by a forecast known as a *trend.*

NB: Metars are also broadcast on volmet services.

Example Metar: EGCC / 0920Z / 210 / 15 G27 / 1000 SW / R24 / P1500 / SHRA / BKN 025 CB / 08 / 06 / Q1013 or 29.92 / RE TS / WS TK OF RWY 24 / NOSIG.

This is decoded as follows:

EGCC	Location identifier (Manchester, United Kingdom)
0920Z	Time the report was taken (09.20 h UTC)
210/15 G27	Wind direction and speed (210° true @ 15 kn gusting to 27)
1000 SW	Horizontal visibility (1000 m to the southwest)
R24/P1500	Runway visual range (runway 24 plus 1500-m visibility)
SHRA	Weather (rain showers)
BKN 025 CB	Cloud (broken @ 2500 ft with cumulonimbus clouds)
08/06	Temperature/dew point (8°C temperature/6°C dew point)
Q1013 or 29.92	QNH (1013 mbar or 29.92 in)
RE TS	Recent weather (recent thunderstorms)
WS TK OF RWY 24	Wind shear (wind shear report on takeoff runway 24)
NOSIG	Trend (no significant change)

Meteorological abbreviations are used in TAFs, metars, trends, etc.

1. Cloud abbreviations

CI	cirrus
CC	cirrocumulus
CS	cirrostratus
AC	alto cumulus
AS	altostratus
NS	nimbostratus
SC	stratocumulus
ST	stratus
CU	cumulus
CB	cumulonimbus (implies hail, moderate/severe icing, and/or turbulence)

2. Amount of cloud coverage (the *okta* is the unit for reporting cloud amount, in eighths of sky covered)

SKC	0 oktas (sky clear)
FEW	1 to 2 oktas (few)
SCT	3 to 4 oktas (scattered)

BKN   5 to 7 oktas (broken)

OVC   8 oktas (overcast)

3. Weather abbreviations (with these prefixes: −, light; +, heavy; VC, within 8 km but not at station, i.e., vicinity):

BC	patches
BL	blowing
BR	mist
CAT	clear air turbulence
COT	@ the coast
DR	drifting
DS	dust storm
DU	dust
DZ	drizzle
FC	funnel clouds
FG	fog
FRQ	frequent
FV	smoke
FZ	freezing
FZ	supercooled
GR	hail
GS	small hail
HZ	haze
IC	diamond dust
ISOL	isolated
LOC	locally
LYR	layered
MI	shallow
OCC	occasionally
PE	ice pellets
PO	dust devils
RA	rain
SA	sand
SG	snow grains
SH	showers
SN	snow
SQ	squalls
SS	sandstorm
TS	thunderstorm
VA	volcanic ash
WDSPR	widespread

4. Storm thunderstorms, i.e., amount of cumulonimbus clouds in an area:

ISOL	isolated (i.e., individual cumulonimbus)
OCNL	occasional (i.e., for well-separated cumulonimbus)
FRQ	frequent (i.e., for cumulonimbus with little or no separation)
EMBD	embedded (i.e., thunderstorm clouds contained in layers of other clouds)

Meteorological Briefings The primary method of preflight met briefing for aircrew, in most parts of the aviation world, is self-briefing. This is done in one of two ways:

1. Using facilities, information, and documentation routinely available or displayed in aerodrome briefing areas.

   NB: Computerized systems providing specific route met conditions are commonplace in most of the major airports around the world.

2. Using a telephone service to call an aviation authority met office. For example, area forecasts (TAFs) or reports (metars) are available directly from a meteorological officer, and/or area forecasts are available from an airmet telephone recording system, especially common in the United Kingdom.

In-flight weather reports that can be accessed include

1. Flight Information Service or Air Traffic (control) Service
2. ATIS (automatic terminal information service)
3. Volmet

Meteorological Forecast A forecast is a prediction, or prognosis, of what the weather is likely to be for a given route, area, or aerodrome. The common types of aviation forecasts are as follows:

1. Area forecasts

   NB: These are for preflight briefings. Also cloud bases in area forecasts are given AMSL (above mean sea level).

2. Aerodrome forecasts, e.g., TAFs and trends
3. Special forecasts

Meteorological Report A met report is an observation of the actual weather at a specific time, either past or present. The common aviation met reports are

- Metars, sigmets, and speci
- ATIS (automatic terminal information service)
- In-flight weather reports, such as volmets, ATIS, and radio communications with an air traffic service unit or flight information service (FIS)

Microburst A microburst is a severe downdraft, or a vertical wind, emanating from the base of a mature cumulonimbus (CB) cloud during a thunderstorm. The microburst downdraft is highly concentrated, typically only 5 km across and often centered in a thunderstorm surrounded on all sides by strong updrafts.

NB: A microburst is a severe form of wind shear.

The mechanics that produce a microburst are found at the height of a mature CB cycle; they only last for a few minutes, up to 10 min, because they are centered in the mature stage of a CB (where updrafts are produced that fuel the downdrafts) which itself only lasts for about 40 min. A mature CB produces strong continuous updrafts around its outer edges (and sometimes outside the cloud itself) which build up to fuel the downdrafts in the center of the CB. The microburst is a result of the downdrafts breaking out of the base of a CB that is colder than the surrounding air, because it has only been warmed at the SALR during the descent within the cloud. Therefore, as it is colder, it is also denser/heavier than the surrounding air, and hence it continues down to the surface where it can often be felt as the *first gust*.

A microburst can be associated with precipitation or "dry" downdrafts. At its most severe, a microburst's downdraft can reach up to 3000 ft/min and have a wind speed of approximately 100 kn. Therefore it is always best to wait for it to pass, because the downdrafts and the rapid reversal of wind direction of a microburst can be fatal to an aircraft.

The typical indications of a microburst are

1. Mature CB clouds with thunderstorm activity, especially rain.

2. Roll cloud formation around a CB.

NB: Roll clouds are formed by the turbulent outflow, associated with microburst downdrafts, from underneath a CB cloud.

3. Virga rain, especially beneath or near a CB cloud.

NB: Rain that evaporates before it reaches the ground absorbs latent heat out of the surrounding air, which creates denser air with an associated higher pressure, causing the downdrafts to fall to the ground at an even greater rate.

Therefore virga rain indicates a possible severe microburst.

4. Flight path and IAS fluctuations, especially on the approach path.

5. Wind direction and speed changes, which can be either reported or measured by aircraft systems, if fitted.

The possibility of a microburst is very high whenever heavy rain is encountered on final approach.

With a microburst out of the base of a CB cloud, the downburst will spread out as it nears the ground and then rebound off the ground as an updraft.

If an aircraft's flight path (approach) is beneath such a cloud, then the aircraft will experience the following characteristics.

1. Initially when entering a zone underneath a CB cloud with a microburst present, an aircraft will experience an updraft (as a result of the downburst spreading outward and rebounding off the ground as an updraft). This has the following effects on an aircraft's flight path:
   a. Energy gain from an increasing headwind will cause the aircraft's nose to rise.

NB: In an extreme case the updraft may be severe enough to cause the aircraft to exceed its stalling angle.

   b. Airspeed (IAS) rises.
   c. Rate of descent is reduced.
   The overall effect in this portion of flight is to cause an *overshoot effect* as the aircraft tends to fly above the desired flight path.

NB: When such an updraft is encountered at the start of an approach underneath a CB cloud, a pilot should be aware of the potential for the initial overshoot effect of the aircraft to be reversed as the aircraft proceeds along the flight path. This is known as the *windshear reversal effect,* i.e., an overshoot followed by an undershoot. Pilots should, of course, observe the overshoot and undershoot effects and react accordingly with changes in attitude and/or thrust. However, pilots should not be tempted to reduce power and attitude too much with an initial overshoot tendency, especially on jet aircraft. This is so because the spool-up time of the engines when you enter the undershoot region (see following points) may be too long to counteract the downdraft and avoid hitting the ground.

2. Approaching the central area underneath a CB cloud with a microburst present, an aircraft would start to experience a strong downdraft, which would be a severe change over a short distance after having previously experienced an updraft. This has the following effects on an aircraft's flight path:
   a. Energy loss from a reducing headwind will cause the aircraft's nose to fall.
   b. Airspeed (IAS) starts to fall (due to the reducing headwind).
   c. Rate of descent increases, with a tendency to go below the glide path.
3. As the aircraft's flight path progresses to the far side of the cloud with a microburst, the aircraft will experience a severe downdraft and a tailwind, directly from the cloud base. This has the following effects on the aircraft's flight path:
   a. Energy is lost from an increasing tailwind.
   b. Airspeed (IAS) continues to fall sharply.
   c. Rate of descent continues to increase with a tendency to go farther below the glide path, with even the possibility of ground contact, if not checked by initiating a missed approach, and even then success depends upon the power height and speed reserves available being sufficient to overcome the downdraft.

The overall effect in this portion of flight is to cause an *undershoot effect* as the aircraft tends to fly below the desired flight path.

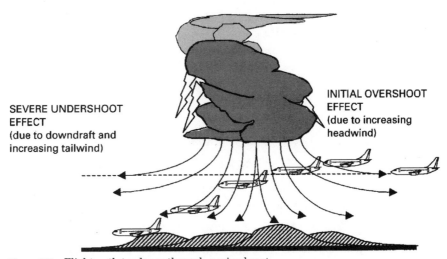

Figure 136 Flight path tendency through a microburst.

It should be clearly understood that microbursts can be of such strength that they can down an aircraft no matter how much power, time, and height are available. Therefore always, always avoid possible microburst areas.

You shouldn't attempt an approach into an area where a microburst is reported or likely. Hold for 10 to 15 min, to allow the storm to move away from the approach path, before you attempt an approach. Continue to monitor for indications of possible microburst activity.

NB: The early identification of a possible microburst, or severe wind shear downdrafts, on an approach is vital to ensure that you have sufficient power, height, and speed reserves to successfully initiate a go-around.

Therefore it is important not to overcompensate for an overshoot tendency by reducing power early in an approach, as this will incur a delay in the delivery of full power when you pass into the downdraft area later on the approach and the go-around has to be initiated. This time delay could be crucial.

If, however, you do encounter an unexpected microburst on an approach, you should initiate an immediate full-power (TOGA) maximum-pitch-up go-around. If you are close to the ground and the landing gear is already down, then you should delay retracting the gear until ground contact is definitely no longer a possibility.

Millibar (hectopascal) The number of feet in a millibar/hectopascal is variable because the rate at which the pressure changes with height is variable. The air pressure (millibar) rate of change is more rapid near the earth's surface and less rapid at high altitudes. By measuring the actual height of a single millibar for a given altitude, the rate of pressure change with altitude can be determined.

The formula used to find the size of 1 mbar is

$$1 \text{ mbar} = 96 \times \frac{\text{absolute temperature (273 K + actual temperature)}}{\text{altitude pressure (mbar)}}$$

To find the size of 1 mbar at sea level, e.g., given an altitude pressure of 1013 mbar and an OAT of +15°C,

$$96 \times \frac{273 + 15}{1013} = 27.3 \text{ ft}$$

Therefore,

$$1 \text{ mbar} = 27 \text{ ft at sea level/1013 mbar}$$

$$1 \text{ mbar} = 40 \text{ ft at 18,000 ft/500 mbar}$$

$$1 \text{ mbar} = 80 \text{ ft at 38,000 ft/200 mbar}$$

However, it is common to use 30 ft for 1 mbar up to an altitude of approximately 5000 ft, and there is 0.0295 inches of mercury (inHg) to 1 hectopascal-millibar (hPa·mbar).

Minimum Approach Breakoff Height (MABH) *See* MABH (Minimum Approach Breakoff Height).

Minimum Control Speed (VMCL) *See* VMCL Speed.

Minimum Decision Altitude (MDA) Minimum decision altitude is the wheel altitude above the runway elevation at which a go-around must be initiated by the pilot unless (1) adequate visual reference has been established and (2) the position and approach path of the aircraft have been visually assessed as satisfactory to safely continue the approach and landing.

MDA is the altitude measured using the local QNH pressure setting, i.e., the height above sea level, or

$$\text{MDA} = \text{airport elevation} + \text{height above the ground}$$

The MDA for an instrument-rated nonprecision approach is calculated as follows. Take the higher of (1) obstacle clearance height (OCH) for the aid and aircraft category or (2) nonprecision approach system minimum.

Localizer only	250 ft
SRA terminates at $\frac{1}{2}$ nm	250 ft
SRA terminates at 1 nm	300 ft
SRA terminates at 2 nm	350 ft
VOR	300 ft
NDB	300 ft

NB: Many operators add another 50 ft to the MDA, especially for large aircraft.

Minimum En route Altitude. *See* MEA (Minimum En route Altitude).

Minimum Height/Altitude Rule To comply with the instrument flight rules both inside and outside controlled airspace, and without prejudice to the usual low-flying rules, the minimum height rule dictates that an aircraft should not fly at a height of less than 1000 ft above the highest obstacle within a distance of 5 nautical miles (nmi) of the aircraft unless

1. It is necessary for the aircraft to do so in order to land.
2. The aircraft is flying on a route notified (Notam, AIP) for the purpose of this rule (this may include controlled airspace such as terminal control areas and airways).
3. The aircraft has been otherwise authorized by the competent authority.
4. The aircraft is flying at an altitude not exceeding 3000 ft above mean sea level and remains clear of cloud and in sight of the surface.

An additional ICAO height rule increases this height to 2000 ft above the highest point in mountainous terrain.

Minimum Sector Altitude *See* MSA (Minimum Sector Altitude).

Mist Mist is simply a parcel of low-level air in contact with the ground that has small suspended water droplets, which have the effect of reducing visibility. It is usual for mist to precede and to follow fog, unless an already formed fog patch is blown in across an area, e.g., sea fog. Mist is reported when suspended water droplets reduce the horizontal visibility to 1000 to 2000 m.

MLS (Microwave Landing System) MLS is a development of the ILS (Instrument Landing System) that uses two beams, one in azimuth and one in elevation as in an ILS, but gives the ability to fly a curved or off-center localizer beam and/or a curved or varied glide slope angle. Both beams use the same frequency in the SHF band, as the beams share the frequency by time-multiplexing the signals that send out azimuth, elevation, and other data signals in a predetermined sequence.

An MLS has a narrow fan-shaped azimuth beam for the localizer, which sweeps to and fro between the limits of its coverage, namely, 40° left and right of the centerline out to a maximum distance of 20 nm.

See Fig. 137 at the top of the next page.

From the point of view of the aircraft on the approach, the azimuth beam starts on the left and sweeps to the right as the to scan and then back to the left as the fro scan. During the complete cycle of the to and fro scan, an aircraft would have

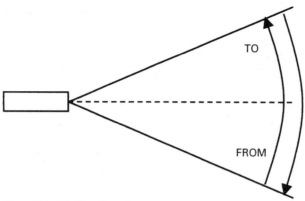

Figure 137   MLS localizer beam scan.

the beam pass over it twice. An aircraft close to the left-hand side of the coverage would have a relatively large time interval between its to and fro sweep passing over the aircraft. And an aircraft close to the right-hand side of the beam's coverage would have a relatively short interval. The time interval between the two scans is therefore a direct indication of the aircraft's position in the covered area. Similarly the localizer path can be set to any line in the covered area in a similar manner.

The elevation beam works in a similar way, but it has a horizontal fan-shaped beam, which sweeps from its lower limit of 0.9° to its upper limit of 15° from the horizontal up to a maximum of 20,000 ft and out to 20 nm within the coverage of the azimuth beam.

The advantages of an MLS over an ILS are as follows:

1. MLS offers a varied approach localizer (azimuth) as either a curved or off-center beam and a glide slope (elevation) beam as either a curved or a varied glide slope angle. Theoretically an incoming aircraft could choose the glide and approach path more suited to its particular requirements; e.g., general aviation or heavy jet aircraft have different optimum approach paths. However, it is more likely to be set to a predetermined path that has been determined by the surrounding terrain, noise abatement sensitive areas, etc. In general it is advantageous because it makes certain airports more accessible which had previous conflicts with terrain or noise abatement areas in using an ILS approach.

2. The MLS beams are not sensitive to reflections from terrain and obstructions near the transmitter, which affected the ILS beams.

3. The MLS frequency range has more transmitting channels—200 available channels—which removes the ILS problem of conflicting signals that occur because of the ILS's limited frequency range.

MLW (Maximum Landing Weight)   Maximum landing weight is the maximum gross weight of the aircraft permitted for landing.

NB: Sometimes a performance-limited MLW (i.e., short runway, obstacle clearance) may restrict the aircraft to a weight less than its structural maximum [certificate of airworthiness (C of A)] weight.

MMO   *See* VMO/MMO Speed.

Momentum The momentum of a body is the product of the mass of the body and its velocity, which enables the body (aircraft) to remain on its previous direction and magnitude for a quantity of time after an opposing force has been applied.

The momentum of a jet aircraft is significantly greater than that of a piston engine aircraft in all its handling maneuvers, climbs, descents, turns, etc., because of its greater weight and velocity. Intelligent anticipation is now necessary to stop unwanted excursions developing to a significant extent on heavy aircraft, especially jets; in the light piston aircraft days, it was acceptable to allow an error in flight path to occur before applying a correction. Now because of the effects of momentum in heavy jet aircraft, we have to control the flight path more closely at a subconscious level to adjust the *required flight path* to maintain the *projected flight path* in a much more preemptive manner. Jet aircraft have been designed to be able to exert greater control forces in both direction and magnitude to keep the effects of momentum to an acceptable level. However, the jet aircraft's momentum is still a significant factor that the pilot should always be aware of, especially on the approach.

Morse Code Letters are as follows:

A	·-	B	-····	C	-·-·
D	-··	E	·	F	··-·
G	--·	H	····	I	··
J	·---	K	-·-	L	·-··
M	--	N	-·	O	---
P	·--·	Q	--·-	R	·-·
S	···	T	-	U	··-
V	···-	W	·--	X	-··-
Y	-·--	Z	--··		

The numbers are as follows:

1	·----	2	··---	3	···--
4	····-	5	·····	6	-····
7	--···	8	---··	9	----·
0	-----				

Mountain (Lenticular) Clouds Airflow rises over mountains due to the orographic uplift and cools adiabatically. If it cools below its dew point temperature, then the water vapor will condense out and form clouds, either *lenticular* clouds (often on the hillside when there is a stable layer of air above the mountain) or as cumulus or even cumulonimbus clouds (when there is unstable air above the mountain).

See Fig. 138 at the top of the next page.

Rotor or roll clouds, particularly common with lenticular cloud formation, may also form at low level downstream of the mountain as a result of surface turbulence.

Mountain/lenticular (lens-shaped) clouds indicate standing (mountain) wave CAT. They are found at height, in rising air, above the downwind side of a range of hills, often extending for up to 100 nm downward of a line of hills and at a height of up to 25,000 ft.

MRA *See* VRA/MRA Speed.

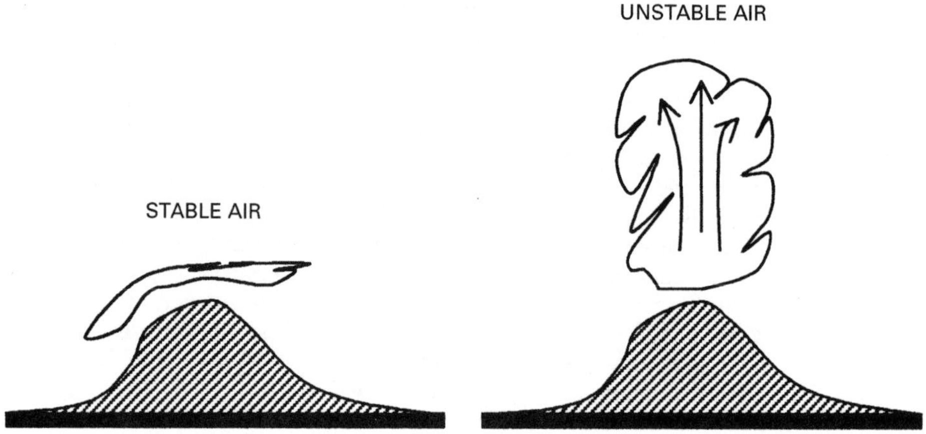

STABLE AIR

UNSTABLE AIR

Figure 138   Mountain cloud.

MSA (Minimum Sector Altitude)   Minimum sector altitudes (MSAs) are published on instrument approach charts and en route charts.

Instrument approach charts   The published MSA provides at least 300-m (1000-ft) vertical clearance within 25 nm of the homing facility for the particular instrument approach. If the aircraft remains at or above the relevant MSA, then it should remain clear of terrain and obstacles as it tracks toward the aerodrome prior to commencing the approach. Some aerodromes have MSAs that apply to all sectors; however, most aerodromes have different MSAs for different sectors, depending upon the direction the aircraft is arriving from and the terrain over which it has to cross. Jeppesen publishes MSAs in a circle at the top of its charts, with the sectors delineated by magnetic bearings from the homing facility. Aerad publishes this information in the same manner, or as four sector safe altitude (SSA) boxes, one in each corner on the plan section of the instrument approach chart. An SSA of 3400 ft AMSL within 25 nm is published as **SSA 25 nm 3$_4$.**

En route charts   These show grid moras (or minimum sector safe altitudes). When determining your current MSA or grid mora altitude from an en route chart, you should take the most restrictive MSA of all the adjacent grid moras to your present position.

MTOW (Maximum Takeoff Weight)   The maximum takeoff weight is the maximum gross weight of the aircraft permitted for takeoff.

NB: Sometimes a performance-limited MTOW (i.e., short runway, obstacle clearance) may restrict the aircraft to a weight less than its structural maximum (C of A) weight.

Multiplex Auto Land System   *Multiplex* is a term used for an autopilot landing system that comprises two or more independent autopilots/channels, which are used collectively to provide a redundant system design.

A multiplex auto land system can utilize one of these:

1. A dual digital control computer channel, known as a duplex system
2. A triple digital control computer channel, known as a triplex system
3. A quadruple digital control computer channel, known as a quadruplex system

The independent autopilot/channel systems are interconnected, and together they provide continuous auto landing control. In the event of a single autopilot/channel failure, or a different reading, that system is outvoted by the others and is automatically disengaged. This multiplex redundancy system allows the autopilot to operate in either a fail-passive or operational condition, depending upon the number of channels engaged, thus allowing the system to perform what are termed *Land 2* or *Land 3* status auto landings.

Myopia Myopia is the condition of shortsightedness. It is associated with a longer-than-normal eye, which results in the image being formed in front of the retina so that images of distant objects are out of focus.

Navigation and Anticollision Lights—on an Aircraft An aircraft basically has the following navigation and anticollision lights:

1. A steady green navigation light on the starboard (right) wing that shows from dead ahead around 110° on the right side.
2. A steady red navigation light on the port (left) wing that shows from dead ahead around 110° on the left side.
3. A steady white navigation tail light that shows through 70° either side of dead astern.

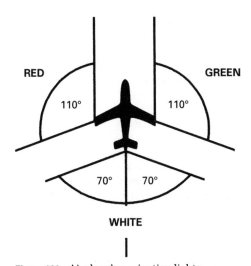

Figure 139 Airplane's navigation lights.

4. A flashing red anticollision light showing in all directions.
5. High-intensity white strobe lights positioned on both wing tips, found on most commercial aircraft. These are used to make the aircraft more visible to other aircraft.

Knowing the arcs of the aircraft's navigation lights helps to ascertain a collision risk. For example, if at night you see a red navigation light of an approaching aircraft out to the left, then your paths will not cross; i.e., red to red is safe. However, a red light out to the right is unsafe as a collision risk exists, i.e., red to green is unsafe. If at night you see a green navigation light of an approaching aircraft out to the right, then your paths will not cross; i.e., green to green is safe. However, a green light out to the left is unsafe as a collision risk exists, i.e., green to red is unsafe. If both red and green lights from another aircraft are visible, then the other aircraft is flying directly toward you. If the steady white light from another aircraft is visible, then the other aircraft is flying away from you.

The red anticollision light should be switched on whenever an engine is running, day or night.

The white strobe lights should be switched on whenever the aircraft is airborne. Many aircraft have a switch mode that automatically switches on the strobe lights when airborne.

NB: However, it is good airmanship to switch on the strobe lights whenever the aircraft enters or crosses an active runway.

It is permissible to switch off or reduce the intensity of the flashing strobe lights if they become a distraction to crew, especially in clouds, or to persons on the ground.

The green and red navigation lights should be switched on during flight at night.

NB: However, it is good airmanship to display the navigation lights during the daytime.

Notwithstanding any individual company minimum equipment list (MEL), the following are generally the limits imposed by most local authorities:

At night on the ground	Failure of the anticollision light or the navigation lights on the ground means the aircraft cannot dispatch as required to be displayed during night flights.
At night in flight	Failure of the anticollision light or the navigation lights in flight requires the aircraft to land as soon as it is safe and possible to do so, unless authorized to continue by ATC.
By day on the ground and in flight	Failure of the anticollision light during daylight does not stop an aircraft from flying, provided that the light is fixed at the earliest opportunity. The navigation lights are not required to be on during daylight hours.

NDB/ADF   A *nondirectional beacon* (NDB) is a medium-range radio navigational aid that sends out a signal in all directions for aircraft to home to. The NDB transmits in the 200- to 1750-kHz, medium- and low-frequency band, and it uses a surface wave propagation path.

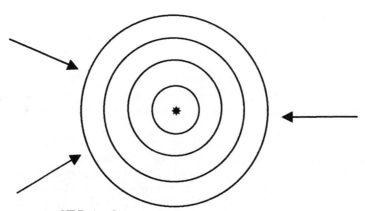

Figure 140   NDB signal.

The *automatic direction finder* (ADF) is a needle indicator fitted in the aircraft that shows the direction to the selected NDB from the aircraft. This is displayed either as a relative bearing (angle between the aircraft heading and the direction of the NDB) on an RBI instrument or as a QDM on an RMI instrument.

The NDB/ADF system is used for en route tracking, holds, and NDB letdown procedures where low-powered NDBs are used as locators. The principle of use is the same for any of these procedures.

Tracking to an NDB Flying to an NDB is similar to following a compass needle to the north pole. If you fly the aircraft toward where the needle points, eventually you will arrive overhead. If, however, you wish to fly to an NDB on a particular magnetic track, which is especially true for an NDB letdown, then you have to maintain the correct ADF tracking by making sensible heading changes, with the "follow the head" technique.

That is, if the head of the ADF needle is 10° left of the desired track, this means the NDB is left of the aircraft's nose, and the relative bearing is 10° left. Therefore you have to alter your heading farther to the left to intercept the desired inbound track. For example,

Desired track	090°	
ADF head indicates	080°	
Alter heading to	070°	until the ADF needle reads the desired track, then alter the heading to hold the 090° track

Tracking from an NDB When tracking away from an NDB, you can either "follow the head" of the ADF needle on the reciprocal heading, as per tracking to an NDB, or "pull the tail" of the ADF needle to maintain the outward track, which is more commonly used.

That is, if the tail of the ADF needle is 10° left of the desired track, this means the NDB is to the right of the aircraft's tail, and the relative bearing is −10° off the tail. Therefore you have to alter your heading to the right to "pull" the needle and therefore the aircraft back onto your desired track. For example,

Desired track	090°	
ADF tail indicates	080°	
Alter heading to	100°	until the tail of the ADF needle reads the desired track, then alter the heading to maintain the 090° track

The NDB/ADF system can also be used for fixing an aircraft's position.

The ADF indication can be displayed on an RBI instrument, or on an RMI instrument, or on an HSI (horizontal situation indicator) instrument.

The errors suffered by an NDB/ADF are as follows:

1. *Interference.* Interference from other stations that are transmitting on a similar frequency is the largest problem associated with an NDB station. Station interference is interference from another NDB station. However, this will not cause an error during daytime if the NDB is used within its protected range (the area in which freedom from interference is guaranteed). At nighttime, however, returning sky waves (see night effect error in item 3) can cause interference within an NDB's protected range.

2. *Static.* All forms of static can affect the ADF system, especially thunderstorms. Electrical thunderstorms, usually in and around CB clouds, cause the ADF needle to be deflected in its direction and away from the selected NDB station. Precipitation static also affects the range and accuracy of ADF bearing information, although to a lesser extent.

In such conditions, VHF aids should be used in preference to ADF signals.

3. *Night effect.* Night effect is the natural phenomenon of strong NDB sky waves being returned to the earth, especially during dusk and dawn, when the ionosphere is less dense and attenuation is least. This causes a particular interference with the NDF sky wave's own surface wave and results in a fading of the signal and a possible wandering ADF needle.

Night effect is a particular form of interference from the selected station. It can be detected by listening for fading of the ident on the carrier wave (BFO on) and by the instrument needle "hunting" for a signal.

4. *Coastal refraction.* The speed of a surface wave is faster over water than over land. This change of speed means an NDB's surface wave is refracted as it passes over a coastline when crossing at an angle. An aircraft receiving such a refracted signal would be given a false indication of the beacon's position, or a "ghost beacon." This effect worsens the farther back from the coast the beacon is sited.

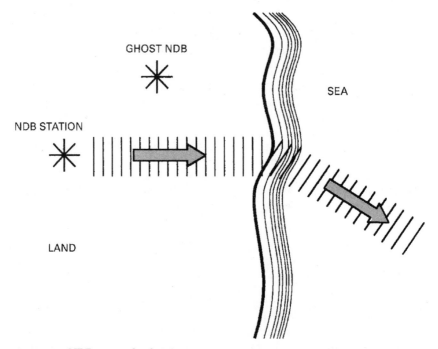

Figure 141   NDB—coastal refraction.

Therefore to lessen coastal refraction error, if possible you should
■ Use beacons closer to the coast.
■ Only take bearings at right angles to the coast.
■ Fly at a higher altitude.

5. *Mountain effect.* Mountain reflections of an NDB signal can produce areas of zero coverage beyond the mountain and in areas affected by signal fading and ADF needle wandering between the station and the mountain.

6. *Quadrantal error.* Aircraft quadrant error is the effect of NDB signals being reflected from the aircraft itself, especially the engines. The aircraft has its own

electromagnetic field which can distort incoming signals, particularly for signals arriving at 45° and 135° left and right of the aircraft's nose and tail, in the four quadrants; hence it is called *quadrantal error*.

NB: Quadrantal error is least on the aircraft's fore and aft cardinal points, that is, 000° and 180°.

Quadrantal error is predictable and can be either manually or automatically compensated for, depending on the sophistication of the equipment.

7. *Synchronistic transmissions.* These occur when an ADF picks up two stations at the same time and at the same power level, and this results in the ADF needle showing the middle direction between the two stations.

The ICAO requirement for an NDB is that it maintain its accuracy to within ±5°. However, not all errors are predictable, and therefore it is possible that the actual error experienced by an NDB/ADF system is greater than its designated maximum.

The following technique is used for ADF interception and tracking:

NDB/ADF homing procedure

1. Tune and identify the required NDB. The ADF needle will now indicate the QDM to the beacon.

NB: Any angle between the aircraft's heading and the ADF needle is the relative bearing of the beacon from the aircraft's nose.

2. Turn the aircraft the shortest way toward the ADF needle pointer to align the needle with the aircraft's heading.

3. Lay off drift to maintain the QDM.

NB: The drift allowance is the angle between the aircraft heading and the QDM indicator.

The aircraft is now heading along a QDM to the beacon.

Maintaining a specified track

To the NDB. If the head of the ADF needle moves away from the QDM (magnetic bearing to the beacon), then you have to adjust the aircraft's track in the *same* direction by approximately twice the track error. That is, use the "follow the head" technique.

From the NDB. If the tail of the ADF needle moves away from the QDR (magnetic bearing from the beacon), then you have to adjust the aircraft's track in the *opposite* direction by approximately the same track error amount from the original track. That is, use the "pull the head" technique.

Intercepting a specified track

1. Tune and identify the required NDB. The ADF needle will now indicate the QDM to the beacon.

2. Turn the aircraft to intercept the required track in the same method as maintaining a track, except that an initial larger adjustment may be required.
    - *To the NDB.* If the head of the ADF needle is 30° left of the desired track, alter your heading farther to the left to intercept the desired inbound track. For example, desired track is 090°; ADF head indicates 060°; alter heading to 030°.
    - *From the NDB.* If the tail of the ADF needle is 30° left of the desired track, alter your heading to the right to "pull" the needle and therefore the aircraft

back onto your desired track. For example, desired track is 090°; ADF tail indicates 060°; alter heading to 120°.

3. When you are steady on an intercept heading, check that the ADF needle is moving toward the required QDM or QDR.

4. Use the appropriate amount of lead time to turn onto a heading to capture the required QDM or QDR track.

After you have been cleared for the procedure by ATC, there are two further requirements that must be met before you descend during an NDB procedure:

- You must be at the descent point as defined on the procedure. This may be expressed as a fix, distance, or time.

- You must be within ±5° of the ADF final approach track.

These requirements relate to any point on the procedure, i.e., the outbound procedure, and not just the inbound track.

NDB Range   The range of an NDB beacon is controlled primarily by its power, with a maximum theoretical range of 300 nm over land and 600 nm over sea, although its range is often restricted to avoid interference with other signals. That is,

$$\text{Range over land (nm)} = 3\,\sqrt{\text{power output (W)}}$$

$$\text{Range over sea (nm)} = 9\,\sqrt{\text{power output (W)}}$$

An NDB in theory has the farthest range, because of its surface wave propagation path, i.e., 600 nm over sea and 300 nm over land, compared to the VOR's line-of-sight propagation path.

Net and Gross Flight Paths/Performance   The difference between the net and gross flight paths/performance is as follows.

The gross performance is the average performance that a fleet of aircraft should achieve if satisfactorily maintained and flown in accordance with the techniques established during flight certification and subsequently described in the aircraft performance manual. Gross performance therefore defines a level of performance that any aircraft of the same type has a 50 percent chance of exceeding at any time.

The net performance is the gross performance diminished to allow for various contingencies that cannot be accounted for operationally, e.g., variations in piloting technique and temporary below-average performance. It is improbable that the net performance/flight path will not be achieved in operation, provided the aircraft is flown in accordance with the recommended techniques, i.e., power, attitudes, and speed.

Normally performance graphs show net performance; however, some performance charts, especially for three-engine aircraft, assume gross performance is equal to net performance. This means that no margin exists between what the graph suggests the aircraft will achieve and what the aircraft will achieve.

Net Takeoff Flight Path—Obstacle Clearance   The net takeoff flight path (NTOFP) is the true height versus the horizontal distance traveled from reference zero, assuming the failure of the critical engine at V_1, and is used to determine the obstacle clearance by a specified minimal amount. Namely, the NTOFP is the gross flight path reduced by 0.9 percent gradient ability, with a guaranteed obstacle clearance safety margin of at least 35 ft.

The NTOFP starts from reference zero, where a jet aircraft is already at a height of approximately 35 ft, or 15 ft in wet conditions. Because obstacle heights are generally referred to the elevation at the end of the takeoff distance available (TODA), i.e., reference zero point of 35 ft when the TOD is limiting, the calculated height gained by the aircraft at a point in its NTOFP is a true height that can be compared to the obstacle height, to ascertain the obstacle clearance margin.

NB: When TODR is less than TODA, then adjustments have to be made to the obstacle height (in the second segment) to compare with the height gained by the aircraft.

The NTOFP is a phase of flight that is divided into segments and is defined by changes in the aircraft's configuration, engine thrust or speed, during which the aircraft must be "cleaned up" from its takeoff configuration.

The plan view of the NTOFP is often referred to as the *obstacle domain,* which is funnel-shaped and commences at the end of the TODA, where the width of the NTOFP, or obstacle domain, is one-half the wing span plus 60 m (European JAR OPS; 75 m in other states including the United Kingdom). The width of the NTOFP domain (on each side of the extended runway centerline) thereafter increases at a rate of the horizontal distance traveled divided by 8 up to a maximum width of 900 m.

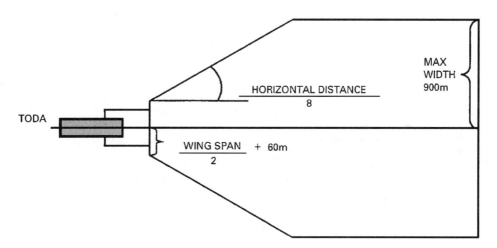

Figure 142   NTOFP, obstacle domain.

Any obstacle that falls within the NTOFP must be cleared by its minimum safety margin of 35 ft. The NTOFP normally ends at a net height of 1500 ft where the aircraft transits to its en route phase of flight, unless obstacle clearance is still restrictive, thus requiring the NTOFP to be extended to the point where the obstacle clearance is no longer restrictive.

The takeoff flight path is affected by the prevailing wind conditions, which either increase (headwind) or decrease (tailwind) the climb gradients and horizontal distances of the flight path. When a flight path has to be calculated segment by segment, the effect of the wind must be considered individually for each segment. Obstacle distances are adjusted to allow for the effect of the wind in the second segment (up to 400-ft gross height), while wind-corrected climb gradients are used for third- and final-segment calculations.

Newton's Three Laws of Motion   These laws address the property of inertia.

1. That which is still will tend to remain still, and that which is moving will tend to continue moving and will resist any change of speed or direction. An example is the mixture of fuel and air in the engine induction system.

2. If it is required to alter the state of rest or uniform motion of a body, or change its direction, a force must be applied to it, and the more sudden the change desired and the greater the mass affected, the greater the force required. An example is the piston in the engine cylinder when the engine is running.

3. This is the best known of Newton's laws. The application of a force upon a body will cause an equal and opposite reaction upon the surface which produces the force. An example is the gas turbine engine.

Night   Night is defined as being

- From 30 min after sunset at the surface
- Until 30 min after sunrise at the surface

   NB: This differs from the definition of night required for night training flights, which is when the sun exceeds 12° below the horizon.

Nighttime—Visual Acuity   Nighttime visual effect is a form of empty field myopia when airborne and a form of blackhole effect when landing. That is, the runway always appears to be farther away in terms of vertical distance than it actually is; therefore you tend to land earlier than expected—so beware.

Noise Abatement   There are two main parameters affecting noise, which in practice can be handled to achieve a noise abatement profile:

1. Noise is proportional to the power being developed by the engines.

2. Noise is inversely proportional to the distance between the noise source and the listener.

Therefore to reduce noise, we must reduce power and get as far away from the noise-sensitive areas as quickly as possible.

   The following basic technique is the only one that could be used if the airfield is completely surrounded by noise-sensitive areas. After the aircraft leaves the runway, the technique is split into two separate parts:

- In the noise abatement first segment, the aircraft adopts a steep climb at full power, to gain as much height as possible before the listening posts' noise-sensitive areas.

Then the flight path changes to the second segment.

- In the noise abatement second segment, the engines are throttled back to climb power while the aircraft maintains its initial high attitude. This profile continues until either a declared height has been reached or the noise-sensitive area has been cleared, at which time a lower climb attitude is adopted.

This leads into the departure profile's third phase, acceleration and cleanup, followed by the fourth phase, en route climb.

Noise—Jet/Gas Turbine Engine The noise generated by a jet/gas turbine engine is from the shear effect of different displaced air velocities. The shear is the difference between the jet's faster displaced air and the slower ambient air around it. The higher the power, the faster the speed of the jet efflux; and the greater the shear effect, the greater the noise.

NB: The development of bypass engines has reduced the speed of the jet efflux and therefore the noise considerably.

Jet/gas turbine engine noise can be reduced with bypass engines; that reduces the shear effect between the displaced slower bypass engine air and the ambient air that reduces noise. Jet/gas turbine engine noise can be controlled by employing a reduced-thrust takeoff and employing a maximum angle climb after takeoff (this allows the aircraft to get above any noise control zones).

Noise—Propeller The noise generated by a propeller aircraft is due to the shear effect of different displaced air velocities. The shear is the difference between the propeller's faster displaced air and the slower ambient air around it.

The following can reduce propeller noise:

1. Increase the number of propeller blades, say, from four blades to three, as a design feature. This increases the mass of airflow, therefore its velocity can be reduced to maintain the same thrust. This results in reducing the shear effect between the displaced propeller air and the ambient air that reduces noise.

2. Employ a reduced-thrust takeoff. To control propeller noise, employ a maximum angle climb after takeoff. This allows the aircraft to get above any noise control zones.

Nondifferential Spoilers The difference between differential and nondifferential spoilers lies in how they provide lateral roll control when already extended as speed brakes. Nondifferential spoilers will extend farther on one side but will not retract on the other side in response to a roll command; and when already fully extended as a speed brake, both sides remain in the extended speed brake position, and so the spoilers do not provide any roll control.

Normally Aspirated Piston Engine A normally aspirated engine works because the breathing of the cylinder is due to the pressure differential below the standard sea-level value of 14.7 psi. In other words, it only uses the atmospheric air density that is available to produce a charge in its cylinders and is not boosted by a supercharger; therefore the normally aspirated piston engine's power output is restricted by its engine cylinder capacity.

Nosewheel The requirements of the nosewheel are as follows:

1. It must carry the aircraft's direct compression load weight.
2. The nosewheel must provide a towing attachment.
3. It must withstand shear loads.
4. To enable the aircraft to be maneuvered about the airfield, the nosewheel must be able to castor freely when subjected to compression and shear loading (i.e., right brake to turn right).

5. Self-centering is required. Whenever the weight is removed from the wheel, it must be centered, to ensure the correct positioning of the nosewheel before retraction.
6. Steering is required to maneuver the aircraft on the ground, using hydraulic power.
7. Antishimmy is necessary. Shimmy is damped by one of the following:
   a. Provision of a hydraulic lock across the steering jack piston
   b. Hydraulic damper
   c. Heavy self-centering springs
   d. Double nosewheels

Nosewheel Shimmy Nosewheel shimmy is the unstable swiveling oscillation of the wheel due to the flexibility of tire sidewalls, especially at high speed. Excessive shimmy can vibrate dangerously throughout the entire aircraft, causing wear in the wheel bearings, low tire pressures, and wear on the undercarriage linkage and mountings.

Notams Notams are *notices* to *airmen*. They contain information on any aeronautical facility, service, procedure, or hazard, timely knowledge of which is required by people concerned with flight operations. It may be given as either a class 1 notam distributed by teleprinter for urgent matters or a class 2 notam distributed through the mail for less urgent matters.

OAT This is the acronym for the ambient outside air temperature.

OBI (Omni Bearing Instrument) The OBI is a navigation instrument sometimes referred to as a first-generation VOR indicator. It is used by the pilot to select the required VOR radial and to display tracking guidance relative to the selected radial. If the aircraft is on the selected radial, the VOR needle or CDI (course deviation indicator) will be centered; and if the aircraft is not on the selected course/track, then the CDI will not be centered.

The OBI is a "track up" display, i.e., the selected track accommodates the top of the dial position, regardless of the aircraft's heading.

NB: Later versions of the OBI also have the capability of displaying ILS information.

NB: *Course* is another term for *track.*

The OBI consists of the following:

1. The *omni bearing selector (OBS)* is a small knob geared to the compass card that is used by the pilot to select or dial in the required course under the course index at the top of the instrument display.

NB: The compass card is actually rotated.

2. *Course deviation indicator or bar (CDI)* is superimposed over the compass card. It is a vertical bar that is fixed under the course index with only left and right freedom of movement which indicates the aircraft's horizontal off-track angular deviation from the selected radial (track). Since the CDI indicates angular deviation, the actual distance for a given CDI will be smaller as the aircraft gets closer to the VOR ground station.

The CDI presentation may differ between different OBIs. It may move laterally as a whole, or it may be hinged at the top and swung laterally over the dot scale.

3. The *deviation dot scale* indicates the amount of angular deviation from the selected track. Normally there are five dots on either side of the center position, although the inner dots on both sides are represented by a continuous circle.

One dot deviation represents 2° VOR track deviation, and full-scale deviation represents 10° or more VOR track deviation.

NB: Some older instruments only have a four-dot scale; therefore on these instruments one dot represents 2.5° VOR track deviation.

4. The *to/from indicator* specifies whether the selected omni bearing will take you to or from the VOR station, i.e., whether you are looking at a QDM (to) or QDR (from). This is important because the CDI can be centered on either a radial or its reciprocal, that is, 100° or 280°, because they are the same position line.

NB: It is important to select a QDM and a to flag on the OBI when heading toward a station and a QDR with a from flag when you are tracking away from a VOR station.

5. The *compass card/scale* is, of course, 360° and is controlled by the OBS which displays the selected course under the course index at the top of the instrument display.

It is important to realize that the aircraft heading has nothing to do with the OBI. Namely, the VOR display is not heading-sensitive. This means that the OBI display will be the same irrespective of the aircraft's heading, as long as the position of the aircraft with respect to the selected track remains unchanged. This is so because the OBI is only an indication of the aircraft's position to the selected VOR radial. However, the VOR cockpit display (e.g., OBI) should be used as a command instrument whenever possible. That is, the aircraft's heading, allowing for wind drift, etc., should roughly be in the same direction as the OBI's selected track, thereby ensuring the correct to/from orientation. Therefore the OBI will act as a command instrument, and by flying toward the deflected CDI, the pilot can center it and thereby regain track.

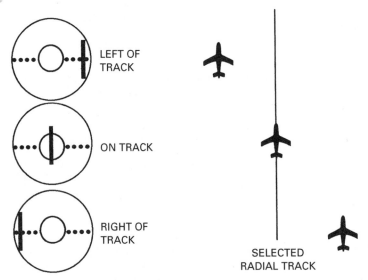

Figure 143 OBI tracking indications.

To track to or from a VOR using an OBI:

- Select the VOR frequency in the correct VHF/NAV box.
- Identify the station.
- Check that the warning flag is not displayed.
- Select the desired track with the OBS ensuring a correct flag and heading orientation. For example, the heading is 280° to a station, after wind drift. The desired track is 280° with a to flag showing.

Oblique Mercator Chart An oblique Mercator chart is a cylindrical projection in which the great circle tangent is neither the equator (normal Mercator chart) nor a meridian (transverse Mercator chart). Oblique Mercator charts are designed for particular routes, i.e., London to Melbourne, Australia. The great circle tangency on an oblique Mercator chart is sometimes also referred to as the *false equator* of the projection.

Scale and convergency are correct at its tangential route and can be assumed to be correct within its usable limits. Inside its usable limits it has the same properties as a Lambert chart and is considered to be orthomorphic, with great circle tracks as straight lines and rhumb lines as curves.

The advantage of this chart is that it is ideal and correct for a particular long-haul route.

Obstacle Clearance   *See* Sloping Runway—Obstacle Clearance Performance; Net Takeoff Flight Path (Obstacle Clearance); and Reference Zero—Obstacle Clearance.

Obstacle Clearance Height/Altitude   *See* OCH(A).

Obstacle Clearance—in the Second Departure Segment   If an aircraft is limited by an obstacle in the second segment because its climb performance is insufficient to clear the obstacle(s), then the following flight procedures or a combination of them can be used to improve the aircraft's climb gradient and thereby clear the close-in obstacle(s).

1. Increase takeoff flap (remaining within the takeoff flap range). This will enable a shorter takeoff run, thus increasing the airborne distance to any very close-in obstacles. This allows for a longer airborne time to the obstacle during which the aircraft can achieve a greater height before reaching the obstacle.

   NB: This improvement has to be greater than the reduced climb gradient due to the airborne drag associated with an increased takeoff flap setting.

2. Reduce takeoff weight to a level that achieves the required climb gradient that clears the obstacles. Reducing takeoff weight improves the aircraft's obstacle clearance because
   a. The horizontal distance between the obstacle and reference zero is increased. Since the field length TOR/D required is less for the lighter aircraft, this allows a longer airborne time, and therefore the aircraft can achieve a greater height before it reaches the obstacle.
   b. The climb gradient is improved because of a better lift/weight ratio, because lift is directly proportional to weight.
3. Increase V_2 climb, maintaining takeoff weight.
4. Use the maximum angle climb profile.

Obstacle Clearance—in the Third Departure Segment   If an obstacle in the third segment limits an aircraft because the obstacle in question is higher than the second-to third-segment level-off height, then any of the following flight procedures or a combination of them can be used to clear the distant obstacle(s).

1. Use the extended V_2 climb profile technique.
2. Reduce takeoff weight to a level that achieves the required climb profile which clears the obstacle. Reducing takeoff weight improves the aircraft's obstacle clearance because
   a. The horizontal distance between the obstacle and reference zero is increased, because the field length TOR/D required is less for the lighter aircraft. This allows a longer airborne time, and therefore the aircraft can achieve a greater height before it reaches the obstacle.

    b. The climb gradient is improved because of a better lift/weight ratio, because lift is directly proportional to weight.

3. Use flight path climbing turns to avoid the obstacles.

   NB: The distance and the height of the obstacle dictate the procedures available. However, these vary according to aircraft type because of different performance capabilities.

Obstacle—Lighting Obstacles greater than 492 ft (150 m) are lit by high-intensity flashing white lights, day and night. Any failed lights are notamed.

   Obstacles less than 492 ft (150 m) but higher than 300 ft are lit when the obstacle is considered significant with medium-intensity flashing red lights at nighttime. These lights are *not* notamed when failed.

   NB: A building or obstacle greater than 300 ft becomes an air navigation obstacle, and it is lighted accordingly.

   NB: An obstacle may have both white lights at 492 ft (150 m) and red lights at 300 ft.

OCA *See* OCH(A).

Occluded Front An occluded front is the combination of both cold and warm fronts.

   NB: A front is shown as a line on a weather chart that represents the front's *surface* position. And an *occluded* front is represented by a combination of semicircles (warm front symbol) and barbs (cold front symbol) along the front line.

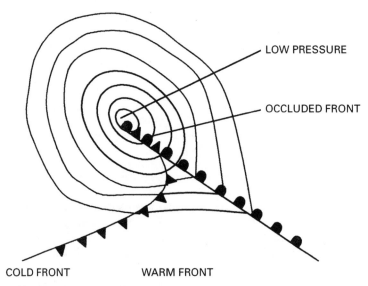

Figure 144 Occluded front shown on a weather chart.

   An occlusion occurs because the cold front moves faster than the warm front, and therefore the cold front will inevitably catch up to the warm front. This occurs first near the center of the depression, where the original distance between the two fronts is the shortest. The characteristic weather of the two fronts also becomes combined. However, because temperatures between the cold air behind the warm sector and the cold air in front of the warm sector are never the same, the type of occlusion will be either a warm or cold occlusion.

Warm occlusion This occurs where the cold air in front of the warm sector is cooler than the cold air behind the warm sector. Therefore the cooler air in front of the warm sector undercuts the cold air behind the warm sector.

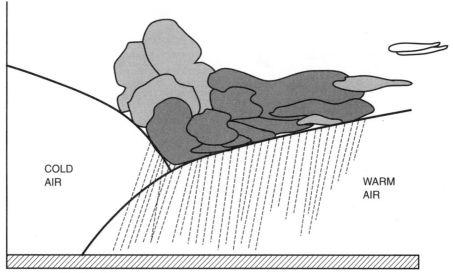

Figure 145 Warm occlusion.

Cold occlusion This occurs where the cold air behind the warm sector is cooler than the cold air in front of the warm sector. Therefore the cooler air behind the warm sector undercuts the cold air in front of the warm sector.

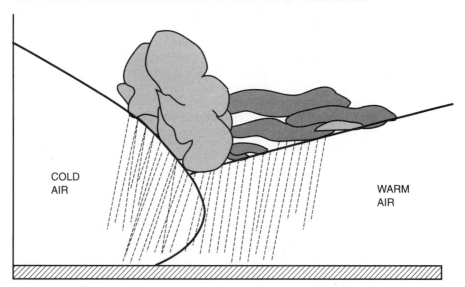

Figure 146 Cold occlusion.

OCH(A)   The OCH is the obstacle clearance height above the aerodrome level (and so has relevance when using QFE). And the OCA is the obstacle clearance altitude above mean sea level (and so has relevance when using QNH).

A typical OCA(H) is written as 850 (170) on Jeppesen charts. The difference between the numbers equals the runway elevation. Different OCA(H) values are published for different categories of aircraft (CAT A, B, C, or D) based on speed and thus maneuverability. The starting point for calculating approach minima is the OCA(H) value published on the instrument approach charts.

Ohm   An ohm is a unit of electrical resistance (expressed by the symbol R), i.e., the degree of resistance or opposition to current flow (or amperes).
NB:

$$1 \text{ kilohm } (1 \text{ k}\Omega) = 1000 \text{ ohms } (\Omega)$$

$$1 \text{ megohm } (1 \text{ M}\Omega) = 1{,}000{,}000 \ \Omega$$

Ohmmeter (Gauge)   An ohmmeter measures and indicates the amount of ohms (electrical resistance) of electric current.

Ohm's Law   Ohm's law describes the relationship between applied emf (measured by V in volts), current (measured by I in amperes), and resistance (measured by R in ohms) in a circuit. The current flowing in a circuit is directly proportional to the applied emf and inversely proportional to the resistance. Given any two of these quantities, the third can be easily calculated from the following formulas:

$$I = \frac{V}{R} \qquad R = \frac{V}{I} \qquad V = IR$$

Omega   Omega is a long-range global navigation system that uses ground-based beacons and operates in the VLF (very low frequency, 3 to 30 kHz) band using surface wave propagation. It is a hyperbolic system that has eight ground-based transmitters located around the world in Norway; Liberia; Hawaii and North Dakota, in the United States; La Reunion, Argentina; Australia, and Japan. Each station is identified by its unique duration of pulses that it sends out on 10.2, 11.05, 11.33, and 13.6 kHz. Only one station transmits a particular frequency at any one time, in a preprogrammed 10-s cycle, during which time all the stations will transmit on all four frequencies. This allows the aircraft's omega equipment to run a phase comparison on the same frequency from different stations without mixing up the signals. With a memory store, it is able to compare the phase of the signal of one station with that of another station transmitted at a slightly different time. Thereby the omega onboard equipment runs phase comparisons between the different station signals to fix a position on a hyperbolic lattice.

NB: A hyperbolic lattice is made up of hyperbolic position lines (lines of equal difference in range between two stations) overlaid with hyperbolic position lines from another pair of stations. It produces approximately 8-nm-wide lanes.

In practice, the 10.2-kHz frequency is the only one used for navigation; the other three are used to prevent beat frequencies for lane ambiguity resolutions.

The computer normally maintains the accuracy of the system in the airborne equipment, which compensates for predictable errors such as the following:

1. Diurnal and seasonal variations in the height of the ionosphere that can affect the propagation path of the VLF radio wave.
2. Ground conductivity changes, e.g., sea and ice caps, that can affect the propagation of the VLF radio wave in different ways.
3. Geographic error, because omega regards the earth as a perfect sphere, which it is not.
4. Latitude error. Omega doesn't allow for different heights of the ionosphere at different places on the earth, e.g., between the pole and the equator.
5. Modal interference. Stations within 600 nm are automatically deselected to prevent confusion between different signal paths.
6. Geometric ambiguity. Stations greater than 7500 nm away are automatically deselected to prevent the possibility that the signal may have come the long way around the earth.

The accuracy of the system is affected by unpredictable errors that the computer does not compensate for, such as

7. Sudden ionospheric disturbances or sporadic intensification of the E layer caused by solar flares, etc.
8. Polar cap absorption, which is caused by solar activity and disturbs the passage of the signal over the pole.

   NB: Notams warn of significant solar activity.

9. Local oscillation drift. When an omega unit has to use a VLF station, it has to make do with its local oscillator rather than the required atomic clock, resulting in a local oscillator drift, which can reduce the accuracy of the system.

Overall the accuracy of omega is said to be better than 3 nm on 95 percent of occasions and is generally better by day than by night.
See also Rho Rho Rho Omega.

Omni Bearing Instrument (OBI)   *See* OBI (Omni Bearing Instrument).

Orthomorphic Chart   An orthomorphic chart must have

1. The same scale in all directions, from a given point
2. A right-angle graticule

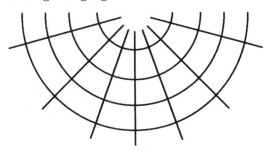

Figure 147 Orthomorphic charts.

Orthomorphic charts show shapes and therefore angles and distances correctly, which is known as orthomorphism.

Oxygen—Passenger (Cabin)   The passengers should get oxygen at a cabin altitude of 14,000 ft.

The passenger oxygen systems are normally chemically generated oxygen systems, comprising sodium chlorate and iron powder which, when mixed chemically, generate oxygen, which is continuously dispensed via a mask to the user.

The passenger oxygen system is activated by one of these:

- Manually by the flight crew.

- Automatically by a barometric pressure controller which releases locking pins that allow the masks to drop from the overhead compartment whenever it senses a 14,000-ft cabin altitude. Pulling on the mask, attached to its individual oxygen generator, causes an electrical firing mechanism to mix the chemical agents that generate the oxygen, which is then continuously supplied to all the attached masks until the system is emptied.

There must be enough passenger oxygen for a 15- to 20-min descent, plus enough oxygen for 10 percent of the required amount for the rest of the flight to the nearest diversion aerodrome, based on the worst-case scenario.

Oxygen—Pilot   A pilot should go on oxygen at a cabin altitude of 10,000 ft. Prior to takeoff there must be enough oxygen in the crew system to allow for a 15- to 20-min descent (i.e., cruise altitude down to 10,000 ft), plus enough oxygen for the rest of the flight to the nearest diversion airport, based on the worst-case scenario.

The "normal" selection on a pilot's oxygen mask will provide a sea-level pressure mixture of oxygen and ambient cabin air.

For life support reasons, 100 percent oxygen on demand should be used. It provides partial pressure of oxygen at or near sea-level (SL) pressure. That is, sea-level atmosphere pressure is 14.7 psi, and 21 percent of the atmosphere is oxygen; ergo, SL oxygen pressure is 3.08 psi.

It is not practical to keep the aircraft at the atmosphere's SL pressure of 14.7 psi, but it is practical to increase the percentage volume of oxygen, thereby maintaining SL oxygen partial pressure.

NB: SL partial pressure of oxygen (3.08 psi) can be maintained up to 34,000 ft when 100 percent oxygen is supplied.

Between 34,000- and 41,000-ft cabin altitude, 100 percent oxygen can maintain an oxygen partial pressure equivalent to 10,000 ft without any adverse side effects.

A 100 percent emergency oxygen provides pure oxygen supplied continuously under positive pressure for

- Life support conditions above 34,000 ft

- Medical conditions, especially suspected hypoxia

- Whenever smoke and/or other harmful gases are present in the cabin

NB: Normal oxygen supply mixes oxygen with ambient air on demand. Therefore any cabin fumes would be mixed and supplied to the crew if a normal oxygen supply were selected. Emergency 100 percent oxygen supplied under positive pressure doesn't mix ambient air and therefore ensures a nonharmful and constant oxygen supply.

Whenever a crew oxygen bottle has been discharged as a result of excess pressure, the following indications will be present: First, a green disk on the aircraft's fuselage is blown out whenever excess pressure occurs in the crew's oxygen cylinder.

NB: Excess pressure in the crew oxygen cylinder causes a ruptured disk to explode, allowing the oxygen to discharge through a pipeline to the outside of the aircraft's fuselage, where it blows out the green indicator disk.

Second, oxygen pressure gauges in the cockpit will indicate below normal or zero on most system types.

Pan Pan   The radio-telephone spoken word *Pan Pan* indicates that the commander of an aircraft has an urgent message to transmit concerning the safety of the aircraft or person(s) onboard or another aircraft, person, or property within sight. A typical Pan Pan call from a pilot would be

1. Pan Pan Pan
2. ATC name, e.g., Atlanta Control
3. Aircraft call sign
4. Message, including relevant points (failures, actions being taken, and requirements)

PAPI Lights   Precision approach path indicator (PAPI) lights are a development of the VASI system which also uses the red/white signals for guidance to maintain the correct approach path. However, the lights are arranged differently.

The PAPI system has a single wing bar, which consists of four light units on one or both sides of the runway, adjacent to the touchdown point. The colors seen by the pilot indicate whether the aircraft is right on slope, too high, or too low.

Figure 148   PAPI slope guidance.

PAR (Precision Approach Radar)   A PAR is similar to a surveillance radar approach (SRA) except it has a different ground-based radar system that provides returns in vertical (slope) navigation as well as horizontal (azimuth) navigation. Therefore the controller can give voice instructions in both horizontal and vertical parameters.

This enables the PAR to be a much more accurate and controlled approach system, and as such it is classified as a precision approach with a corresponding lower MDA. The PAR system minimum is 200 feet MDH.

Parallel Circuit   A parallel circuit is one in which there are two or more alternative paths which subsequently reunite. A feature of this arrangement is that the total current is shared between the parallel circuit paths at its first junction and then restored to a single value at a second junction. The word *parallel* is also applied to the way electrical sources, resistors, and loads are connected to different circuit paths (branches). In this way, the failure of one device is somewhat isolated and does not affect them all.

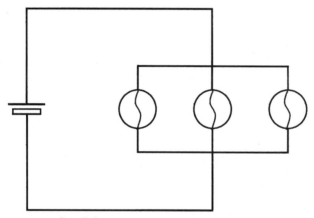

Figure 149   Parallel circuit.

Parallel Yaw Dampers   Parallel yaw dampers (early type) apply rudder control through the same control run as the pilot, and their activity is reflected in the rudder bar activity. (It moves the rudder pedals.) While this provides a visual indication of the yaw dampers' serviceability, it also increases the rudder loads experienced by the pilot. To prevent this making matters worse in the event of an engine-out failure on takeoff, or a crosswind landing, the damper can be switched off for takeoffs and landings.

As this damper in effect parallels the pilot's actions, it has become known as the parallel yaw damper.

Parallels of Latitude   Parallels of latitude run east/west on the earth. They are described by their angle above or below the equator. The units of measurement used to describe the position of a parallel of latitude are degrees, minutes, and seconds (1 degree = 60 minutes, and 1 minute = 60 seconds). The maximum possible latitude is 90° north or south (i.e., north and south poles).

See Fig. 150 at the top of the next page.

A change of longitude or east/west distance (along a parallel of latitude), has a varied distance (nm) per degree at different latitudes. The distance along the equator is measured as 1 nm equal to 1 minute. As we move away from the equator, the

Figure 150   Parallels of latitude.

east/west distance decreases for the same change in longitude, until eventually at the poles the distance between the meridians has reduced to zero.

Therefore the change-of-longitude distance decreases as latitude increases, and this is known as *departure*.

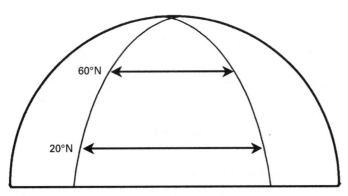

Figure 151   East/west distance with an increase in latitude.

Payload   Payload comprises passengers and/or cargo.

Penalties   The penalties experienced at very high altitudes are
1. Restricted operating speed range
2. Reduced maneuverability
3. Reduced aerodynamic damping
4. Reduced stability

Performance A Aircraft   For an aircraft to be categorized as a performance group A aircraft, it must be able to suffer a total loss of power from its most critical power unit at any stage of flight without this loss resulting in a forced landing. In essence, it must be able to retain directional control and climb to MSA. Multiengine aircraft greater than 5700 kg are normally categorized as performance group A aircraft.

Performance group A is the most stringent of all the various aircraft performance group categories.

Performance C Aircraft   Multiengine aircraft less than 5700 kg are normally categorized as performance group C aircraft. Performance C aircraft have to meet the following performance levels:

- When in VMC conditions, in a 75-m domain (NTOFP) on either side of the extended centerline, with an engine failure, the aircraft would use a see-and-avoid obstacle clearance principle.
- When in IMC conditions, an aircraft with an engine failure must be able to stay with a 75-m plus one-eighth horizontal distance traveled domain (NTOFP) on either side of the extended centerline and climb up to 1500 ft AAL, so that the aircraft would clear any obstacle.

Performance—Stopping Distance   *See* Stopping Distance—Performance.

Piston Engine   The piston engine converts the heat/pressure energy of the fuel it burns to mechanical energy. The conversion of heat (pressure) energy to mechanical energy is accomplished by inducing an air/fuel mixture (known as a *charge*) into a cylinder, which is then compressed by a piston. An electric spark causes the mixture to ignite, and this combustion causes a rapid rise in the temperature and pressure of the gases, which forces the piston down the cylinder. The linear movement of the piston is converted to a rotary movement of the crankshaft by the connecting rod. The burned gases are then exhausted into the atmosphere.

An aero piston engine suffers from three main disadvantages:

1. A piston engine suffers from a lack of performance/power output with increased altitude because the normally aspirated piston engine works on the principle of constant volume and not constant pressure.

It can be seen that the lower the pressure in a constant-volume cylinder, the lower the power output. The pressure experienced in the cylinder depends upon the amount of the air/fuel mixture (charge) that can be induced into the cylinders during the induction stroke/cycle, and the air charge is dependent upon the atmospheric pressure experienced, which decreases with altitude. For example, at an altitude of 20,000 ft AMSL the air pressure/density is approximately one-half that of sea level. This results in a corresponding reduction in the weight of charge in the normally aspirated piston cylinder, which results in a decrease in the overall power output. This lack of power output with increasing altitude renders the aircraft unable to produce the required power to overcome its own friction at a relatively low altitude, e.g., at approximately 25,000 ft. This results in a lower TAS and therefore a shorter range.

2. Produced airspeed is low due to propeller rpm limitation. (*See* Propellers.)

3. A piston engine is a mechanically inefficient engine, especially when compared to the gas turbine, because it is a complex, multipart moving piece of engi-

neering. This complexity lends itself to high maintenance demands. Also it produces a comparatively low power output which cannot be increased with larger cylinders because the increased engine weight is more penalizing than the increased power output obtained. That is, the ratio of power output performance to engine weight is limiting. This results in high maintenance demands and low power output/unit weight ratio.

Piston Engine—Engine Power Monitored/Indicating System *See* Engine Power Monitored/Indicating Systems—Piston Engine.

Pitching Moment—Lift/Weight Couple A pitching moment around the lift/weight forces occurs if they are not acting through the same point (line). This will cause either a nose-up or a nose-down pitching moment, depending on whether the lift is acting in front of or behind the center of gravity (cg) point.

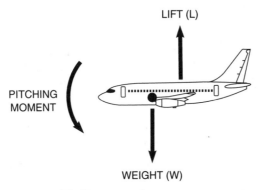

Figure 152 Pitching moments.

NB: A cg forward of the center of pressure (cp) has a nose-down pitching moment. A cg aft of the cp has a nose-up pitching moment. A cp moves if the angle of attack changes, and the cg moves as the weight changes (due to fuel being used). Therefore their positions will vary during a flight.

Whenever the pitching moment of the lift/weight couple is not perfectly balanced, extra forces are provided by the horizontal tailplane to center the aircraft's pitching moment.

See Fig. 153 at the top of the next page.

NB: Lift forward of weight has a nose-down pitching moment, which is counterbalanced by the downward deflection of the horizontal tailplane, which creates a nose-down counterpitch. Ergo, lift aft of weight requires the opposite balance.

The tailplane force has a turning moment in the pitching plane (nose up or nose down) about the lateral axis at the cg point. Its effectiveness depends on its size and the length of its moment arm from the cg.

Any movement of the cp on the main lines (through which the lift force acts) or of the cg (through which the weight acts) will require a different balancing force from the horizontal tailplane. The balancing lift force that the tailplane can produce depends upon its airspeed. Therefore it determines just how much movement

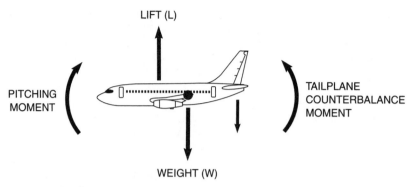

Figure 153 Pitching moment balanced by the horizontal tailpipe.

of the cg and therefore the length of its moment arm to the cg can be balanced, thus establishing a forward and aft cg limit.

Pitching Moment—Thrust/Drag Couple If the forces of thrust and drag are not acting through the same point (line), then they will set up a moment causing either a nose-up or nose-down pitch, depending on whether the thrust is acting above or below the dragline.

Figure 154 Thrust/drag couple pitching moment.

Therefore a change in thrust (increase or decrease) in straight and level flight can lead to a pitching tendency of the aircraft. Or likewise an increase or decrease in drag can lead to a pitching tendency of the aircraft. For example, an increase in thrust on an aircraft with engines mounted under the wing, with a higher dragline, will cause a nose-up pitch, as thrust is increased.

It is possible that the lift/weight couple and the thrust/drag couple will balance each other and that there will be no residual pitching moment. For example, thrust/drag causes a nose-up pitching moment, which balances the nose-down moment due to the lift/weight couple.

See Fig. 155 at the top of the next page.

Pneumatic Systems A pneumatic system uses compressed air as a fluid/gas to exert a pressure/force to move a surface.

The advantages of a pneumatic system over a hydraulic system are that

1. Air weighs less than hydraulic fluid, which is beneficial to the aircraft's overall weight limitations.

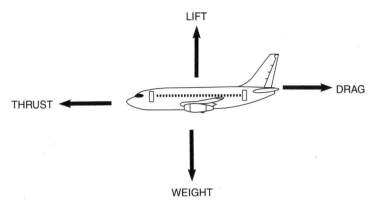

Figure 155   Thrust/drag and lift/weight couple pitching balance.

2. There are no problems of availability and cost with air, which could exist with hydraulic fluids.

PNR (Point of No Return)   The PNR is the last point on a route at which it is possible to return to the departure aerodrome with a sensible fuel reserve.

Normal PNRs are based on the aircraft's safe endurance. The all-engine PNR formula is

$$\text{Time to PNR} = \frac{EH}{OH}$$

where E = safe endurance time
$\quad\ H$ = ground speed home
$\quad\ O$ = ground speed out

$$\text{Distance to PNR} = \text{time to PNR} \times \text{ground speed out } O$$

One-engine PNR distance is calculated as follows:

1. Calculate a 1-nm round-trip fuel flow (all-engine ground speed out and one-engine inoperative ground speed home).
2. Divide safe endurance by 1-nm round-trip fuel flow.

The one-engine PNR time is distance divided by ground speed out.

PNR calculations are important for aircraft where diversion airfields are not readily available, i.e., over large water areas such as the Pacific Ocean. It is crucially important to have a PNR point if you elect to carry island holding fuel instead of diversion fuel, because you become solely committed to landing at your destination once you pass the PNR.

NB: PNRs are not really required on a route over land with diversion airfields available en route.

Polar Stereographic Chart   A polar stereographic chart is a flat projection tangent to the pole. Therefore scale and convergency are correct only at the pole, and the farther you move away from the pole, the greater the error in both scale and convergency.

Rhumb lines are curved away from the poles, akin to parallel lines of latitude, and great circle tracks are assumed to be straight lines. This chart is used for polar navigation.

Position Report   An aircraft under IFR which flies in or intends to enter controlled airspace must report its time, position, and level at such reporting points or at such intervals of time as may be notified or directed by ATC.

A typical position report is

November Four Six Five Foxtrot Tango	aircraft identification
Haven—Two Six	position and time
Flight Level Three Five Zero	level
Key West—Five Eight	next position and estimate

Precipitation   Precipitation refers to falling water, in either a liquid or ice state, that reaches the ground. This includes the following types:

1. *Rain* is defined as water drops from 0.2 to 5.5 mm in diameter.

2. *Drizzle* is defined as fine water drops less than 0.2 mm in diameter.

3. *Snow* is defined as precipitation consisting of ice in the form of crystals, grains, needles, or snowflakes.

4. *Hail* is defined as precipitation consisting of small balls or pellets of ice.

5. *Sleet* is defined as a mixture of rain and snow precipitation.

6. *Freezing rain or drizzle* is defined as water droplets which freeze on contact with a cold surface, i.e., an aircraft or the ground.

It is possible to use precipitation as a means of identifying the cloud type and its associated hazards, e.g., showers/intermittent precipitation, which generally falls from cumuliform cloud types, and continuous precipitation, which generally falls from stratiform cloud types.

Precision Approach Radar (PAR)   *See* PAR (Precision Approach Radar).

Pressure (Air)   Air pressure is the weight of a column of air or the gravity force of air molecules. The air is made up of molecules of gases and water vapor; and like all types of molecules, they have a mass and therefore occupy a space. Gravity pushes these molecules down to the earth's surface, which results in a greater weight of air and therefore a greater pressure toward the earth's surface. Therefore air pressure decreases with a rise in altitude. Air pressure is influenced by gravity and is proportional to density.

Air pressure can be measured by using a simple mercury barometer, and the pressure experienced can be expressed as either inches of mercury (inHg, as used in the United States) or millibar (mbar)/hectopascal (hPa). These are both a unit of pressure force of 1000 dynes per square centimeter (dyn/cm^2).

NB: The hectopascal is the standard unit of pressure measurement recognized by the ICAO.

Pressure Altimeter   *See* Altimeter—Pressure.

Pressure Altitude   Pressure altitude or pressure height is the ISA height above the 29.92-in/1013-mbar pressure datum at which the pressure value experienced represents that of the level under consideration.

Pressure altimeters are calibrated to ISA conditions, and they measure air pressure values, which they interpret to a corresponding altitude/height. Therefore a particular ISA calibrated pressure always corresponds to a particular height, and when this altitude is measured above the 29.92-in/1013-mbar pressure datum, this is known as a *pressure altitude*. Pressure altitudes are often described as flight levels, where the final two zeros are omitted; e.g., a pressure altitude of 6500 ft is expressed as a flight level and written as FL65.

Any difference in the actual mean sea-level (MSL) pressure from the ISA 29.92-in/1013-mbar pressure datum will give rise to a difference between the pressure altitude and the actual altitude. Therefore, when flying at a pressure altitude (flight level) the pilot needs to know what the actual height is for (1) ground clearance and (2) calculating the performance capabilities of the aircraft.

NB: Traffic vertical separation should not be affected by a difference between the pressure altitude and the actual height because all aircraft should be on the same pressure subscale setting.

To calculate a pressure altitude's actual height, you have to calculate the difference between the regional QNH and the 29.92-in/1013-mbar pressure datum and convert this pressure difference to a height. Then add or subtract this height from the pressure altitude, depending upon whether the region QNH is above or below the 29.92-in/1013-mbar pressure datum. (Assume that 0.0295 inHg/1 mbar = 30 ft unless otherwise instructed, and that pressure decreases with height.)

The best explanation of this would be an actual example. Given a cruising altitude of FL40 and a regional QNH 29.24 inHg/990 mbar, what is the actual height of the aircraft?

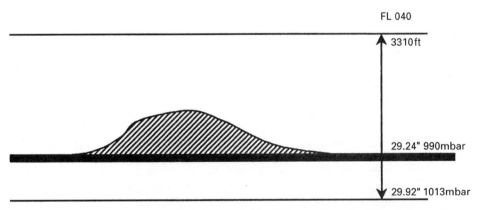

Figure 156   Pressure altitude actual height calculation. Difference in inches/mbar between the region QNH and pressure altitude datum: 29.92 in − 29.24 in = 0.68 in/1013 mbar − 990 mbar = 23 mbar. Difference in feet between the region QNH and pressure altitude datum: 0.68 in ÷ 0.0295 × 30 ft = 23 × 30 = 690 ft. Actual AMSL height: 4000 ft − 690 ft = 3310 ft.

Pressure Altitude—Error   The pressure altitude error experienced on the altimeter is called the *barometric pressure error*. Barometric pressure error is the effect of flying into an area with a different sea-level pressure that gives rise to an incorrect altimeter reading. An aircraft would normally fly at a standard 1013-mbar/29.92-inHg pressure datum, and therefore when it flies into an area with a lower actual sea-level pressure, it has the effect of overreading the aircraft's altitude. Namely, the aircraft is at a lower height than indicated. Hence the adage "High to low beware below."

Conversely, the opposite is true when flying from a low- to a high-pressure area.

Adjusting the pressure subscale setting to the correct regional QNH will indicate the correct actual height, but this is not possible if you are flying on the standard 29.92-in/1013-mbar setting.

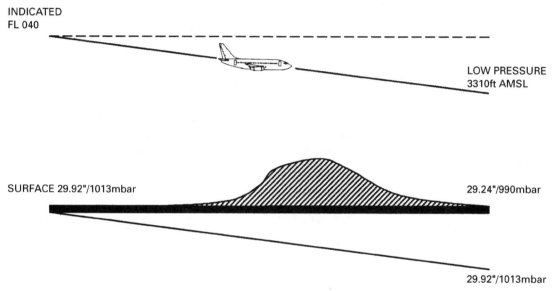

Figure 157   Pressure altitude error.

Pressure Gradient Force   A pressure gradient force is a *natural* force generated by a difference in pressure across a horizontal distance, i.e., gradient, between two places. It acts at right angles to the isobars, and it is usually responsible for starting a parcel of air to move from an area of high pressure to an area of low pressure.

In the simplest case, a parcel of air movement is represented as a wind that blows across the isobars from an area of high pressure to an area of low pressure. The greater the difference between their pressures, the greater the wind strength. Thus, if the isobars are very close together, rapid changes in pressure occur, the pressure gradient is said to be *steep,* and the resultant wind is *strong.* If they are widely spaced, the pressure changes are less marked, and so the pressure gradient is *flat* and the resultant wind is *weak.*

NB: This is a *simple* pressure-driven wind, and *not* the gradient wind.

Pressure System   A pressure system is a circulating air mass that is classified as either a low or a high, which relates to the direction of pressure change toward the center of the air mass at the surface. A low-pressure system will typically have more than one air mass that includes a warm air mass incorporated into a cold unstable air mass, with fronts in between. A high-pressure system is typically made up of a single stable air mass.

Pressure System—Airflow   The airflow between low- and high-pressure systems is pressure-driven. Remember, a pressure system is classified as high or low in relation to its surface pressure. However, the upper-level pressure system is reversed, and in describing the airflow between a low- and a high-pressure system we have to examine both the surface and upper-level pressure systems and related pressure-driven airflow.

As a random starting point, at the surface of a high-pressure system the air flow outward (divergence) at low level into the center of a surface low-pressure system (convergence). Here it rises to an upper-level high and flows outward (divergence) and into an upper-level low (convergence), where it subsides down to where it started, the surface high pressure.

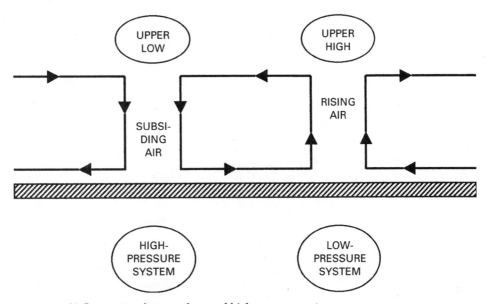

Figure 158   Airflow pattern between low- and high-pressure systems.

Primary Cell Battery   *See* Battery.

Primary (Pulse) Radar   *See* Radar—Primary (Pulse).

PROB   PROB, or probability, is used in weather forecasts when the forecaster is uncertain if the weather conditions will exist and therefore assesses the probability of their occurring as less than 50 percent. (If it is greater than 50 percent, then it would be listed as a tempo.) For example, PROB 30 means a 30 percent chance of the weather conditions occurring.

Procedure Turn   There are two types of procedure turns
1. The 45° procedure turn consists of
   - An outbound track from the fix (usually the reciprocal of the approach track) to a set distance or timed point

■ An outbound turn of 45° away from the outbound track for either 1 min (or 1 min 15 s) from the start of the turn or for 45 s from wing level once the turn has been completed (plus or minus wind correction in terms of drift and time)

NB: Left or right in a description of the procedure turn refers to the direction of this initial turn.

■ A 180° turn in the opposite direction to intercept the inbound track

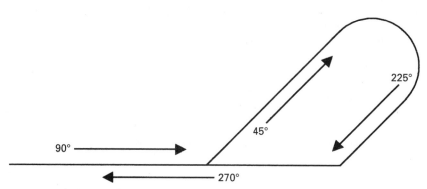

Figure 159 A 45° procedure turn.

2. The 80°/260° procedure turn consists of
■ An outbound track from the fix (usually the reciprocal of the approach track) to a set distance or timed point
■ An outbound turn of 80° away from the outbound track
■ Followed almost immediately by a 260° turn in the opposite direction to intercept the inbound track

NB: If the initial turn is into a strong headwind, then the heading can be held for a brief time before the 260° inbound turn is commenced. For a tailwind, stop the turn slightly before completing the 80° turn, and gently roll immediately onto the inbound turn.

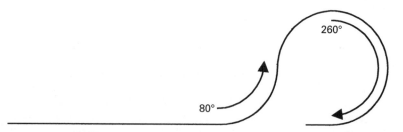

Figure 160 The 80°/260° procedure turn.

Procedure turns are used when no suitable fixes permit a direct entry onto an instrument approach procedure.

Propeller(s)   The propeller produces a forward thrust (horizontal lift) by employing Newton's third law: The application of a force upon a body will cause an equal and opposite reaction.

The propeller is connected to the engine via a prop shaft that rotates the propeller, which generates an accelerated mass of air rearward, thereby converting the shaft horsepower of the engine to a thrust force. The thrust force developed by the propeller is equal to the mass of the air times the rate of change in momentum given to the air, which has the reaction of driving the aircraft forward

$$\text{Thrust force} = \text{air mass} \times \text{velocity}$$

Air mass There are three interactive features of the propeller which directly determine the size of the slice of air made by the propeller blade, which determines the amount of the air mass's rearward displacement and indirectly its velocity, and therefore the thrust force produced: propeller blade angle, angle of attack, and pitch.

1. *Propeller blade angle*

   The propeller blade angle is similar to the angle of incidence of an aerofoil section. It is the angle between the chord line of the blade and the plane of rotation (usually vertical) of the propeller.

   The blade angle determines the angle of attack and the pitch of the propeller blade, which determines the mass of air accelerated rearward and therefore the thrust force produced. There is a reduction of blade angle with increasing diameter toward the blade tips in order to maintain a constant angle of attack along the blade and a constant geometric pitch.

2. *Angle of attack*

   The propeller angle of attack is the angle between the chord line of the blade and the relative airflow, as with an aerofoil. Therefore it is initially set by the blade angle of the propeller. Like the wing, the propeller aerofoil operates more efficiently at a small angle of attack, approximately 3° to 5°. It produces its maximum thrust force/torque force ratio at this optimum angle of attack. A higher angle of attack will produce greater thrust but also more resistant torque for the same forward velocity.

   Angle of attack decreases with an increase in forward airspeed (due to resistant forces experienced); therefore the efficiency of a fixed-blade/pitch propeller will decrease with an increase in forward airspeed away from its optimum setting.

3. *Pitch*

   The pitch is the setting of the blades (the blade angle) with reference to the plane of rotation, which determines the amount of forward distance covered during one complete revolution, and thereby the amount of the air mass generated rearward. The pitch has two components.

   a. *Forward travel geometric / effective pitch*

      The geometric pitch of the propeller is the distance the propeller would travel forward in one complete revolution if it were moving through the air at its given blade angle.

      For a fixed-pitch propeller the geometric pitch is a constant value, whatever the operating conditions.

      The effective pitch (or advance per revolution) of the propeller is the actual distance the propeller would travel forward in one complete revolution. In flight the propeller does not move through the air at the geometric pitch, and the difference between the geometric pitch and the effective pitch is called the *slip*. The effective pitch is a shorter distance than the geometric pitch, due to the resistance of the air experienced.

The greater the forward travel, the greater the new high-energy air mass displaced and therefore the greater the resultant thrust.

b. *Blade angle of attack / pitch turning effort*

Alteration of the blade angle of attack/pitch will affect the forward distance traveled as well as the torque or turning effort required. That is,

Fine pitch (low blade angle)
= low turning effort required but a low forward travel
Coarse pitch (high blade angle)
= high turning effort required but a higher forward travel

Therefore in general terms, to increase thrust/airspeed the propellers should be moved to a coarse pitch position, which increases the amount of the air mass sliced rearward.

Also a low air density provides less resistance to the rotating blades, which allows the blade rotational speed to increase for the same engine power. Therefore a coarser blade angle will maintain the rpm and thrust produced. Remember that a blade is at its most efficient just below its maximum rpm setting. Above its maximum rpm, the blade tip speed is supersonic and compressibility loss (drag) occurs. Therefore a variable pitch presents this.

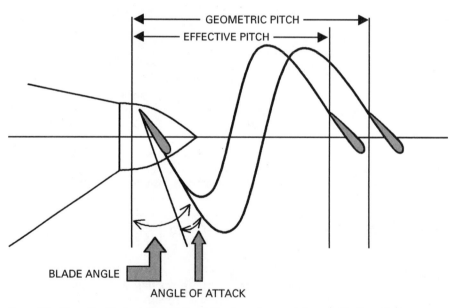

Figure 161 Propeller blade angle, angle of attack, and geometric and effective pitch.

Air velocity The velocity at which the air mass is displaced rearward is also critical in the thrust force equation. The air's velocity is a function of the rpm speed, with the propeller at its most efficient when the speed of the tips is just below sonic speed. The rotational speed is governed

■ Directly by the engine power supply.

■ Indirectly by the blade angle/pitch. That is, with a coarse blade angle/pitch the larger air mass displaced puts a greater load on the propeller, thereby causing the propeller rpm to be reduced, which therefore requires a higher engine power supply to maintain a given rpm. With a fine blade angle/pitch the opposite is true.

The blade angle/angle of attack for a fixed-pitch propeller is only correct for one specified engine speed—aircraft forward speed and altitude, where the propeller will convert the power of the engine to thrust with maximum efficiency. Therefore for a fixed-pitch propeller a compromise must be made between the blade angle (and therefore the angle of attack and pitch) desirable for takeoff and that required for cruise condition. Generally, the angle best suited to high-speed cruise conditions (high blade angle/coarse pitch) is chosen, because it will spend the greater part of its working life here. Consequently the propeller efficiency is poor for take-off performance, where it requires a low blade angle/fine pitch.

The thrust force produced by the propeller can be resolved as follows:

1. The forward thrust component/force in the direction of flight is produced by the propeller moving through the air. The propeller produces forces in the same way as a wing, and this is manifested as pressure forces on the propeller blades in the direction of flight. This pressure force is determined by
   a. Angle of attack (more efficient at small angle of attack)
   b. The relative airflow velocity, which is made up of two components:
      (1) The forward speed of the aircraft
      (2) The rotational speed of the propeller
         The velocity at which the relative airflow meets any section of the propeller blade determines the size of the force produced, again as for a wing.
2. The torque component/force, which acts about the propeller axis in the plane of rotation and opposes the rotation of the propeller, is a combination of the following various effects on the propeller:
   a. Centrifugal force
   b. Torque bending force/torque reaction
   c. Thrust bending force
   d. Aerodynamic twisting force
   e. Centrifugal twisting moment/force
3. The direction and the relative size of the resultant force components (forward thrust and torque) depend upon the propeller blade angle of attack, which produces the greatest ratio of thrust force to torque force at low angles of attack and determines the path of the blade through the air.

See Fig. 162 at the top of the next page.
The amount of force created by the propeller is governed by two factors:

■ The output of the engine, which drives the prop shaft
■ The blade angle, angle of attack, and pitch of the propeller

The degree to which the propeller design can be increased to absorb engine power is restricted in the following areas.

1. Blade length is restricted because of
   a. The need for adequate ground and fuselage clearance
   b. The need to maintain subsonic blade tip speeds

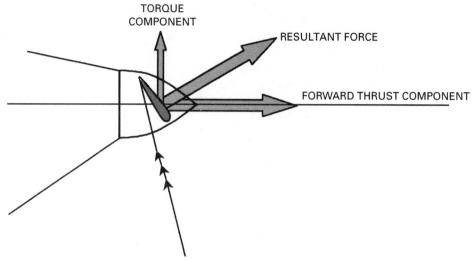

TORQUE
COMPONENT

RESULTANT FORCE

FORWARD THRUST COMPONENT

Figure 162   Propeller forces.

(The longer the propeller blade, the greater the tip speeds; and supersonic propeller speeds are inefficient.)

2. Blade chord size is restricted because
   a. An increase in the chord size will reduce the aspect ratio (blade diameter/chord) and, as with the wing, this gives a lower-efficiency output than the optimum ratio.
   b. Larger chord lengths also increase the centrifugal twisting moment, which tends to twist the blade to a finer pitch, causing higher loads on the root fittings and higher torque values, which give a lower resultant force/lower thrust output. (Forward thrust force minus torque force equals resultant force.)
3. The number of blades can be increased to a certain value. But with more blades the hub diameter and weight become excessive, and blade interference begins to reduce efficiency.

The advantages of a propeller are that

1. The propeller creates a high-energy slipstream, which has three main effects on the aircraft.
   a. The slipstream creates extra lift over the wing.
   b. The slipstream suppresses the stall speed of the aircraft.
   c. The slipstream makes the fin/rudder more effective.
2. The propeller/piston engine has a quick response rate to a throttle input, which gives an early application of the slipstream effects on the aircraft.

Therefore a propeller-driven aircraft has good recovery qualities from slow airspeed situations. The disadvantages of a propeller is that the aircraft suffers from a lack of airspeed due to propeller rpm limitations, as a result of propeller compressibility losses. This is so because the propellers suffer the effects of compressibility when the speed of the propeller blade tips becomes sonic. Therefore the propellers have a revolution speed limited to just below sonic speed, which limits the thrust force produced by the propeller and therefore the aircraft's airspeed. This results in a lower TAS and therefore a shorter range.

Propeller Blade   The propeller blade is twisted along its length to maintain a constant blade angle of attack. This is required because due to the length of the blades, the outer extremities, the tips cover a greater distance than the center when the propeller is rotating. Thus although the propeller rpm is the same throughout the length of the blade, the distance traveled rapidly increases toward the tips, resulting in a greater rotational speed in this area. An increase in revolution speed increases lift on the aerofoil propeller blade section (centrifugal forces). Therefore it becomes necessary to reduce the blade angle increasingly toward the tips by twisting the blade in order to maintain the same angle of attack along the blade length. This ensures that every section of the blade meets the relative airflow at its most efficient angle of attack.

If the blade angle were not twisted to maintain a constant angle of attack along its length, then the angle of attack would increase toward the tips. There would be a tendency for the tips to have a greater forward travel and a greater torque value than the remainder of the blade with a consequent reduced thrust/torque ratio. This would result in a loss of efficiency, raising the level of stress and damage experienced on this section of the propeller blade.

Propeller(s)—Braking Thrust   It is possible to increase the braking effect on some aircraft types by reversing the pitch of the propeller to a negative angle, causing it to produce thrust in the opposite direction.

Propeller Efficiency   The propeller efficiency in producing thrust to propel the aircraft forward is the ratio of the useful work done by the propeller (propeller thrust) in moving the aircraft to the work supplied by the engine (engine brake horsepower).

$$\text{Propeller efficiency} = \frac{\text{prop thrust (air mass} \times \text{velocity)}}{\text{engine brake horsepower (bhp)}}$$

The highest efficiency obtained by a propeller is usually 85 to 88 percent. The brake horsepower is proportional to the torque of the propeller; so the efficiency depends upon the ratio of the thrust force to the torque force, and this ratio depends on the blade angle of attack. Propeller efficiency varies with angle of attack, with optimum propeller efficiency occurring at the optimum angle of attack of approximately $3°$ to $5°$.

For a fixed-pitch propeller, the blade angle of attack for any rotational speed will depend directly on the aircraft's forward airspeed. The efficiency can therefore be shown to vary with speed at a constant rpm; i.e., blade angle of attack decreases with an increase in forward airspeed. The propeller's maximum efficiency will only occur at one rpm speed, and the blade angle will usually be set so that the rpm speed for maximum efficiency is close to the cruising speed (i.e., coarse blade setting). At other speeds, for example, at takeoff speed, the propeller efficiency will be relatively low; i.e., only a small proportion of the engine power delivered is used to propel the aircraft. The takeoff performance therefore suffers if the blade is set to coarse for the cruise condition, and vice versa.

Propeller—Feathering   *See* Feathering.

Propeller—Fixed-Pitch   *See* Fixed-Pitch Propeller.

Propeller—Noise   *See* Noise—Propeller.

Propeller—Variable-Pitch   *See* Variable-Pitch Propeller.

Q

QDM QDM is the magnetic bearing (radial) to the station.

QDR The magnetic bearing (radial) from the station is the QDR.

QFE This zeros the altimeter on the airfield elevation datum. There are two types of QFE.
1. Airfield QFE is measured at the highest point on the airfield.
2. Touchdown QFE is measured at the touchdown point of the runway in use for precision approaches.

_Q_ Feel Q feel is a sophisticated computer-based artificial feel system based on $^1/_2\,\rho V^2$, that is felt by the pilot through the control column and rudder pedals and is commonly used on aircraft with powered flight controls (i.e., elevator, rudder, and ailerons). It meets the requirements of progressive feel to match variable control surface deflection at a constant speed and/or for a constant angle of deflection at varying speeds.

The inputs to a Q feel system are the static and dynamic pressure and the control surface angle of deflection.

QFF This is similar to QNH except that it uses the *actual conditions* (not ISA) to find the sea-level pressure, not ISA. It is more commonly used by meteorologists than by pilots.

QGH Letdown A QGH is a type of VDF letdown, usually available from military controllers. The controller passes headings to steer and descent instructions to the pilot, similar to a radar letdown. The pilot does not have the responsibility of allowing for drift to bearings, etc., which is the case for a normal DF letdown procedure. A QGH letdown procedure has no approach chart, although minimum descent heights still apply.

QNE This is not an altimeter setting, but it is the height shown at touchdown on the altimeter with 29.92 in or 1013 mbar (hPa) set on the subscale. It is used at very high aerodromes where QFE pressure is so low that it cannot be set on the altimeter subscale.

QNH This is a local altimeter setting that makes the altimeter indicate the aircraft's altitude above mean sea level (AMSL) and therefore the airfield elevation. There are two types of QNH: airfield QNH and regional QNH, which is the lowest forecast QNH in an altimeter setting region. QNH is QFE reduced to sea level using ISA values for the calculation.

QTE This is the true bearing from the station.

QUJ This is the true bearing to the station.

Radar The name *radar* was devised from *ra*dio *d*etection (direction) and *r*anging.

Radar Advisory Service The radar controller will use radio communications to provide traffic information and advisory avoiding action necessary to maintain separation from other aircraft.

Radar Control Service This is available wherever radar coverage exists in controlled airspace, i.e., airways, terminal control zones, aerodrome traffic zones, and control areas whereby ATC is responsible for

1. Monitoring and separation from other aircraft

2. Radar vectoring

3. Controlled airspace crossing

4. Navigation assistance

5. Weather information

6. Hazard warnings

7. Emergency assistance

NB: However, it remains the responsibility of the aircraft commander, *not* ATC, to maintain terrain clearance even when under radar control.

NB: Instructions from ATC under a radar control service have to be adhered to; i.e., ATC instructions are mandatory, unless in exceptional circumstances the commander believes the ATC instructions will create a danger to the safety of his or her aircraft, say for CFIT or TCAS RA warnings.

Radar Information Service The radar controller will use radio communications to provide traffic information only (no avoiding action will be offered).

Radar—Primary (Pulse) Primary (pulse) radar works on the reflected signal principle. When the transmission of electromagnetic radio energy encounters an object, it is reflected back to the point from where it was originally transmitted; therefore it is used to detect aircraft. It does not require any specific equipment in the aircraft. Primary radar uses UHF/VHF frequencies, which have minimal static and atmospheric attenuation, and a line-of-sight propagation path. Primary radar is also known as pulse radar and reflective radar.

Primary radar is solely a ground-based system that consists of a combined transmitter and receiver parabolic dish, which detects airborne objects in terms of direction and range. The direction of the reflected signal can be determined by slowly rotating the radar dish, during which time it will transmit millions of pulses in a narrow radar beam and will receive almost instantaneously any reflected returns. The angle of the radar antenna, relative to north at the time the echo is received, indicates the horizontal direction (azimuth) of the object, i.e., the QDR to the object.

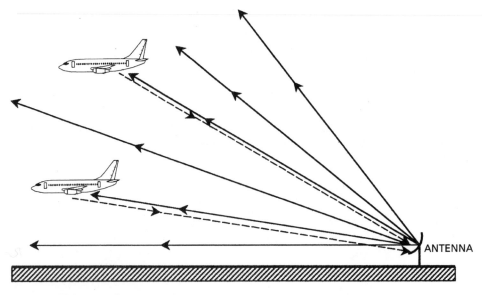

Figure 163 Primary radar returns.

The range can be determined by measuring the elapsed time of a reflected echo, between the transmission of a pulse of radio energy and its reception back at the source. It is a simple calculation to determine the distance or range of the object causing the echo.

$$\text{Distance} = \frac{\text{speed} \times \text{time}}{2}$$

NB: Given that, electromagnetic energy travels at the speed of light, or 300,000 km/s, or 162,000 nm/s.

The lapsed time represents a return journey (once out, once back). Therefore the time needs to be halved to equate to a one-way journey time.

Primary radar returns, i.e., aircraft, are shown on cathode-ray tube displays as echoes. This information can be used for a number of aviation purposes, such as surveillance radar that provides a ground radar controller with an overview of the area of responsibility, radar vectoring maneuvers, and surveillance radar approach (SRA). For example, ATC provides the pilot during a SRA landing with azimuth tracking guidance (e.g., steer heading) and desired height information (e.g., 4 nm to touchdown, should be at 1200 ft AGL).

NB: SRAs are flown on QFE.

Atmospheric attenuation, antenna power, and the height of the aircraft all affect the range capabilities of a primary radar system. Atmospheric attenuation reduces the range of a primary radar system.

NB: The greater the atmospheric attenuation, the more power required to achieve the same range.

The greater the power output of the antenna. the greater the range of the transmitted radio energy.

NB: To double the range, you have to quadruple the power.

Because of the curvature of the earth, the higher the aircraft, the greater the distance at which it can be detected by radar.

$$\text{Radar range (nm)} = \sqrt{1.5 \times \text{aircraft height (ft AGL)}}$$

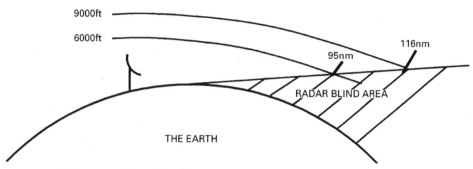

Figure 164 Radar range/aircraft height.

Primary radar has the following operational disadvantages:

1. *Clutter.* Return echoes from high ground and/or heavy precipitation clutter the radar screen and may be similar to aircraft returns. Therefore during periods of heavy rain, primary radar may be significantly degraded. However, some primary radars incorporate an electronic moving-target indicator that shows only moving targets on the radar screen. Therefore this system helps to eliminate return echo clutter from stationary objects, e.g., high ground.
2. *Similar returns.* Returns from different aircraft are indistinguishable from each other. Therefore it can be very difficult for the radar controller to identify one particular aircraft versus another. A controller may request one aircraft to carry out a maneuver to distinguish its radar blip.

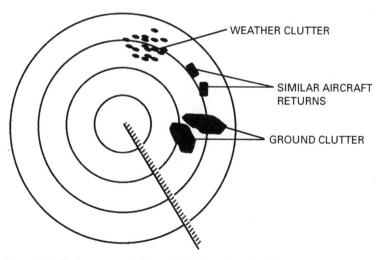

Figure 165 Radar screen clutter and similar aircraft returns.

3. Blind spots
   a. *Tilt of the antenna.* For an aircraft to be detected, the radar beam must be directed roughly toward it. If the radar is tilted up, then it will miss lower aircraft at a distance; or if the radar is tilted down, then it will miss higher aircraft nearer to the radar dish.
   b. *Curvature of the earth.* Because the radar uses a line-of-sight propagation path, it does not follow the curvature of the earth. Therefore distant aircraft at low altitude can be hidden from the radar by the curvature of the earth.
   c. *High terrain.* High terrain will reflect radar signals. This creates a blind spot behind any such terrain or manmade objects.

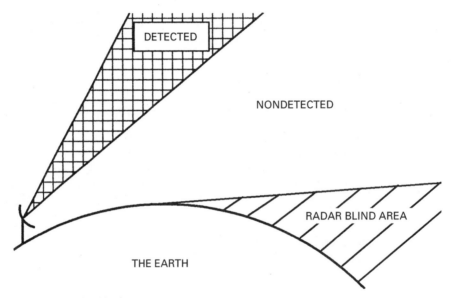

Figure 166 Radar blind spots.

Radar—Secondary (Surveillance) Secondary radar works on the respondent/reply signal principle. It requires the active participation of an aircraft that transmits a new return/reply signal back to the ground station, whenever it receives an interrogation signal. This is accomplished by using an active onboard device called a *transponder.* The transponder also allows the return signal to carry additional information, such as identification code and even the aircraft's altitude, which allows the radar controller to easily identify a particular aircraft on the screen. Therefore secondary radar is in fact two sets of radar talking to each other.

The secondary radar equipment consists of the following:

1. A ground antenna is a highly directional rotating ground radar aerial that transmits a coded interrogation signal, on a 1030-MHz frequency, asking an airborne transponder to respond. An individual station is recognized by its own unique set of transmission pulses.

NB: Because only a small amount of transmission energy is required from the ground to trigger the airborne transponder, the ground antenna can be quite small. Often it is mounted on a primary radar dish.

2. An airborne transponder is triggered by the ground interrogator signal to send an automatic secondary responding/reply pulse signal, on a 1090-MHz frequency, back to the ground station. The transponder reply carries coded information such as the aircraft's identity (i.e., its assigned numerical code).

NB: The secondary radar's response signal is much stronger than the simple reflected primary radar signal.

3. The ground receiver is in fact the ground antenna that also operates as a receiver. It receives any secondary responding signals and passes them back into the interrogator system. A decoder then accepts the signals from the interrogator, decodes them, and displays them on the radar screen. It displays an aircraft identification code, the altitude, and the ground speed.

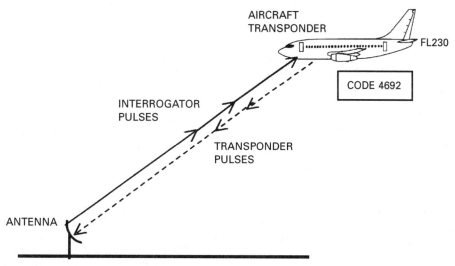

Figure 167 Secondary radar signals.

Secondary radar displayed information can be used for surveillance radar, providing the radar controller with an improved system of identification and return clarity, as opposed to the primary radar display, thus allowing radar vectoring maneuvers and surveillance radar approaches (SRAs). (ATC provides the pilot during an SRA landing approach with azimuth tracking guidance, e.g., steer heading, etc., and desired height information, e.g., 4 nm to touchdown, should be at 1200 ft AGL.)

NB: SRAs are flown on QFE.

The main advantage of secondary radar is that the reply pulse from the aircraft's transponder can carry coded additional information, such as the following:

1. *Positive identification.* ATC uses secondary radar to distinguish a particular aircraft from others in its vicinity by assigning a special code, such as 1800 or 4200, which the pilot dials into the transponder, known as mode A—Alpha.

2. *Abnormal situations.* Abnormal situations such as radio failure, aircraft hijacking, and emergencies are universally assigned special transponder codes, which are used to communicate the aircraft's situation to ATC. These codes are dialed into the transponder by the pilot in place of any specific identity code.

3. *Altitude/ground speed reporting.* Most modern transponders are equipped with altitude and/or ground speed reporting, known as mode C—Charlie. For this information to be displayed on the ground radar screen, the pilot has to select the mode C or ALT function on the transponder unit. It is a special encoding altimeter that feeds the altitude, which is always calibrated to 1013 mbar/29.92 inHg to the transponder for transmission. Likewise the ground speed is fed into the transponder and then transmitted to the ground radar screen.

4. *Ident.* Identing an aircraft's responding signal provides a positive identification of the aircraft on a screen when the pilot presses the transponder's ident button.

   The second main advantage of secondary radar is the improved display on the controller's radar screen, because the system is not degraded to the same extent as primary radar.

5. The improved radar screen display includes the following:
   a. There is minimal weather or ground clutter.
   b. There are minimal blind spots.
   c. Targets are shown as the same size and intensity, thereby making smaller aircraft returns easier for the controller to follow.
   d. The radar controller is able to select specific displays.
   e. Mode C altitude and reporting displayed information is used:
      (1) Ident code, e.g., 3828
      (2) Flight level
      (3) Aircraft ground speed
      (4) The flight's designator code, say, QF801

Radar Vectoring   Radar vectoring occurs when a radar controller passes to an aircraft a heading to steer, e.g., Delta 204, steer heading two seven zero.

NB: Bear in mind that the radar controller is trying to get you to achieve a particular track over the ground. However, not knowing precisely what the wind drift is, the controller will occasionally request a modification to your heading to achieve and/or maintain the desired track.

No radio navigation instruments are required in the aircraft for it to be radar-vectored, but radio communication is necessary. To terminate radar vectoring the controller uses the phase "Resume own navigation."

Radar vectoring is a very useful procedure in busy terminal areas where an aircraft may be vectored quite efficiently onto a final approach to land. Radar vectoring by the controller may cease once the aircraft is established on a visual approach or on an instrument approach such as an ILS; or it may continue down to an SRA or PAR minima.

Radar—Weather   *See* Weather Radar.

Radio Altimeter   Radio altimeters provide an accurate height measurement from 2500 ft down to 50 ft AGL for pulse radar beams, or 0 ft for continuous wave radar beams. They are usually fitted alongside barometric altimeters in most commercial aircraft. The basic principle of the radio altimeter is that a wide conical beam is directed vertically down toward the ground, and the time taken for the reflected signal to return corresponds to its height.

A radio altimeter can use one of two beams:

1. A *pulse radar beam* measures the time taken by the reflected pulse signal as a height.

NB: Pulses are transmitted at time intervals, and this limits the ability to accurately update its height measurement when the aircraft is close to the ground. This is true because the time between pulse transmissions can be large compared with the very small time delay for the returning signal. This means the height of the aircraft could have changed dramatically, and therefore may be momentarily inaccurate, before a new radio pulse is sent out to remeasure the aircraft's height.

2. A *continuous wave radar beam* varies the modulation of the transmission frequency. The difference between the frequency being transmitted and the one returning is the beat frequency, which is converted to a time, given the sweep rate that is converted to a height measurement.

NB: The radio altimeter's central frequency is 4300 MHz, and its frequency sweeps can be up to 100 MHz on either side. The frequency sweep rate is automatically adjusted by the radio altimeter to produce the most accurate results. A higher sweep rate is used at low heights to cope with the quick signal returns. Therefore a continuous wave (CW) radio altimeter is very accurate close to the ground, and a lower sweep rate is needed at greater heights to prevent ambiguous returns (i.e., the same frequency being transmitted before its previous transmission has been returned). Radio altimeters are accurate to within ±1 ft or 2 percent up to 500 ft AGL, whichever is greater; or ±1 ft or 3 percent up to 2500 ft AGL, whichever is greater.

Radio altimeter displays are either a separate dial or incorporated on an EFIS ADI. The radio altimeter height information is also fed into other systems such as GPWS and flight data recorders.

The radio altimeter normally becomes active at 2500 ft above the ground, for both separate dial and EFIS radio altimeter instruments.

NB: Some types of radio altimeter may become active at a different height.

On an EADI instrument the radio altimeter is often displayed during the approach as a digital countdown from 2500 ft. On some types at 1000 ft AGL the display may automatically change to a white circular scale, calibrated in increments of 100 ft, with the indicated height being displayed as a magenta color marker on the outside of the circular scale. As the aircraft descends, the circular scale is simultaneously erased so that the scale diminishes in an anticlockwise direction, leaving the magenta height marker at the end of the remaining circular scale. The radio height is also shown as a digital readout in the center of the circular scale. On most systems, a marker shows the decision height, either as an RA circular scale or as a digital number. At reaching the DH plus 50 or 100 ft an aural alert chime sounds, and at the decision height the circular scale and/or the decision height digital display and/or the marker flashes and changes color.

On the B737, pressing the DH reset button on the EADI control panel will cancel the alert chime, stop the DH marker flashing, and change the display back to its normal colors.

Radio Magnetic Indicator (RMI) *See* RMI (Radio Magnetic Indicator).

Ramp Weight Ramp weight (RW) is the gross aircraft weight prior to taxi.

NB: RW must be within its structural maximum [certificate of airworthiness (C of A)] weight limit.

$$RW = TOW + \text{fuel for start and taxi}$$

Range—Headwind Range is increased when flying into a headwind because the best range speed will be a little faster, because the airspeed represents the rate of distance covered. Therefore range will be increased with a headwind.

The increased fuel flow is compensated for by a higher speed, allowing less time en route for the headwind to act.

Range—Maximum *See* Maximum Range Cruise.

Range Maximum (Loading) *See* Loading—Maximum Range.

Rapid *Rapid* is a weather term that relates to the rapid change, in less than one-half hour (30 min), of the original weather to a new and different prevailing weather condition.

Rate of Climb/Descent This is the vertical component of the velocity of an aircraft. It determines the time it will take to either climb to or descend from a given height. Normally it is expressed in terms of feet per minute.

The heavier the aircraft, the greater its rate of descent. This is so because a heavy aircraft will fly at a higher airspeed for a given angle of attack, and so its rate of descent will be increased.

Rate 1 Turn *See* Turns.

Rate 1 Turn Angle of Bank *See* Turns.

Rate 2 Turn *See* Turns.

Rate 3 Turn *See* Turns.

RBI (Relative Bearing Indicator) The relative bearing indicator is a simple ADF instrument that is used to display NDB navigation/information. The RBI comprises the following:

1. A fixed 360° compass card. (The 000° is at the 12 o'clock position, and so is the aircraft heading.)

2. An ADF needle, which seeks out and shows the relative bearing of the nondirectional beacon from the aircraft's heading.

The pilot in conjunction with the directional indicator uses the RBI, by adding the relative bearing (ADF needle) to the aircraft's magnetic heading (DI), to determine the QDM to the nondirectional beacon.

See Fig. 168 at the top of the next page.

If the QUJ bearing is required, then add the aircraft's true heading to the relative bearing. With each change of magnetic heading, the ADF needle will indicate a different relative bearing. Relative bearings between 090° and 270° indicate that the NDB is behind the aircraft.

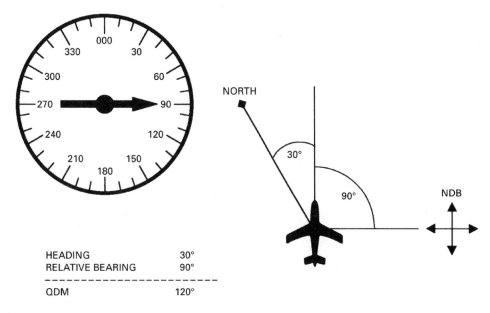

HEADING	30°
RELATIVE BEARING	90°
QDM	120°

Figure 168   RBI to an NDB.

A further development of the RBI instrument is the moving card ADF that can be manually oriented, slaved, to the aircraft's heading. This subtle change means the needle head now indicates the QDM to the NDB.

Real Wander   Real wander occurs whenever the direction of a gyro's spin axis actually moves from its alignment in space. If this occurs, it gives rise to inaccurate instrument readings. Real wander can be either deliberately induced by applying an external correcting force (e.g., as in the alignment of tied gyros) or caused by imperfections in the gyroscope (e.g., unbalanced gimbals or bearing friction).

NB: This is also known as *mechanical drift.*

Rectified Airspeed (RAS)   *See* Calibrated Airspeed (CAS).

Reduced-Thrust Takeoff   *See* Variable-Thrust Takeoff.

Reference Zero—Obstacle Clearance   When reference zero moves away from the obstacle, i.e., back into the clearway, this changes the horizontal distance to the obstacle, which increases the height gained by the aircraft at the obstacle, thus improving obstacle clearance. Reference zero moves back into the clearway, thus improving obstacle clearance, with the following scenarios:

1. TORA is limiting rather than TODA, and therefore TODR < TODA. Remember, reference zero is the ground point at the end of the TOD.

   See Fig. 169 at the top of the next page.

2. Sloping runway.

3. Reduced aircraft takeoff weight reduces the TORR and TODR.

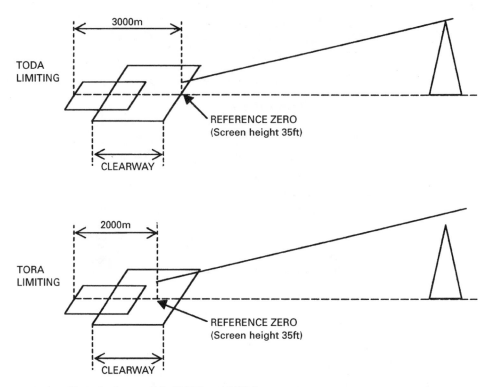

Figure 169 Obstacle clearance for TORA and TODA.

NB: In addition, the climb gradient is improved which provides an additional benefit to obstacle clearance.

Refueling—Precautions *See* Fueling—Precautions.

Relative Bearing Indicator. *See* RBI (Relative Bearing Indicator).

Relight Engine (Flight Start) Boundaries The purpose of engine relight boundaries is to ensure that the correct proportion of air is delivered to the engine's combustion chamber to restart the engine in flight.

For this reason, the aircraft's flight manual outlines the approved relight envelope of airspeed against height. This ensures that within the limits of the envelope, the airflow ingested into the engine will rotate the compressor at a speed that generates and delivers a sufficient volume of air into the combustion chamber to successfully relight the engine.

NB: The approved flight envelope will usually be subdivided into a starter assist and windmill boundaries.

See Fig. 170 at the top of the next page.

To relight an engine, it is necessary to have

- The correct volume of air
- Fuel delivery

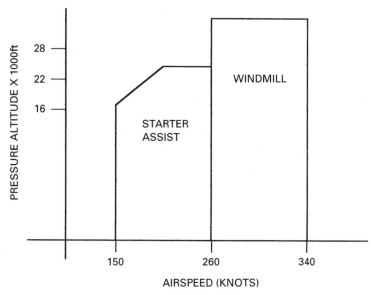

Figure 170 Generic engine relight (flight start) envelope.

- Ignition system on

The air and fuel have to be delivered in roughly the correct proportions so that a defined mixture is obtained. Therefore because the fuel is delivered at an approximately constant volume, the air delivery has to be adjusted/maintained to obtain the correct air/fuel ratio and thereby create the correct mixture for a stable combustion. However, air volume ingested into the engine decreases with altitude, due to a decrease in air density, and decreases with a decrease in forward airspeed. An observance of the restart height/speed envelope is required to ensure that the correct volume of air to restart the engine, either windmilling or starter assist, is delivered to the engine's combustion chamber.

Let us examine when and where the air volume delivered to the engine is or is not correct to allow the engine to relight.

1. Extreme boundary limits
    a. The engine will not relight at extreme high speeds, that is, >340 (IAS), because the volume of air is so great that it will blow out the relight ignition.
    b. The engine will not relight at very low speeds or high altitudes because the volume of air is too low and therefore does not maintain the combustion.
2. The engine will relight satisfactorily within the windmilling envelope. (A windmilling engine is one that is turning over solely under the influence of the ambient airflow into the engine.) Within the windmill speed and altitude envelope, the correct proportions of air are delivered to mix with the fuel delivery that ensures a stable relight.
3. The engine will relight satisfactorily within the starter assist envelope. (At low airspeeds where the wind milling engine creates an insufficient volume of air to relight the engine, a starter assist relight uses the engine's starter to rotate the engine compressors to generate an adequate air intake into the engine.) This

ensures that a correct proportion of air is delivered within the starter assist lower-speed envelope, to mix with the fuel delivery to ensure a stable relight.

Different aircraft engine types have different relight boundaries and nonnormal relight drills, which the pilot should be aware of.

REM   Rapid Eye Movement (REM) is a sleep cycle.

Remote Indicating Compass   The remote indicating compass is a combination of the directional indicator and the magnetic compass instruments, as a single instrument. It utilizes the rigidity of the gyro to avoid the compass turning and acceleration errors, and a magnetic north sensing input to prevent DI gyro wander to maintain its correct orientation at all times without any external influence.

The remote indicating compass is made up of the following:

1. The *detector unit* is used to sense the earth's magnetic field and thereby magnetic north. It is therefore a replacement for the direct reading compass magnet.

The detector unit is essentially a small magnetic sensor fixed to the aircraft, which has three arms. Each arm has windings that produce an induced current when they are positioned in the earth's magnetic field, in the following manner. If an arm is lined up with magnetic north, then a maximum electric current is induced in this arm; and if an arm lies at a right angle to magnetic north, then a minimum current is induced. By feeding the induced current readings of all three arms into a Selsyn system (self-synchronizing), the direction of the detector and hence the aircraft can be calculated.

Figure 171   Remote indicating compass, detector unit.

2. The *gyro unit* is a tied gyro with an axis of rotation in the horizontal plane, which indicates the aircraft's heading by moving the compass card on the instrument display, just as a directional indicator does. However, the gyro is aligned not to north, but to the aircraft's heading by the feedback system.

3. The *feedback system* is best regarded as an error detector system. It keeps the gyro aligned to the detector unit–derived aircraft magnetic heading. It does this by an electrical feed from the signal self-synchronizing system to the gyro rotor,

which generates a current in the rotor. This produces an electromagnetic force which turns, or *precesses,* the gyro and hence the gimbal which moves the compass card to indicate the aircraft's magnetic heading.

Reverse Thrust The most efficient system for stopping at high speed is the *reverse thrust system.* Heavy jet aircraft have a high kinetic energy (momentum) during a landing roll or an aborted takeoff; and the most efficient method of stopping that maintains the initial deceleration rate, when the aircraft is at high speed, is best achieved by reverse thrust for two reasons.

1. The net amount of reverse thrust increases with speed because the acceleration imposed on the constant mass flow is greater. This is so because the aircraft's forward speed is additional when using reverse thrust as opposed to subtracting when in forward thrust.

2. The power produced is greater at high speeds because of the increased rate of work done. This means that the kinetic energy of the aircraft is being destroyed at a greater rate at higher speeds.

Low wing-mounted engines have the additional benefit when in reverse thrust at high speeds of suffering a general blockage of airflow under the wing, causing a higher velocity over the upper surface and hence higher drag.

Therefore to get the best value from reverse thrust, it is important to use it as early as is prudent after the touchdown and to open up to full power as soon as possible. It is also important not to cancel reverse thrust too early. Otherwise your deceleration will be compromised; then either too much strain will be placed on the brakes, or, worse, your landing distance could be compromised.

NB: The attraction of reverse thrust is that, provided it is used symmetrically, a slippery runway does not compromise the aircraft.

NB: At low speeds, the brakes take over in an auto brake system, and they are used to maintain the constant deceleration rate of the aircraft. Sometimes an overlap of reverse thrust and brakes may be used to maintain the constant deceleration at the middle speeds. Most modern aircraft also employ a spoiler system all the way through the landing roll to produce drag and hence create a braking force, in tandem with the other systems (e.g., brakes and thrust reverses).

You should cancel the reverse thrust before you reach low speeds, i.e., approximately below 70 kn, for the following engine handling considerations:

1. On a four-engine or greater aircraft, the reversed flow of the inner engines, being ahead of the outer engines, tends to upset the intake flow of the outer ones if reverse thrust is left engaged at low speeds.

2. Any engine in isolation will tend to breathe its own reversed exhaust, at low speeds, which may blow up surface contamination into the engine.

In addition, if reverse thrust is maintained at a low speed on a contaminated surface (e.g., snow or dust), then forward visibility is compromised as a cloud is blown forward by the reverse thrust.

Reverse thrust is not certified for in flight deployment on most aircraft types, except for a very few exceptions, which use it as an air brake.

Rho Rho Rho Omega Some omega systems use omega stations in a phase measurement mode (like VLF) which creates circular position lines with lanes a whole wavelength wide (that is, 16 nm). This principle is based on range and is called *rho*

rho fixing for a two-station position fix or *rho rho rho* fixing for a three-station position fix.

Rhumb Lines   Rhumb lines are tracks with a constant track direction between two points on a sphere; therefore, it must be a longer distance than a great circle track.

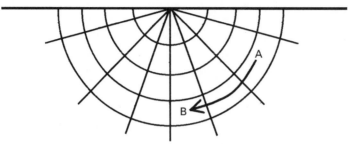

Figure 172   Rhumb line.

Rhumb lines are always found on the equatorial side of the great circle track. Meridians and the equator are both great circle tracks and rhumb lines, and parallels of latitude are solely rhumb lines.

Ridge   A ridge is a U-shaped extension to a high-pressure system. Stable air subsides into a ridge, similar to a high-pressure system, and therefore weather similar to that found in an anticyclone will occur in a ridge.

Risky Shift   Risky shift is the tendency for a group of people to decide upon a riskier, even more dangerous decision than an individual who is part of the group would make by herself or himself. It is human nature that groups of people are more likely to decide to take a greater risk than an individual would. This is generally thought to occur because in a group we take confidence in other people and their decisions more than we do in our own.

RLW   The runway or restricted landing weight is the maximum landing weight the runway can carry.

RMI (Radio Magnetic Indicator)   The radio magnetic indicator can be used to display ADF or VOR navigation information, and it is regarded as an advanced development of the RBI.
 The RMI comprises the following:

1. A remote indicating 360° compass card that is continuously and automatically being aligned with magnetic north. Therefore the RMI displays the aircraft's magnetic heading at the top of the dial.

2. Either a single or dual needle. The needle seeks out the direction of the station to which it is tuned and is superimposed onto a compass card that is oriented to the aircraft's magnetic heading. This means the needle's head indicates a QDM, and the needle's tail indicates a QDR.

3. Selection button, sometimes known as rabbit ears, that enables the pilot to change between ADF and VOR needle indications.

Selected as an ADF needle, it seeks out the direction of an NDB station as a QDM and displays the relative bearing of the beacon from the aircraft, which is the sum of the QDM minus the aircraft heading.

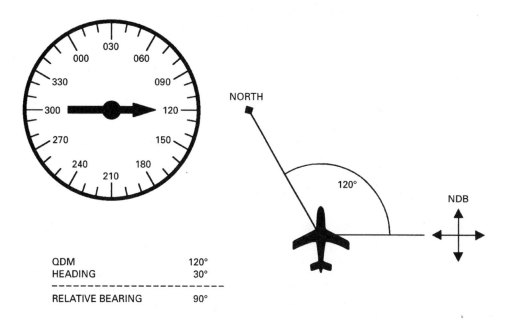

QDM	120°
HEADING	30°
RELATIVE BEARING	90°

Figure 173 RMI, QDM to an NDB.

Selected as a VOR needle, the head indicates the QDM and the tail indicates the QDR, which is itself an indication of the VOR radial the aircraft is on.

Usually the RMI will have two needles, which can display two separate ADF or VOR systems; or a combination of an ADF and a VOR signal at the same time; or the same ADF or VOR signal, which enables the pilot to verify the signal. ADF/VOR system 1 is represented on the RMI by a single-bar needle, and ADF/VOR system 2 is represented on the RMI by a double-bar needle.

RNAV This form of onboard area navigation aircraft equipment utilizes either a basic VOR/DME system or other position sensors, such as DME/DME, INS/IRS, Omega/VLF, or Loran C.

A simple area navigation system allows the operator to input the bearing and distance of a geographic location with reference to a given station (entered as VOR/DME bearing and range, or an INS/IRS latitude and longitude) to position a way point, thereby making a way point a specified geographic location. A series of way points is used to define an area navigation route or the flight path of the aircraft.

NB: A way point is sometimes likened to a *phantom* station because it provides the RNAV user with the same navigation information that a real VOR/DME installation would provide.

Thereafter, the pilot has access to deviation information from the track between way points by displayed information on the primary navigation instrument and distance to go from the DME reading. The sophistication of the equipment determines the amount of information displayed and the ability of the system to automatically select and deselect the primary and alternative ground-based navigation aids. The advantages of the system allow for more direct routing; it is not restricted to established airways, which is more efficient.

The accuracy of the system depends upon the following:

1. The selection of the waypoints. If a flight path is selected to coincide with a radial of a VOR, then the along and cross-track errors are minimal. However, as the way point is offset, the greater the offset distance, the greater the expected error.

2. Ground/slant range errors are not corrected for in some early RNAV systems, but the error is negligible at ranges greater than 3 times the aircraft's altitude.

NB: RNAV is now sanctioned for navigation guidance on some departure or approach procedures.

Roll Dampers The purpose of a roll damper is to

- Damp/remove Dutch roll
- Provide roll damping in turbulence
- Provide spiral stability

The roll damper works through the aileron controls and is normally associated with providing spiral stability, by taking a term from the yaw damper and feeding it back into the rudder power control system. This ensures that a positive aileron angle is always needed to maintain a turn. When roll damper occurs, the spiral stability is bound to be improved because when the ailerons are released, they center and the aircraft naturally recovers from the turn.

Rotation Rate On modern aircraft the importance of the correct rotation rate at the correct speed VR ensures that the aircraft leaves the ground within the correct distance (runway performance) and achieves the correct initial climb speed $V2$.

NB: The correct pitch attitude after liftoff also has to be achieved to maintain the $V2$ speed.

There is a *natural* rotation rate appropriate for each aircraft type. However, under high-performance conditions (low weight, low altitude, and low temperature) the aircraft will have a high rate of acceleration, and the rotation rate will need to be that much faster. However, under these conditions it is easy to overrotate and tail-strike the aircraft, so beware. Similarly under a low-performance condition (maximum weight, high altitude, and high temperature) a lower rate of rotation will be required, reflecting the slower acceleration of the aircraft.

Historically early jet aircraft suffered from ground stalls and failed to leave the ground if overrotated. This was so because early jet transport aircraft had a rather symmetric wing section, which was designed primarily for good high-speed qualities. This section was a critical producer of lift, but a sharp nose profile caused the wing to stall at an incidence, which could be reached on the ground. A significant contribution to this condition was the absence of propeller slipstream. This event has been removed on later designs with the introduction of leading-edge flaps, slats, or droops, which enable the wing to retain its lifting ability at higher incidences, even at VMU.

Rotation Speed *See VR* Speed.

R/T Communication Failure *See* Communication Failure.

R/T Emergency Frequency *See* Emergency Frequency.

R/T Position Report *See* Position Report.

Rudder The rudder is a hinged control surface at the rear of the fin (vertical tailplane) that is controlled by the pilot's rudder pedals. As the rudder control surface is deflected, the airflow and thereby the aerodynamic force around the rudder (vertical tailplane) change. Moving the left rudder pedal deflects the rudder to the left, causing an increase in the airflow speed, thus reducing the static pressure on the right-hand side of the rudder control surface. In addition the left side of the rudder becomes presented more face on to the relative airflow, which causes an increase in the dynamic pressure experienced. These effects create an aerodynamic force to the right on the rudder (vertical tailplane) that rotates (yaws) the aircraft about its vertical/normal axis at its cg point to the left.

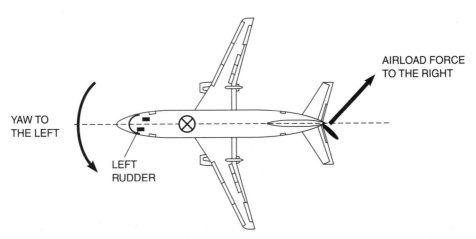

Figure 174 Rudder effect on aircraft.

Left rudder pedal movement moves the rudder control surface to the left, producing an aerodynamic force to the right, which yaws the aircraft to the left. Ergo, the opposite is true. Right rudder pedal movement moves the rudder's control surface to the right, producing an aerodynamic force to the left, which yaws the aircraft to the right.

Runway Lights

Runway centerline lights Precision approach runways usually have flush white runway centerline lights. The color of these lights changes over the last 600 m (caution zone) to alternate red and white lights. Then over the last 300 m they change to all red lights. However, some runways, especially non-precision-approach runways, have no centerline lights, just edge lighting.

Touchdown zone lights   These consist of rows of flush white lights on either side of the runway centerline during the first 900 m. They are commonly used for CAT II and III runways.

Runway edge lights   These are white. They may be directional-only lights (i.e., only visible to aircraft aligned with the runway) or they may be equally visible in all directions. Some are flush with the runway, and others are elevated. For precision approach runways, the color of the runway edge lights changes over the last 600 m (caution zone) to alternate red and white (sometimes yellow) lights. Then over the last 300 m they change to all red lights.

Threshold lights   Green threshold lights cross the width of the runway threshold.

NB: Extra green threshold wing bar lights are sometimes extended outboard of the runway edge threshold lights or at the displaced threshold for poor-visibility approaches.

The last 600 m (2000 ft) of the runway for U.S. ASLF runway lights, or the last 900 m (3000 ft) of the runway for British Calvert-style lighting systems, will have alternating red and white edge/centerline lights, which change to red-only edge/centerline lights over the last 300 m (1000 ft) of the runway.

Runway Threshold—Height of Aircraft   Given a 3° glide path and a 1000-ft touchdown, the height of the aircraft over the runway threshold fence is 50 ft. Use the 1:60 rule:

$$\text{Height} = \frac{\text{angle} \times \text{distance along track}}{60}$$

$$= \frac{3 \times 1000}{60} = 50 \text{ ft}$$

However, larger aircraft will have a reduced clearance height, due to their extended wheels.

Runway Visual Range (RVR)   *See* RVR (Runway Visual Range).

RVR (Runway Visual Range)   RVR is a highly accurate instrument-derived visibility measurement that represents the range at which the runway's high-intensity lights can be seen in the direction of landing along the runway. Its readings are transmitted to the air traffic controller who informs the pilot. RVR is used, when available, as a visibility minimum for low-visibility precision-approach landings, in preference to the general visibility measurement, which may also be reported.

RVR values are measured at three points along the runway:

- Touchdown point
- Midpoint
- Endpoint

RVR is a more precise measurement than general visibility reports. This is true because a visibility report is the worst condition in all directions and the RVR is the visibility down the runway, which will never be less than the general met visibility reported and could be considerably better, particularly if the runway is not into the sun.

Usually the RVR is measured by a ground transmissometer at the side of the runway. This device measures the intensity of a light beam after it has passed through the atmosphere. Data from each transmissometer are computed every 1.5 s and passed to the tower to give an automatic RVR readout. These data are refined to be equivalent to the RVR as seen from a cockpit, that is, 5 m above the runway centerline.

RVR is reported at up to three points on the runway, whenever it is detected as being less than 1500 m. Midpoint and stop end values are reported only if they are less than the touchdown zone (TDZ) and less than 800 m or if they are less than 400 m. When all three values are given, the names are omitted. RVR is measured in steps of 25 m up to 200 m, in steps of 50 m up to 800 m, and thereafter in 100-m steps. The table below shows how the actual RVRs would be reported.

TDZ	Midpoint	Stop end	Reported as
1600	1800	2000	Not reported
1200	900	1000	RVR 1200
1000	700	9000	RVR 1000 midpoint 700
600	500	550	RVR 600 500 550
600	500	800	RVR 600 midpoint 500
400	450	300	RVR 400 stop end 300

NB: The quoted RVR may be followed by the runway designator; for example, RVR 300/25 means RVR 300 m on runway 25.

S

SALR (Saturated Adiabatic Lapse Rate) The SALR is the adiabatic change in temperature for saturated air, as it rises. The SALR commences at a height where a parcel of air's temperature is reduced to its dew point temperature, 100 percent relative humidity, and its water starts to condense out to form a cloud.

Above this height, the now saturated air will continue to cool as it rises; but because it releases latent heat as the water vapor condenses to a liquid form, i.e., clouds, its cooling rate is reduced to a rather regular drop of 1.5° per 1000 ft of height/altitude gained.

SAT The SAT is the ambient static air temperature. It is commonly used as a different name for outside air temperature (OAT).

Saturated Adiabatic Lapse Rate *See* SALR (Saturated Adiabatic Lapse Rate).

Scale Scale on a chart is defined as the ratio of chart distance to earth distance.

Screen Height Screen height relates to the minimum height achieved over the runway before the end of the clearway, should an engine failure occur on takeoff. The screen height also marks the end of the takeoff distance.

The screen height for propeller engine aircraft in dry conditions is 50 ft.

NB: Most propeller aircraft have an increased accelerate/stop distance in wet conditions but no change in $V1$ or screen height.

The screen height for jet aircraft in dry conditions is 35 ft.

NB: The screen height is less than that for propeller aircraft due to the lower coefficient of lift of the jet aircraft's swept wing.

In wet conditions the jet aircraft's screen height is reduced to a minimum of 15 ft in most cases. The screen height is reduced for a jet aircraft using a wet $V1$, due to a proportion of the airborne distance being added to the ground run, between the wet $V1$ and VR, if an engine failure occurs at the worst point, i.e., just after the wet $V1$ and prior to VR.

NB: The $V2$ speed will be achieved only at 35 ft; therefore at a reduced screen height of 15 ft the aircraft speed will be less than $V2$.

Ergo screen height relates to engine failure scenarios and changes with runway conditions for jet aircraft; that is, 35 ft for one engine inoperative/dry conditions and 15 ft for one engine inoperative/wet conditions.

Sea Breezes *See* Land/Sea Breezes.

Secondary Cell Battery *See* Battery.

Secondary (Surveillance) Radar *See* Radar—Secondary (Surveillance).

Separation—on the Approach *See* Approach—Separation.

Separation—on the Departure   *See* Departure—Separation.

Series Circuit   A series circuit has no junctions or branches, and just one circuit path. The word *series* is also applied to the way in which electrical sources, resistors, or loads are situated, i.e., in a chainlike connection through which the same current passes, and their effect on the electric current is additional. Thus it is connected in series.

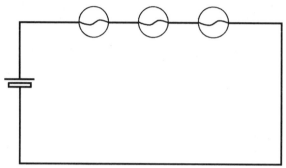

Figure 175   Series circuit.

Series Yaw Damper   The series yaw damper is a development of the parallel yaw damper and is commonly found in modern jet aircraft. This system is attached to the rudder control circuit at the back of the aircraft, and as such it *does not move the pilot's rudder pedals* when it moves the rudder. This means that the rudder foot forces are not increased and therefore allow the series yaw damper to be used for engine-out takeoffs and landings.

Servo-Assisted Altimeter   *See* Altimeter—Servo-Assisted.

SID (Standard Instrument Departure)   A SID details a specific initial route or track from a particular aerodrome runway, often with altitude and occasionally speed constraints at specific points along the track.

SIDs are published (Aerad/Jeppesen) named charts, e.g., Conga 2A, and are used by ATC to simplify the departure clearance, without having to describe the track and altitude details. For example, "Delta 345, you are cleared for a Conga 2A departure."

SIDs are generally designed to separate departing traffic from arriving traffic, to provide an efficient interception of the en route track, and to meet any local constraints, such as noise-sensitive areas, other aerodrome zones, and even danger and restricted military areas. Also SIDs considerably reduce radio transmissions, thereby reducing workload, and vary in their degree of difficulty.

NB: SIDs are generally only available to instrument-rated pilots.

Sideslip   A sideslip is an out-of-balance flight condition that is used to increase the rate of descent and steepen the descent path at a constant airspeed. A sideslip employs *crossed controls,* i.e., control column one way and the rudder the other, without using flap. It uses the opposite rudder to prevent the nose of the aircraft from yawing toward the lower wing. This maintains the aircraft slipping sideways, around its roll axes only. The greater the bank angle (aileron) and more opposite rudder used, the steeper the descent.

NB: Sideslipping is not an approved procedure for some aircraft.

You perform a sideslip maneuver as follows:

Enter the sideslip   Ensure that you have adequate height to recover, since a high rate of descent can be achieved in a sideslip.

■ Close the throttles.

■ Bank the aircraft, using the ailerons.

■ Apply opposite rudder to stop any yaw.

■ Backpressure may be required to prevent the nose from dropping.

Do not trim the aircraft, since a sideslip is a transient maneuver.

Maintain the sideslip   The greater the bank angle with ailerons and rudder used, the steeper the flight path.

■ Maintain the bank angle with the ailerons.

■ Control the heading with the opposite rudder.

■ Maintain airspeed with the elevator, but the airspeed indicator is not totally reliable in this maneuver.

Recovery from a sideslip   To recover into a normal, in-balance descent:

■ Level the wings with ailerons.

■ Simultaneously remove any excess rudder pressure, and centralize the slip "ball" indicator.

Sigmet   A sigmet is a meteorological report that advises of significant meteorological (sig/met) conditions that may affect the safety of flight operations in a general geographic area, i.e., en route or at the aerodrome.

The criteria for raising a sigmet include

1. Active thunderstorms

2. Tropical revolving storms

3. Severe line squalls

4. Heavy rain

5. Severe turbulence

6. Severe airframe icing

7. Marked mountain waves

8. Widespread dust or sandstorms

Signals—Lights and Pyrotechnics from an Aerodrome to an Aircraft

Description of signal	Meaning to an aircraft in flight	Meaning to an aircraft on the aerodrome
1. Continuous red light	Give way to other aircraft and continue circling.	Stop.
2. Red pyrotechnic lights, or red flare	Do not land, wait for permission.	—

3. Red flashes	Do not land; aerodrome not available for landing.	Move clear of landing area.
4. Green flashes	Return to the aerodrome and wait for permission to land.	To an aircraft: You may move onto the maneuvering area and apron.
5. Continuous green light	You may land.	You may take off.
6. White flashes	Land at this aerodrome after receiving a continuous green light; then after receiving green flashes, proceed to the apron.	Return to starting point on aerodrome.

Signals—Lights and Pyrotechnics from an Aircraft to an Aerodrome

Description of signal	Meaning from an aircraft in flight to an aerodrome
1. Red pyrotechnic light, or red flare	Immediate assistance is requested.
2. Continuous green light	By night: May I land?
	By day: May I land in direction different from that indicated?
3. White flashes	I am compelled to land.
4. White pyrotechnic lights, or switching on and off of navigation lights, or switching on and off of the landing lights	I am compelled to land.

Sink Rate (on the Approach)   *See* Approach—Sink Rate.

Sleep Patterns   Pilots should manage their sleep patterns so that they awake from a sleep a few hours before reporting for duty. This ensures that they arrive for their flight fresh and awake, as if it were the beginning of a new day at the office. Obviously this is difficult for commercial pilots, especially long-haul, who have to sleep at times when their bodies are saying they should be awake.

Slip   Slip is the difference between geometric pitch and effective pitch.

Sloping Runway—Obstacle Clearance Performance   The height of an obstacle is measured against the reference-zero elevation, that is, 35 ft, assuming that reference zero is at the end of the TODA; that is, TODA = TODR. Thereby on a level runway the height of the obstacle and the height gained during the climb-out are measured from the same datum and therefore are true comparisons. If the runway is sloped downhill, the TODR is reduced, or TODR < TODA, and the reference-zero point (35 ft) is reached earlier, i.e., a shorter horizontal distance from the brake release point; or it moves back into the clearway. Therefore the reference-zero ground point is at a higher ASL height than any obstacles, which results in the effective height of the obstacle (i.e., height of the obstacle compared to the revised reference-zero point) being reduced, thus improving obstacle clearance. However, if the runway is sloped uphill, the reference-zero ground point and flight path begin at a lower ASL height, even if TODR = TODA; therefore the effective height of the obstacle is increased, thus reducing obstacle clearance.

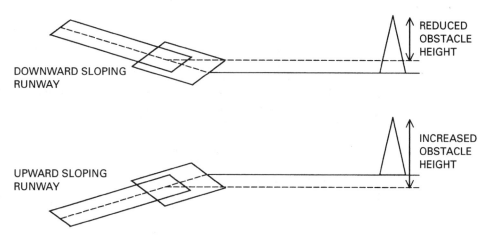

Figure 176 Obstacle clearance with a sloping runway.

Smog Smog is the combination of smoke (and/or other airborne particles) and fog. Smog is usually found under an inversion layer that acts as a blanket, stopping vertical convective currents.

Any particles, especially smoke, suspended in these lower layers beneath the inversion layer will be trapped there, and so a rather dirty layer will form, which is particularly common in urban industrial areas. These small particles may also act as *condensation particles* and therefore encourage the formation of fog. This combination of smoke, fog, and an inversion layer is known as smog. The most well-known example of smog is seen in Los Angeles, California, which has all the ingredients for smog: clear, cold nights that form an inversion layer and moist air that creates fog and dirty urban smoke. Similar effects can be seen in rural areas if there is a lot of pollen, dust, or other foreign matter in the air.

Spanwise Airflow Spanwise airflow over a wing

1. Creates wing tip vortices
2. Reduces aileron (wing control surface) efficiency
3. When reversed increases disturbed airflow on the wing's upper surface at the tip, contributing to a wing tip stall

Reversed spanwise airflow is caused because as the angle of attack is increased on a swept wing, the boundary layer spanwise airflow changes direction to flow outward to the tip. This occurs because air is flowing at an angle to the isobars on the wings, and the spanwise component of the air velocity is not slowed down to the same extent as the chordwise component. This results in a change of direction of the boundary layer on the upper surface of the wing outward to the wing tip.

The designer prevents spanwise airflow on a wing, especially a swept wing, with *fences and vortex generator* surfaces. They are designed to direct the airflow over the wing's upper surface perpendicular to the leading edge.

SPECI   A SPECI is an aviation selected special weather report for an aerodrome. It is generated whenever a critical meteorological condition exists, e.g., wind shear or microbursts. It is similar in presentation to a metar.

Special Forecast   A special forecast is a meteorological forecast for flights over long routes, outside the coverage of the local country area forecast.

NB: Special forecasts can also include forecasts (TAFs) for the departure, destination, and up to three alternative aerodromes.

Special forecasts may be specifically requested from the met forecast office for an individual flight via facsimile, AFTN (Aeronautical Fixed Telecommunications Network), telex, phone, and presumably in the future via the Internet. However, the met forecast office often requires prior notice to compile a special forecast. In most countries usually a 2-h notice is required for routes less than 500 nm or a 4-h notice for routes greater than 500 nm.

Specific Fuel Consumption   The specific fuel consumption (SFC) is the quantity/weight of fuel consumed per hour, divided by the thrust of the engine:

$$\text{SFC} = \frac{\text{fuel burn, lb/h}}{\text{engine thrust, lb}}$$

Specific Gravity   The specific gravity (SG) of a substance is the ratio of the weight of a unit volume of the substance to the weight of the same volume of water under the same conditions of temperature and pressure. For example,

$$1 \text{ L of freshwater weighs } 1 \text{ kg} = \text{SG } 1.0$$

$$1 \text{ L of jet A1 fuel weighs } 0.8 \text{ kg} = \text{SG } 0.8$$

As such, specific gravity is used to determine the weight (measure) of a quantity of fuel.

The specific gravity of a substance varies with its density, which in turn varies inversely with its temperature; i.e., as temperature increases, the SG decreases.

See also Fuel—Measurement.

Specific Heat Capacity   Specific heat capacity is the ability of a material to hold thermal or heat energy.

Land, especially bare rock and concrete, has a low specific heat capacity. This is true because land loses its heat very quickly, although it heats up just as easily and quickly when solar radiation is present, because it requires only a small amount of insulation to raise its temperature. This means in the summer season land becomes hot and in the winter it becomes cold.

NB: Furthermore, the reflective quality of the type of land plays an important part in determining the amount of thermal/heat energy initially absorbed by the land and thus its temperature. That is, white surfaces reflect and dark surfaces attract.

Water, especially the sea, has a high specific heat capacity. This is so because water requires a lot of heat energy to achieve a small temperature rise, and because it is also translucent, the radiation can penetrate its surface to heat its lower layers. This makes water a good heat storage material, which warms up, and more importantly cools down slowly, to make it a good specific heat capacity material. This means that water, and in particular the sea, maintains a rather stable temperature throughout the year.

NB: The specific heat capacity difference between land and sea is an important factor in many weather phenomena such as land/sea breezes and maritime air masses.

Speed Control (on the Approach)   *See* Approach—Speed Control.

Speed Margins   A speed margin is the difference between the aircraft's normal maximum permitted operating speed and its higher certified testing speed.

The piston/propeller aircraft enjoys a relatively large margin between VNO and VDF and has very few overspeed tendencies. Therefore the speed margin for a piston/propeller aircraft is not very significant.

For the piston engine/propeller aircraft

VNO   Normal operating maximum permitted speed

VNE   Higher, never-exceed operating speed

VDF   Maximum demonstrated flight diving speed, established during design certification flight trials

The jet aircraft's margin between VMO/MMO and VDF/MDF is relatively small. Because of its low cruise drag and the enormous power available from its jet engines, especially at low altitudes, the jet aircraft has a distinct overspeed tendency.

1. At 16,000 ft there is usually enough power available to achieve VDF without exceeding maximum continuous thrust.

2. At 37,000 ft there is usually still enough power available to achieve a speed greater than MMO but not VDF.

3. And of course any significant descent allows speeds to build up quickly, and the aircraft can easily go beyond MDF and well beyond VDF at low altitudes.

Therefore the speed margin on a jet aircraft is very significant.

For the jet aircraft:

VMO/MMO   Maximum indicated operating speed, in knots or Mach number. This is the normal maximum operating speed, which ensures an aircraft's structural strength integrity and adequate handling qualities.

VDF/MDF   Maximum demonstrated flight diving speed in knots or Mach number, established during the design certification flight trials. This flight diving speed incurs reduced aircraft structural strength integrity and often a lower level of handling qualities.

Speed Stability   Speed stability is the behavior of the speed after a disturbance at a fixed power setting, which is a consequence of the drag values experienced by the aircraft frame. Speed is said to be stable if, after it has been disturbed from its trimmed speed, it returns naturally to its original speed. For example,

- An increase in speed leads to an increase in drag, thereby returning to the original speed.

- A decrease in speed leads to a decrease in drag, thereby returning to the original speed.

Speed is said to be *unstable* if, after it has been disturbed from its trimmed state, the speed divergence continues resulting in a negative speed stability. For example,

■ A decrease in speed leads to an increase in drag, which causes a further decrease in speed, thereby causing a negative speed divergence.

■ An increase in speed leads to a decrease in drag, which causes a further increase in speed, thereby causing a positive speed divergence.

Speed stability is of great importance at low speeds on the approach where the jet aircraft's speed is only just above VIMD and therefore only just in the stable speed region. This is a matter of concern because the jet aircraft could easily slip below VIMD and therefore into an unstable speed region during the most critical phase of flight. Here the aircraft is sliding up the back end of the drag curve, where thrust required increases with reducing speed. This is true because the drag values increase faster than the lift values, resulting in poor speed stability and a deteriorating lift/drag ratio which, if not balanced with increased thrust, will result in a tendency to progressively lose speed and to enter into a sinking flight path profile.

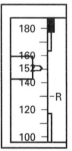

Speed Tape Symbols (on the EADI)   The speed tape consists of a graduated scale with a typical range of 80 kn that moves relative to a fixed reference pointer to indicate the current airspeed. It has several distinctive symbols to illustrate speed-related flight conditions, including the following:

1. High-speed, red and black barber's pole is seen at the top of the speed tape, and the bottom end of the barber's pole indicates one of these (whichever is the lowest, most limiting speed)
   a. VMO/MMO
   b. Maximum operating speed, for extended gear
   c. Maximum operating speed, for current extended flap

2. High-speed, yellow hollow pole is seen at the top of the speed tape, below the high-speed, red and black barber's pole; and the bottom end of the yellow hollow pole indicates one of these:
   a. The speed that provides a 0.3g maneuver margin to the VMO/MMO high-speed buffet
   b. The speed for the next normal flap position to be extended during an approach

3. Low-speed, red and black barber's pole is seen at the bottom of the speed tape, and the top end of the barber's pole indicates the speed at which the stall warning device is activated, i.e., the stall speed.

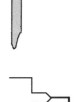

4. Low-speed, yellow hollow pole is seen at the bottom of the speed tape, above the low-speed, red and black barber's pole; and the top end of the yellow hollow pole indicates:

    a. The aircraft's present minimum maneuver speed, which is one of these:

        (1) A 0.3*g* maneuver margin above the low-speed buffet at high altitudes

        (2) A 0.3*g* maneuver margin above the stall warning device, i.e., the stall speed, at low altitudes

5. *Trend arrow.* The tip of the green trend arrow superimposed on the speed tape shows the predicted speed within the next 10 s based on the current airspeed and acceleration or deceleration trend.

    Fast/slow indicator. Less sophisticated systems may have a fast/slow indicator instead of a trend arrow. The fast/slow indicator is located on the side of the (E)ADI and provides the pilot with a speed command, based on the actual speed versus the required speed.

6. The command speed symbol is usually represented by double magenta horizontal lines or a triangle, when the command speed is within the displayed speed tape range.

    NB: When the command speed is not within the displayed speed tape range, it is displayed in digital form either above or below the tape, depending on its numerical value relative to the speed tape range displayed.

7. This represents the *V*1 speed when it is within the displayed speed tape range.

    NB: When the *V*1 speed is not within the speed tape range, it is displayed in digital form at the top of the speed tape scale.

8. During takeoff this symbol represents the rotate speed. During landing this symbol represents the MCP or FMC VREF (VAPP) landing speed for the configuration of the aircraft.

9. This represents the minimum flap retraction speed, i.e., the minimum safe maneuver speed for the next flap position. It is only displayed during the takeoff and go-around stages of flight.

10. This represents the flap-up maneuvering speed during a climb or driftdown for an aircraft in a clean configuration.

    NB: It is usually displayed only when the flaps are fully retracted.

Spin   A spin is a condition of stalled flight in which the aircraft follows a spiral descent path. As well as being in a stalled condition, the aircraft has a yaw component, which induces one wing to produce more lift than the other, resulting in a roll. Greater drag from the stalled lower wing results in further yaw, which leads to further roll, which leads to further yaw, etc.

There are three stages in a spin.

1. The incipient spin, or the beginning of the spin, is an unsteady maneuver in which the entry path of the aircraft is combined with the phenomenon of *auto rotation.*

2. In the fully developed spin, the aircraft

- Is rolling
- Is yawing
- Is pitching (up to a high angle of attack) } in motion about all three axes
- Is stalled
- Is sideslipping
- Has a steady rate of rotation, autorotation
- Is rapidly losing height at a steady rate of descent, at a low airspeed

NB: In a spin the wings will not produce much lift, because they are stalled. The aircraft will accelerate downward until it reaches a vertical rate of descent, where the greatly increased drag balances the weight. The height loss will be rapid as the aircraft spins downward about its vertical axis.

3. The recovery from the spin is initiated by the pilot.

NB: Do not confuse a spin with a spiral dive. In a spiral dive the nose attitude is low, the wing is not stalled, the airspeed is high and rapidly increasing, and the rate of descent is high.

A spin develops from a stall, probably as a result of the nose being raised too high and the speed being allowed to drop too low. A wing drop is then essential. This may occur by itself due to a natural yawing of the aircraft at the stall, or the pilot yawing the aircraft with rudder may induce it. This yaw causes the outer wing to speed up and generate more lift, causing it to rise; its angle of attack decreases, taking it farther from the stalling angle. The inner wing slows down and generates less lift, causing it to drop. Its angle of attack increases, and this wing stalls; or if it has already stalled, it goes farther beyond the stalling angle.

The wing drop after the stall may also occur because of misuse of the ailerons just prior to the aircraft's stalling.

NB: Trying to raise a dropped wing with the opposite aileron may have the reverse effect when the aircraft is close to the stall. If, as the aileron goes down, the stalling angle is exceeded, instead of rising the wing may drop quickly, inducing a yaw.

Autorotation will now commence as a result of the dropping wing becoming further stalled, with a consequent decrease in lift and increase in drag. The aircraft will roll, a sideslip will develop, and the nose will drop and yaw further. If no corrective action is taken, the rate of rotation will increase and a spin will quickly develop.

NB: The shallower the spin, the lower the rate of the rotation and therefore the less the rate of descent. A rearward cg will encourage a flatter spin, and a forward cg will result in a steeper spin.

Remember that it is the yaw that turns a stall into a spin. An aircraft would not usually go straight from stall to a spin. There is usually a transition, typically taking a few turns in the unsteady and steep autorotation mode before settling into a fully developed and stable spin.

A spin is recognized by the following characteristics:

1. A steep nose-down attitude

2. Continuous rotation, indicated by a high rate of turn on the turn indicator

NB: Ignore the slip reading, as it can be stuck hard on either side in some modern aircraft.

3. Buffeting, in some cases, and a lot of banging and clattering noise

4. Low airspeed, almost constant

5. A rapid loss of height

6. General sloppiness of manual controls

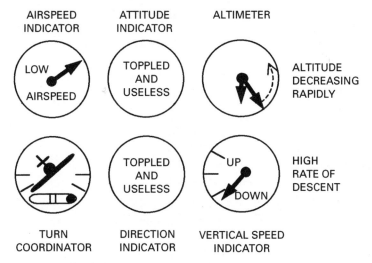

Figure 177 The flight instruments in a spin.

NB: The gyroscopes may topple in a spin, so information from the attitude indicator will be of no value.

A typical recovery technique from a spin is as follows:

1. Close the throttles, keep the ailerons centralized and neutral, and retract the flaps.

2. Verify the direction of the spin on the turn indicator.

3. Apply opposite rudder.

4. Pause, to allow the rudder to become effective and stop the yaw.

   NB: Remember that it is the yaw that turns a stall into a spin.

5. Ease the control column forward to lower the aircraft's nose, which reduces its angle of attack and "unstalls" the wings. Apply full forward control column if necessary.

   This is a vital part of the spin recovery.

   NB: In the process of unstalling the wings, the nose attitude will become steeper and the mass of the aircraft will move closer to the spin axis. The result will be a noticeable increase in the rate of rotation just before the recovery.

6. As soon as the rotation stops, centralize the rudder.

7. Level the wings and ease out of the ensuing dive.

8. As the nose comes through the horizon, add power and climb away to regain height.

An *incipient spin* means the beginning or onset of a spin. To recover from this stage of a spin:

1. Ease the control column forward sufficiently to unstall the wings.

2. Apply sufficient rudder to prevent further yaw.

3. Apply maximum power.

   NB: If the nose has dropped below the horizon, do not apply power until after the recovery is complete and the nose rises above the horizon.

4. As the airspeed increases and the wings become unstalled, level the wings with co-ordinated use of rudder and ailerons, ease out of the descent, and resume the desired flight path.

Spiral Dive   A spiral dive is a nose-low unusual attitude with the wings banked and an increasing airspeed.

An excessive bank angle maneuver will often lead to the nose dropping while the wings are banked, resulting in a spiral dive. A spiral dive is usually entered as a result of a mishandled maneuver from the pilot.

A dive is recognized by the following characteristics:

1. A low pitch attitude on the artificial horizon

2. Descent indications on the altimeter and vertical speed indicator

3. An increasing airspeed on the airspeed indicator

NB: This distinguishes a dive from a spin, which does not have an increasing airspeed.

4. A high *g* loading

A spiral dive is recognized by the following additional characteristic:

5. A high rate of turn on the turn indicator, and not much slip

To recover from a spiral dive:

1. Reduce power (close throttles).

NB: Removing the power is the commonly used technique; this protects against the speed increasing excessively. However, a technique of leaving the cruise power set allows you to maintain the relationship of stick force to speed, and thereby keeps the transition to a level flight attitude more straightforward.

2. Roll the wings level by applying aileron against the indicated turn until it is substantially central. It is permissible to use the ailerons as firmly as needed to roll the wings level. Coordinated rudder may be used on aircraft where it is standard practice to balance the aircraft in a turn with the rudder.

NB: You must recover laterally first. If the control column is simply pulled back to raise the aircraft's nose while it is still steeply banked, then the spiral dive will be tightened without the descent being stopped. Therefore it is very important to level the wings first.

This removes the spiral element and leaves you in a steep dive.

NB: Reducing power and rolling the wings level can and should be done simultaneously.

You will now be left in a steep dive, and the recovery technique from here on is the same as the recovery from a steep dive.

Spiral Stability   Spiral stability (or a spirally stable aircraft) is defined as the tendency of an aircraft in a properly coordinated banked turn to return to a laterally level flight attitude on release of the ailerons. Spirally stable aircraft have dominant lateral surfaces (e.g., wings).

Spiral instability or a spirally unstable aircraft will see a banked turn increase fairly quickly, followed by the nose falling into the turn, leading to the aircraft entering into a spiral dive, when the ailerons are released in a coordinated turn. After the release of the ailerons in a steady turn, stability is

Stable (positive)   if the bank angle decreases

Neutral   if the bank angle remains constant

Unstable (negative)   if the bank angle increases

Spirally unstable aircraft have a dominant (too large) fin/tailplane area.

What happens is that as the aircraft starts to slip into the turn on release of the ailerons, and before the rolling moment due to the sideslip can take effect, the rather dominant fin jumps into play. This is true because the fin/tailplane area (outside) becomes exposed to the relative airflow, which exerts two forces on the aircraft;

1. Around the vertical axis, which straightens the aircraft directionally

2. Around the longitudinal axis, which increases the bank

This accelerates the outer (upper) wing and causes the bank to be increased further; the increased bank causes another slip, which the fin again straightens. This sequence repeats, and the turn is thus made steeper. Once the bank angle exceeds a given, type-specific, amount (say 30°), the nose falls into the turn and the speed increases as the roll increases and the aircraft enters into a spiral dive.

An aircraft's deficiency in spiral stability can be artificially corrected by taking a term from the yaw damper and feeding it back into the rudder power control system. This ensures that a positive aileron angle is always needed to maintain a turn. When roll damper occurs, the spiral stability is bound to be improved because when the ailerons are released, they center and so the aircraft recovers naturally from the turn.

The relationship between oscillatory and spiral stability can now be better understood, because what is good for one is bad for the other, in terms of fin and rudder size, putting them in conflict with each other. Namely, a large fin improves oscillatory stability but degrades spiral stability.

Spoiler Blowback   Spoilers are limited by very high speeds (VDF/MDF) which cause them to blow back. At very high speeds, the spoilers will be blown back to or near to their fully retracted position. This occurs because the high air load experienced on the spoiler's surface at high speeds is greater than the design limit. Obviously the force experienced is a function of airspeed and angle of deflection.

Spoilers are designed not to blow back in the normal operating speed range of the aircraft. Therefore correct speed management of the aircraft will prevent the spoilers from blowing back. This is ensured by either of the following design/operating restrictions placed on an aircraft type.

1. A lower Mach speed limit (MDF) for aircraft types with differential spoilers is imposed that ensures adequate roll rates exist with the speed brakes fully extended; this also ensures that the air loads experienced, within its speed range, will not cause the spoilers to blow back. MMO is reduced proportionally to provide substantially the same speed margins and still remains comfortably above the highest likely cruise Mach number speed.

2. For nondifferential spoiler designs, the maximum spoiler angle in the speed brake (in flight) mode is set to a value below the blowback angle, usually via a restrictive collar on the speed brake lever in its flight detent position. Therefore always allow some slight upward movement of the spoilers in response to a roll command that provides sufficient rates of roll at high speed.

3. For nondifferential spoilers, the maximum speed brake angle can also be fixed automatically as a function of speed. This is accomplished by a speed-sensitive device calibrated at approximately VMO + 15 kn, which restricts the maximum angle of deflection if the speed brakes were selected above the calibrated speed, VMO + 15 kn. It will also retract the brakes to the lower restricted angle of deflection if the speed increases past the calibrated speed, VMO + 15 kn, after the spoilers have been set to the fully extended speed brake position. Again the restricted angle of deflection always allows for some upgoing spoiler movement to provide adequate rates of roll.

All these types will provide adequate roll rates and prevent spoiler blowback for all combinations of speed, Mach number, and altitude up to the design limits of VDF and MDF.

To correct for spoiler blowback in flight, you should reduce speed by removing thrust to a speed where the spoilers will operate normally, then recycle the speed brake lever.

NB: Spoiler blowback will occur only when the aircraft's speed is excessive (that is, VDF/MDF) which itself should only be experienced in a nonnormal flight condition, e.g., spiral dive, when the recovery drill incorporates reducing speed by closing the throttles.

It is very important to restrict the aircraft to speeds below which the blowback occurs; otherwise the control of the aircraft could be severely compromised because of the following:

1. The ailerons lose effectiveness at high speeds due to a wing twisting effect, to a degree where the ailerons lose all lateral control. Therefore both spoilers and ailerons are ineffective, leaving the aircraft with no lateral control at all.

2. At high Mach number speeds, jet aircraft are also liable to suffer a reversed rolling motion due to sideslip; e.g., left rudder causes a sideways force on the tailplane to create an adverse rolling motion to the right. This effect compounds the control difficulties of an aircraft that already has no lateral control because of ineffective aileron and spoiler blowback at high speeds.

These behaviors do not occur within the aircraft's normal operating speed range of VMO/MMO, but it can occur at the design limit VDF/MDF that an aircraft can reach when it finds itself in a nonnormal condition. First, the pilot would experience the blowback of the spoilers, then find he or she had no aileron roll control, and finally would normally apply rudder in an attempt to turn the aircraft, only to find the aircraft rolling in the opposite direction. Altogether this is not a very healthy situation to be in. So first thrust must be used to reduce speed to recover control of the aircraft. Therefore clearly the safe control of the aircraft's speed is the first and main responsibility, in terms of both prevention and recovery.

Spoilers   These are panels that extend from the upper surface of the wing and have the effect of spoiling or disturbing the airflow over the wing (drag), thereby reducing the lift. They are used for the following.

1. *Roll control.*   The spoilers are raised on one wing and not the other, which creates an imbalance of lift values that produces a rolling moment. The spoilers are connected to the normal aileron controls, and they work in tandem with each other for roll control.

Spoilers are in fact a more efficient roll control surface than the ailerons for the following reasons:

a. Spoilers cause less wing twisting than ailerons.

b. Spoilers are more effective than ailerons at high speeds.

c. Spoilers cause a yaw in the direction of the turn.

The disadvantage of roll control spoilers is that they cause an overall loss of lift, which may cause a loss of height, which is particularly undesirable when flying close to the ground.

In general terms, the spoilers and ailerons will control the aircraft's rolling moment equally, that is, 50 percent control each. Therefore failure of either the aileron or the spoiler system would still leave the aircraft with 50 percent of its roll control. And if the aircraft is handled within certain speed margins and excessive crosswinds are avoided for landing, then the aircraft should not experience any undue difficulties.

Spoilers are normally (hydraulically) powered on heavy/fast aircraft because of the heavy operating forces experienced at high speeds.

2. *Air/speed brakes.* The spoilers are raised symmetrically on both wings to a flight detent position (using the speed brake lever) which causes a large increase in drag and slows down the speed of the aircraft.

NB: Buffet is usually experienced with spoiler (speed brake) deployment.

3. Ground/lift dumpers. The spoilers are raised systematically on both wings to the ground detent position (greater angle than the flight detent position) which causes a large increase in drag that

a. Decreases lift over the wing, causing the aircraft to sink to the ground

b. Acts as ground speed brakes to slow down the speed of the aircraft

Normally an aircraft is limited in its operations to the use of spoilers above 1000 ft AGL. This imposed limit protects against ground contact.

There are six reasons why spoilers have to be used for roll control, as speed brakes, and as lift dumpers. Four of the reasons are to provide a degree of roll control, because the ailerons have the following inadequacies:

1. Spoilers are used to supplement the ailerons because the ailerons are limited in size, and therefore effectiveness, as most of the trailing edge of the wing is occupied with flaps, which are needed to generate the greater lift required at low speeds.

2. On a thin swept wing, ailerons which are too large will experience a high degree of air loading/lift, resulting in the wing twisting at high speed that can produce aileron reversal (removes aileron roll control), which is very detrimental. Therefore spoilers are used to provide a proportional roll control that allows for the design of smaller ailerons and thus removes the in-flight experience of aileron reversal if the aircraft is operated within its defined speed range.

3. Ailerons tend to lose effectiveness at high speeds due to the spanwise diagonal airflow across the aileron, which is less effective than a perpendicular airflow. Therefore spoilers are used to provide roll control, especially at high speeds where they are still effective.

4. High-speed swept-wing aircraft cause a strong rolling moment with yaw, known as *adverse rolling moment with yaw.* Therefore additional roll control in the form of spoilers is needed to balance this tendency.

Other than for roll control, the spoilers are needed to counteract (brake) the aircraft's high speed in the air and on the ground.

5. Because the aircraft has low drag and the engines have a slow "lag" response rate, there is a need for high-drag devices in flight to act as a brake when the aircraft is required to lose speed and/or height quickly.

This is achieved by spoilers on both wings being raised simultaneously to the flight detent position, which creates a drag force opposing thrust and therefore reduces the aircraft's speed and/or height. Some aircraft may have additional air brake spoilers mounted on the rear fuselage.

The use of spoilers as air brakes causes buffeting on the airframe and is not recommended below 1000 ft AGL as a safety measure against ground contact.

6. On landing or during a rejected takeoff there is a need to "dump the lift off the wing" and onto the wheels to assist in stopping the aircraft. This is achieved by the spoilers on both wings being raised simultaneously to the "ground or up" detent in a similar manner as the in-flight speed brake. This position has a greater angle of deployment than the flight detent and/or uses more spoiler panels, and therefore it creates a greater drag force.

SRA (Surveillance Radar Approach)   A surveillance radar approach is a nonprecision approach used at certain aerodromes where a radar controller can provide tracking guidance and desired height information down a final approach.

NB: SRA radar only provides the controller with returns showing the aircraft's lateral position, not its vertical position.

A surveillance radar approach is carried out by the pilot under guidance from the radar controller, who passes the following on the VHF-COM radio:

1. Radar vectors to steer the aircraft onto a final approach

2. Advice to the pilot about when to start a (3°) descent

3. Horizontal tracking instructions in the form of turns to make and headings to steer toward to maintain the inbound track (e.g., "Turn right five degrees, heading two six five").

4. Vertical navigation (descent) guidance in the form of range to touchdown and the desired height at that range. The heights given are for QFE set on the altimeter subscale; e.g., if range is 4 mi, height should be 1230 ft.

An SRA is achieved by a ground radar unit being aligned with a specific runway. Therefore it is only available on certain runways at certain aerodromes.

The minima for a SRA procedure are as follows:

SRA terminating at 1/2 nm   250 ft MDH
SRA terminating at 1 nm   300 ft MDH
SRA terminating at 2 nm   350 ft MDH

Stability—at High Altitude   Stability for longitudinal, lateral, directional, and oscillatory modes is, in general, reduced at high altitudes, in terms of dynamic stability, mainly because aerodynamic damping decreases with altitude. The aircraft will feel, and is, less stable except for spiral stability, which improves with altitude, while oscillatory stability deteriorates very rapidly with altitude. This is so because for a constant IAS, the fin suffers a smaller angle of incidence and therefore has a smaller restoring force as the altitude increases. Therefore the fin is less dominant, which is detrimental for oscillatory stability but as a consequence means that the aircraft's lateral surfaces (wings) become more dominant, which improves the aircraft's spiral stability qualities. (Spiral stability always opposes oscillatory stability, and vice versa.)

Stability Modes   There are the five modes of aerodynamic stability:

1. Oscillatory stability (Dutch roll)
2. Directional stability
3. Spiral stability
4. Lateral stability
5. Longitudinal stability

Stabilizer   A stabilizer is an all-moving horizontal tailplane control surface; i.e., it is not fixed in one position. Normally an all-moving horizontal tailplane is called a *stabilizer* when it is solely responsible for longitudinal balancing and it has a separate elevator with its own controls and movement range for pitch maneuverability. A stabilizer is normally moved by its own independent stab trim system which can be either a manual or an automatic device.

NB: An all-moving tailplane is normally called a *variable incidence tailplane* when it does not have an elevator surface. Therefore the variable incidence tailplane provides pitch maneuverability, by the control column, and longitudinal balancing, by the trim system.

The purpose of the horizontal stabilizer is to provide a longitudinal balancing force to the aircraft. Thus the elevator range and the aircraft's pitch maneuverability are not compromised and remain available to be used solely to control the pitch of the aircraft. The stabilizer covers large and small pitching moments; e.g., a single person moving from the aft to the forward cabin will displace the overall weight balance.

There are four reasons for the use of a stabilizer or variable incidence tailplane, especially on a jet aircraft. And while any of the above requirements in isolation might not demand a variable incidence tailplane, in combination they certainly do. Once the need has been established for one of the requirements, then advantages are also enjoyed in the other areas.

The four reasons for the use of a stabilizer or variable incidence tailplane are as follows:

1. A large center-of-gravity range is required on a large jet aircraft because its considerable fuel burn can drastically reduce the aircraft's weight during a single flight. This is compounded by the fact that the fuel (and therefore its weight) is stored in tanks housed in its swept wings, which because of the sweep cover a distance along the longitudinal axis of the aircraft. Therefore as the fuel is burned, not only does the aircraft's weight decrease, but also the cg point moves along the aircraft's longitudinal axis.

As the cg is moved forward or aft (from its neutral position), a balancing force from the tailplane is required to maintain the aircraft's attitude. However, if this balancing force is provided by the elevator alone, a condition could exist, for very large changes in the cg position, whereby the elevator is fully deflected to maintain the balancing force alone, and therefore no further elevator control would remain to pitch the aircraft.

See Fig. 178 at the top of the next page.

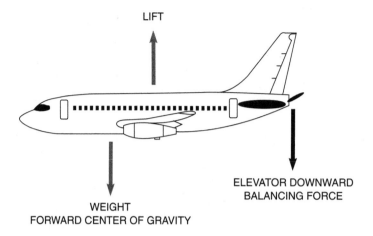

Figure 178   Balancing by a trimmable elevator.

However, maneuverability in pitch would have become very difficult long before full elevator deflection was needed to hold a longitudinally stable condition. With a stabilizer, however, as the cg moves over a comparatively large distance, the incidence of the larger stabilizer is altered a small amount (compared to the elevator) to provide a balancing force, and the elevator remains in a streamlined position. Therefore the elevator remains available over its full range of pitch maneuverability at all times. With a variable incidence tailplane, which is also much larger than the elevator, it, too, can be moved through a smaller angle to produce the required balancing force, while it still retains a large range of available further movement for the aircraft's full range of pitch maneuverability.

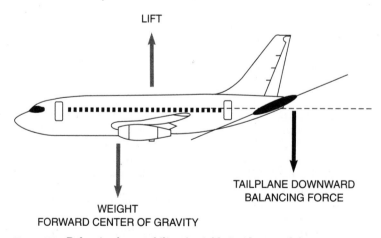

Figure 179   Balancing by a stabilizer/variable incidence tailplane.

The large increase in balancing forces available from a variable incidence tailplane, together with good pitch control from an unrestricted elevator, makes a large cg range and therefore a heavy swept-wing aircraft with a large weight change in flight a practical proposition.

2. A large speed range requires a large balancing force because a function of lift is the airflow speed over the lift-producing wings. High speeds create greater lift [through the center of pressure (cp)], and low speeds produce less lift, because of $\frac{1}{2}\rho V^2$. Also the change in speed will create a change in the distribution (section) of the lift over the wing, which in turn results in a movement along the chordwise line of the cp. Therefore speed has an effect similar to the cg in terms of requiring a tailplane balancing force. On a fixed-tail aircraft, low speeds need an up elevator movement to create a downward force on the tail, giving a nose-up tendency; for high speeds the opposite is true, in basic terms.

Similarly if the speed range is extended considerably, the elevator deflection reaches a point where no further (or insufficient) elevator capability remains to handle the pitch commands of the aircraft. This effect can be made much worse in combination with a cg not in the neutral position.

The variable incidence tailplane/stabilizer erases the speed range problem in the same manner as it solves the cg range problem.

3. Large trim changes are required because of large configuration changes (trailing- and leading-edge flaps especially). These large configuration changes are required because the swept-wing jet aircraft has poor lift capabilities from a clean wing. Therefore flaps are employed to produce a marked increase in the coefficient of lift (CL) maximum by increasing the wing area, chordline, and chamber, thereby moving the cp point on the aircraft. Depending on the variables this can lead to a nose-up or nose-down attitude change.

Therefore the jet/swept-wing aircraft has a large lift-producing configuration range, and the elevator alone is insufficient to produce the balancing force and retain further its ability/range in pitch control of the aircraft. A variable incidence tailplane or stabilizer is therefore employed to handle the trim changes similarly to what it does for cg and speed range balancing forces, and likewise leaves the elevator streamlined to the tailplane with its full up-and-down range of movement remaining.

4. Trim drag is produced when a fixed-tail aircraft requires a deflected (trimmed) elevator to longitudinally balance the airplane, and the drag produced from this deflected elevator can be quite high. (The greater the drag, the greater the thrust required, and the worse the aircraft's performance and endurance.) With a variable incidence tailplane or stabilizer, cruise drag is reduced because its greater area allows it to produce the same balancing for a lower angle of deflection (compared to the elevator). Therefore the variable incidence tailplane or stabilizer and attached elevator are more streamlined to the relative airflow and therefore induce less cruise drag.

In conclusion, the variable incidence tailplane or stabilizer is a very powerful control surface. A pilot should be aware of his or her aircraft type, especially its settings at critical stages of flight, such as takeoff. This is true because the autopilot often trims the stabilizer, and the pilot should monitor its correct automatic position on a frequent basis to guard against incorrect stabilizer trim, which could make the aircraft very unstable.

Stabilizer—Runaway A stabilizer is typically held in its trimmed position by a series of brakes, for both manual and autopilot modes. If a series of these brakes should fail, then the stabilizer will experience backpressure from the airflow which will rotate the stabilizer to its maximum upward or downward mechanical stops, thereby inducing a marked out-of-trim, unstable condition.

Note that it is for this reason that pilots are taught to "subconsciously" monitor the auto trim against what they perceive should be the trimmed amount, so that they can disengage the runaway before it gets excessive. However, pilots should not jump in too early, as an autopilot tends to trim excessively more than a pilot does.

A stabilizer runaway, whether auto or manual trim, should not go undetected for too long; and when it is detected, decisive action should be taken.

Each aircraft type has its own drill, but in general terms

- Hold the control column firmly
- Autopilot (if engaged)—disengage
- Stab trim cutout switches to cut out

If the runaway continues

- Stabilizer trim wheel—grasp and hold

Continue the flight, using manual trim, and adopt early airspeed and configuration conditions. To a lesser extent some autopilots cause the auto trim to change the tail setting significantly in long vertical draughts. This isn't strictly a fault condition, but it should be arrested. Take out the autopilot; be prepared for the stick force as the autopilot comes out, and trim manually.

Stabilizer—Stuck The condition that an aircraft experiences from a stuck stabilizer—other than when the aircraft's equilibrium is matched to the position of the stuck stabilizer—is a degraded longitudinal balancing ability, which is due to the backup employment of the less powerful elevator in providing this balancing force. This condition has the following effects:

1. The longitudinal balancing force that the elevator experiences is much higher than when it is used for its normal pitch control duties, and this translates to heavy stick forces felt by the pilot.

2. Because the elevator surface is smaller than the stabilizer, the response to the pilot's inputs is slower.

3. The elevator's pitch control capability is reduced because the majority of its range is being used to provide a longitudinal balancing force.

These effects could be further compromised by the position of the stuck stabilizer, e.g., if we require a tailplane down force and the stabilizer is stuck giving an up force, or vice versa. This condition would require this effect to be reduced to a level within the elevator's limited range. However, you would not fly in this condition for too long; but if you did, you would find that the higher drag penalty incurred would require higher thrust levels, resulting in greater fuel consumption and therefore a decrease in range.

If a stabilizer has stuck while the aircraft is in a substantially trimmed-out condition, then as long as you maintain the speed at which the tail jammed, you will remain substantially in trim and therefore stable. However, because the aircraft will have to depart from the cruise speed in preparation to land, the following steps will reduce the main reasons for having a variable incidence tailplane in the first place. They also reduce the effects of a stuck stabilizer, which thereby helps to maintain the aircraft's longitudinal stability and maneuverability in pitch, during its approach and landing. Generally the following procedure would be followed:

1. Divert to a nearby airfield (so that the cg movement due to the aircraft's change in weight is not excessive over a prolonged flight).
2. Move the cg to an aft position. This will
   a. Longitudinally balance the aircraft at low speeds and thereby reduce the stick forces.
   b. Reduce the elevator demand for the landing flare.

NB: With a stuck stabilizer the best cg position is *aft*. This can be accomplished by moving passengers to the rear of the aircraft and/or by moving fuel to outer wing tanks, if possible. This is true because in the cruise the aircraft's stabilizer is in a trimmed state, and therefore no immediate problem exists if it becomes stuck in this position. However, as speed is reduced in preparation to land, a higher angle of attack is required to maintain lift, which requires a greater balancing force from the tailplane, which now only consists of the elevator. The elevator, however, has only a limited longitudinal balancing ability, and it is primarily needed to cope with the pitch control of the aircraft. Therefore, moving the weight to the rear of the aircraft will move the cg aft, which will naturally give the aircraft the higher nose attitude it requires for slower flight (especially on the approach). This minimizes the need for the elevator to produce the balancing force, minimizes stick forces, and therefore allows the elevator to retain its pitch control capability, especially for the landing flare. Therefore it is necessary to have an aft cg when the stabilizer is stuck in the cruise position, to allow the aircraft to retain its natural approach attitude, thereby leaving the elevator free for its pitch control during the approach and landing phases of flight.

3. Reduce speed as late as possible, to minimize the length of time a balancing force, with its associated high stick forces, is required from the elevator.
4. Plan a long final approach, and make configuration changes, gear, and flaps earlier than usual to give time to sort out the aircraft before the next change is due.
5. Use a reduced flap setting for landing which will reduce the landing flare required. This allows you to maintain a higher approach speed, which reduces the divergence of the aircraft from its cruise trimmed speed and therefore reduces the balancing force required from the elevator and so reduces the stick forces experienced.

Having exercised the above-mentioned measures in response to this failure, the aircraft will retain enough scope to maintain a longitudinally stable condition and enough elevator pitch maneuverability to adjust the aircraft's approach and landing attitude.

Stall(ing) An aircraft stalls when the streamlined/laminar airflow (or boundary layer) over the wing's upper surface, which produces lift, breaks away from the surface when the critical angle of attack is exceeded, irrespective of airspeed, and becomes turbulent, causing a loss in lift (i.e., the turbulent air on the upper surface creates a higher air pressure than on the lower surface). The only way to recover is to decrease the angle of attack (i.e., relax the backpressure and/or move the control column forward).

The stall is a function of the boundary layer's losing energy because of the skin friction drag it experiences, and as a result the layer increases in thickness as it moves rearward. It separates eventually and becomes disturbed/turbulent airflow when it is faced with an adverse pressure gradient (i.e., pressure increasing in the direction of the flow). Normally the airflow over the upper surface remains attached until it has almost reached the trailing edge; but as the angle of attack is increased, the point of separation or minimum pressure/maximum suction moves nearer to the leading edge, and the magnitude of the suction to pressure increases. This results in the air having a greater distance to travel against an increasing pressure and pressure gradient. These factors cause the point of separation to move forward toward the leading edge. The airflow is no longer following the shape of the wing section, and the pressure variation of the upper surface is altered to a condition such that the turbulent air aft of the point of separation is dominant and the lift force forward of the point of separation is reduced. This results in the wing's having *stalled* and causes the following:

1. Buffeting of the airframe
2. A marked increase in drag
3. Turbulent airflow behind the wing
4. Movement of the cp point
5. A marked decrease in lift, resulting in the aircraft's losing height

It is important to remember that in practice the whole wing does not stall at the same time, and one wing may stall before the other. Therefore where the stall occurs, first determine the response of the aircraft to the stall, i.e., wing drop, nose-down, etc. It is worthwhile to know your aircraft's stall characteristic.

The plan shape of the wing is the main influence on how the stall develops. Thus the design of the wing plan section is such that the wing will normally stall at the root first, causing a nose-down stall characteristic. This is preferred for the following reasons:

▪ There is less chance of one wing stalling before the other and creating a *wing drop stall*.

▪ Aileron control is retained up to the stall.

▪ Separation of the airflow from the wing root section causes a natural buffet effect on the tail, which acts as a stall warning.

The following wing design areas delay the breakup of the airflow and therefore the stall:

1. Wing slots are the main design feature that delays/suppresses the stall speed. A slot is a form of boundary layer control that reenergizes the airflow to delay the airflow over the wing from separating at the normal stall speed. The wing therefore produces a higher CL and can achieve a lower speed at the stall angle of attack.
2. There is a lower angle of incidence and a greater chamber for a particular wing section, e.g., wing tips.

The stall is a function of the boundary layer's losing energy and as a result increases in thickness as it moves rearward due to the skin friction drag experienced. When it is faced by an adverse pressure gradient, it breaks away from the wing's surface as disturbed/turbulent air, and the airflow is said to have stalled.

Therefore to delay the wing stall, we need to design a wing feature that reenergizes the boundary layer. This is achieved with the following boundary layer control features:

1. Slots work as a venturi does and introduce accelerated high-pressure air into the boundary layer. This has the effect of speeding up the airflow in the boundary layer. This helps to
   a. Reduce the effects of skin friction drag

  b. Remove slow boundary-layer air near the wing's trailing edge, which in effect reduces its thickness and increases its energy, thereby helping to prevent the separation of the boundary layer (effectively moving the point of separation rearward) and so obtain a higher CL and depress the aircraft's stall speed

2. The wing tip section experiences higher air loads, causing the wing's tip to reach its stall angle of attack prior to any other section of the wing. To delay this, the wing tip is designed with a lower angle of incidence, twist, or *washout* that requires a greater loading before it reaches its stall angle. This therefore ensures that it does not stall prior to any other wing section.

   Also a greater chamber accelerates the boundary airflow which thereby retains its energy longer and also helps to delay the breakup of the airflow and the development of stall.

NB: A wing that stalls at the tip first will roll over into a dive and possibly a spin, which is clearly undesirable. Therefore the wing is designed to stall at the root first.

An aircraft will stall at a constant angle of attack (known as the *critical angle of attack*). However, because most aircraft do not have angle-of-attack indicators (except "eyebrows" on some EFIS displays) the pilot has to rely on airspeed indications. However, the speed at which the aircraft will stall varies according to the effects of the following properties:

1. *a.* Actual weight. The heavier the aircraft, the higher the indicated speed at which the aircraft will stall. If an aircraft's actual weight is increased, the wing must produce greater lift (remember that the lift force must equal the weight force); but because the stall occurs at a constant angle of attack, we can increase lift only by increasing speed. Therefore the stall speed will increase with an increase in the aircraft's actual or effective weight. The stall speed is proportional to the square root of the aircraft's weight.

   b. Load factor and effect of *g* in a turn (*g* is normal acceleration due to gravity/load factors). The indicated stalling speed will also be greater if the aircraft is performing a maneuver (e.g., banked turn) which increases the aircraft's *g* and thereby increases the effective weight of the aircraft, i.e., load factor. Therefore the airspeed must be higher to generate the greater values of lift required to maintain the lift/weight balance. The stall speed will change in proportion to the square root of the number of *g*.

   See Fig. 180 at the top of the next page.

30° turn = 1.15*g*	increases stall speed by 7 percent
45° turn = 1.4*g*	increases stall speed by 19 percent
60° turn = 2.0*g*	increases stall speed by 41 percent

   Therefore it is obvious that if the aircraft is maneuvering, the speed must be kept well above the normal stalling speed, to allow for the increase in stalling speed during the maneuver.

   c. Effective weight/cg position. A forward cg position requires a tailplane balancing down force which effectively increases the aircraft's weight and therefore increases its stalling speed. For an aft cg position the opposite is true.

2. Altitude. The jet aircraft's stall speed will increase at very high altitudes, albeit only slightly because of

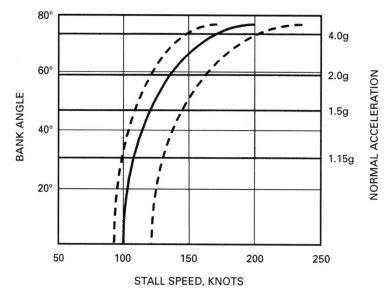

Figure 180 Variation of stall speed for (1) different weight and (2) effective different weight due to increased *g*.

    a. The Mach number compressibility effect on the wing
    b. Compressibility error on the airspeed indicator
3. Wing design/lift. The stall speed is determined by the maximum CL value of the wing, and the CL value is determined by the shape of the aerofoil section. The higher the CL, the lower the stall speed. Contamination, e.g., ice formation, on the wing can disturb the laminar airflow and increase the weight of the aircraft, causing the stall to occur at a higher unknown speed, sometimes even before the stall warning operates.
4. Configuration. Flaps and/or slots extended increase the wing's CL and generally produce a better lift/drag ratio. This can do one of the following:
    a. Reduce the angle of attack for a constant speed to produce the same lift value.
    b. Reduce speed for a constant angle of attack to produce the same lift value.
    Therefore the stall speed is lowered with flaps and/or slots extended.
5. Propeller engine power. The stall speed is reduced (suppressed) with propeller engine power on for two reasons:
    a. The vertical component of the thrust increases as the aircraft's angle of attack increases because the thrust line becomes inclined upward. This helps to support some of the aircraft's weight. The wing therefore has less weight to support, and so at the stalling angle of attack it has to produce a lower lift force, which allows for a lower speed.
    b. The propeller slipstream/airflow adds kinetic energy to the airflow over the wing, thereby having an increased speed value with the power on. Velocity is part of the lift formula, $\frac{1}{2}\rho V^2$. This allows one of these to occur:
        (1) The wing generates greater lift for a given angle of attack and airspeed.
        (2) The wing gives the same lift value at the same angle of attack for a lower airspeed.

An increase in lift helps to support some of the aircraft's weight, thereby reducing the aircraft's effective weight. So because the stall speed is based on the aircraft's weight, the stall speed is reduced with the propeller engine power on; however, the stall angle of attack remains the same.

Not a property but a factor in the stall recovery is momentum. Heavy aircraft have greater momentum, and the effect of this momentum is evident in the stall recovery. When a heavy aircraft stalls, the aircraft begins to sink fast and the initial sink rate is a consequence of momentum, which is almost independent of pitch attitude. Therefore the aircraft's horizontal or forward motion and thus the airflow over the wing are minimal. This means that a stall recovery in a heavy, momentum-charged aircraft needs to be carried out in a methodical manner to reintroduce laminar airflow to unstall the wings. This is done by lowering the nose, applying power, and flying the aircraft around the recovery path, with a minimum height loss, while guarding against a second stall.

Any movement of the cp point at the stall causes a change in the aircraft's angle of attack. Normally a simple swept or tapered wing is designed so that the cp will move rearward at the stall. This is so because the stall is normally induced at the wing root first, where the cp is at its farthest forward point across the wing's span. Therefore the lift produced from the unstalled part of the wing, toward the tips and therefore aft, is behind the root with an overall net result of the cp moving rearward, which results in a stable nose-down change in the aircraft's angle of attack at the stall. If, however, the wing has been designed without any inboard wing stall properties, then the "virgin" wing will usually stall at the tips first, causing the opposite cp movement and effects. This is not a desired stall characteristic, and therefore it is normal for the wing to be designed to stall at the root first.

At very high altitudes the stall speed increases, because to get the same lift force and speed at the constant stalling angle of attack, the dynamic pressure $\frac{1}{2}\rho V^2$ must always be the same. But at very high altitudes, where compressibility effect/errors occur, the dynamic pressure is not the same and therefore the IAS at the stall is not the same for two reasons:

1. Mach number compressibility effect on the wing. At very high altitudes the actual EAS stall speed increases, because the Mach number compressibility effect on the wing disturbs the pressure pattern that increases the effective weight on the wing, resulting in a higher EAS stall speed.

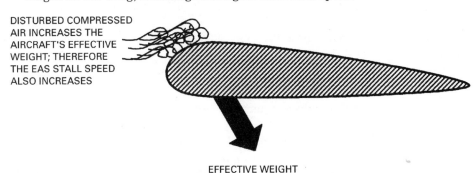

DISTURBED COMPRESSED AIR INCREASES THE AIRCRAFT'S EFFECTIVE WEIGHT; THEREFORE THE EAS STALL SPEED ALSO INCREASES

EFFECTIVE WEIGHT

Figure 181 Compressibility effect on the effective weight.

2. Compressibility error on the IAS/ASIR. The compressibility correction which forms part of the difference between the IAS/ASIR (airspeed indicated reading which is uncorrected) and EAS (equivalent airspeed which is IAS corrected for

compressibility and position instrument error) is larger in the EAS to IAS/ASIR direction due to the effect of the Mach number, resulting in a higher IAS stall speed.

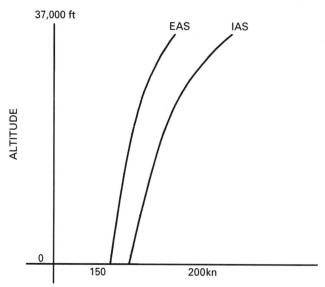

Figure 182 Variation of EAS and IAS/ASIR stall speeds with altitude.

The stall for a given wing section always occurs at a constant angle of attack, but the stall speed can vary due to the effect of a number of properties on the aircraft's effective weight and lift capabilities.

A stall is recognized by its prestall characteristics:

1. *a.* Decreasing airspeed and nose-high attitudes (pitch) with an increasing altitude
   b. Decreasing control effectiveness with a "sloppy" feel.
2. Airflow buffet
3. Activation of any stall warning devices

Stall characteristics are

■ The aircraft nose dropping, which is caused by the center of pressure moving rearward

■ A decreasing altitude normally with a high sink rate

To recover from a stall:

1. Ease the control column forward to unstall the wings by regenerating laminar airflow over the wing.

2. Apply sufficient rudder to prevent further yaw, if required.

NB: Be careful of using ailerons near the stall because a dropping wing can be picked up by moving the control column in the opposite direction. This causes the aileron on the dropping wing to deflect downward, increasing the angle of attack and producing greater lift on that wing. If the wing is near the stalling angle, the aileron deflection could cause the critical angle to be exceeded on that wing and instead of rising, could cause the wing to drop farther. With any yaw a spin could develop. On some aircraft types the ailerons may lose effectiveness, whereas the tail fin elevator and rudder by design will remain effective throughout the stall.

3. Power can be applied to increase forward airspeed which creates extra lift to maintain height; otherwise maintain the best glide speed for a shallow descent.

   NB: On a propeller-driven aircraft any power increase causes a slipstream effect over the wing, which contributes to restoring the laminar airflow over the wing, thus suppressing the stall to below the critical angle of attack. Therefore the slipstream effect provides a larger margin of stall and height recovery; i.e., large pitch-down attitude is counteracted by the slipstream effect. Jet engine aircraft, however, do not produce any slipstream over the wings, and therefore the margin for stall and height recovery is much smaller.

4. When the airspeed increases to at least the cruise speed and as the wings become unstalled, level the wings with the ailerons, ease out of the descent, and gently enter into level flight or into a shallow climb, applying engine power as necessary. Be careful not to restall the wings.

   NB: An aircraft that is longitudinally stable will recover from a stall naturally. If the aircraft's cg is within its range, then the aircraft will always be able to recover from a stall.

To recover from a stall in a turn, follow the same standard recovery technique:

1. Release control column backpressure.

2. Apply rudder to prevent yaw, if necessary.

3. Apply power to maintain height if desired.

4. When the wings are unstalled, use the ailerons to roll the wings level.

However, be aware of a higher stall speed in the turn. This occurs because the load factor (*g* load or effective weight of the aircraft) is increased in the turn, and the stall will therefore occur at a higher airspeed.

NB: The stall speed varies with the square root of the load factor (*g* load or effective weight); i.e., lift has a direct relationship to the effective weight or load factor *g*.

How much the stall speed increases depends on the load factor experienced in the turn:

A 30° bank turn increases the stall speed by approximately 7 percent.

A 45° bank turn increases the stall speed by approximately 19 percent.

A 60° bank turn increases the stall speed by approximately 41 percent.

NB: Stalls at higher speeds than normal, i.e., in turns, are called *accelerated stalls*.

A stall recovery technique near the ground should be the same as a normal stall recovery technique, except for the following:

- Do not change the aircraft configuration until clear of the ground and a positive recovery is established, e.g., stable speed and a positive rate of climb.

- Prompt recovery actions are essential.

The term *incipient stall* means the beginning stages of the stall. It precedes the actual stall. If ever an unwanted stall appears imminent, then recover at the incipient stage, especially if the aircraft is close to the ground. The recovery technique from an incipient stall is as follows:

1. Relax the backpressure (or move the control column forward).
2. Simultaneously apply power smoothly.
3. Use the controls normally, i.e., the ailerons, since the wing is not stalled.

Stall(ing) of a Swept Wing   *See* Swept Wing.

Stall Protection Systems   Certain systems (stall warner and stick pusher) protect against a stall. Stall warners are either an artificial audio warning and/or a stick shaker, which are usually activated at or just before the onset of the prestall buffet. Stick pushers are normally used only on aircraft with superstall qualities and usually activate after the stall warning but before the stall, giving an automatic nose-down command. Both systems normally receive a signal from an incidence-measuring probe.

Stall Speed   *See* V_S Speed.

Stall Speed Variation with cg Position   *See* Center-of-Gravity Effect on Stall Speed.

Standard Setting (Altimeter)   The 29.92-inHg or 1013-hPa (mbar) standard setting will give altimeter readings as a pressure altitude or flight level and is used for traffic-controlled airspace above the transition layer.

STAR (Standard Instrument Arrival)   A STAR details a specific final route or track onto a particular runway approach, often with altitude and occasionally speed constraints at specific points along the track.

STARs are published (Aerad/Jeppesen) named charts, say, WYP 4D, and are used by ATC to simplify the arrival clearance, without having to describe the track and altitude details. For example, "Speedbird 345, you are cleared for a WYP 4D arrival to runway 23 left."

STARs are generally designed to separate arriving traffic from departing traffic, to provide an efficient interception of the runway extended centerline, as well as meet any local constraints, such as noise-sensitive areas, other aerodrome zones, and even danger and restricted military areas. Also STARs considerably reduce radio transmissions, thereby reducing workload, and vary in their degree of difficulty.

NB: STARs are generally only available to instrument-rated pilots.

Static Electricity   Static electricity has the potential to create a spark, inducing a fire risk and/or creating radio interference, if it is allowed to build up in a section of an aircraft. *Bonding* reduces the risk of static electricity buildup on an aircraft surface. Bonding is the joining, by flexible wire strips, of each part of the aircraft's metal structure or metal components to each other, thus providing an easy path for the electrons to move from one part of the aircraft to another. Static wicks are fitted to the trailing of the aircraft surfaces, to dispense static electricity into the atmosphere, thereby reducing static electricity from the aircraft which reduces the risk of flash fires and radio interference.

Static Pressure   Static pressure is the equal weight of all the air molecules at any one point in the atmosphere. As the name implies, static pressure does not involve any motion of the body relative to the air, but it is the ambient pressure around an object.

Steep Dive   An erect steep dive is recognized by the following characteristics:

- An extremely low-pitch attitude on the artificial horizon
- Extremely high descent indications on the altimeter and vertical speed indicator
- A rapidly increasing airspeed on the airspeed indicator

The recovery from an erect steep dive is achieved as follows:

1. Reduce power (close throttles).

    NB: Removing the power is the commonly used technique, as this protects against the speed increasing excessively. However, a technique of leaving the cruise power set allows you to maintain the relationship of stick force to speed and so keeps the transition to a level flight attitude more straightforward.

2. Ease out of the dive by gently pulling up on the elevator with light backpressure on the control column.

    NB: In a steep dive there is a danger of overstressing the airframe at high airspeeds with large and sudden elevator control movements. Therefore only use firm elevator pressure to ease the aircraft out of the dive.

    Maintain up elevator until the airspeed stops increasing; this means that you are passing through the horizon, i.e., level flight.

3. Reapply power for the climb-out, and wait a few seconds for the altimeter to steady; then adjust the pitch attitude, using both the altimeter and airspeed indicators to maintain a stable climb airspeed.

    NB: If you maintained cruise power during the dive, as you pass through level flight push forward, because otherwise the much increased airspeed will cause a strong nose-up pitch.

4. The scan rate should be increased at this point to ensure that there is no over-controlling.

    NB: A high climb rate with a decreasing airspeed and even stalling the aircraft is a distinct possibility if the heavy control forces initially employed in the dive recovery are not released sufficiently during the following climb-out. If this is the case, adjust the pitch attitude downward until the speed stops decreasing.

5. Maintain lateral level flight during the maneuver with the ailerons.

Steep Turn   A steep turn is a turn in which the bank angle exceeds 45° and is normally considered to be at least 60°. In a steep turn you must aim to maintain a constant height and airspeed.

NB: The load factor, drag, and stalling speed all increase significantly in a steep turn.

When entering a steep turn, trim the aircraft for straight and level flight at the desired airspeed and height. Roll into a normal turn, except that as the bank angle increases through 30°:

- Smoothly add power to maintain airspeed, on the ASI.
- Progressively increase the backpressure on the control column to raise the pitch attitude on the AI to maintain height.
- Adjust the bank angle and backpressure to place the nose in the correct position relative to the horizon.
- Balance the turn with rudder in light aircraft (not heavy jet aircraft).

To maintain a steep turn, increase your scan rate as the turn gets steeper, to cope with all the variables.

- Height on the altimeter and vertical speed indicator. Correct with pitch attitude on the attitude indicator and/or on the horizon if in VMC.

- Airspeed on the airspeed indicator. Correct with power.

- Bank angle on the attitude indicator.

- Balance the yaw on the slip indicator, on light aircraft (not heavy jet aircraft).

NB: Steep turns require increased lift, which is generated by increased backpressure on the control column, which increases the angle of attack. The backpressure required to maintain height is quite significant in a steep turn.

When exiting a steep turn, the aim is to roll out onto a particular heading. Then you should

- Allow a greater lead time to roll out because of the high rate of turn, by monitoring the directional indicator compass.

- Lower the pitch attitude on the attitude indicator, by releasing backpressure, at a greater rate than for a medium turn, to maintain height by monitoring the VSI and altimeter.

- Reduce the power gradually to cruise power to maintain the desired airspeed by monitoring the airspeed indicator.

NB: If in VMC, a good look out at the horizon should be included in your scan, to back up your instrument readings, especially for maintaining height and for picking a reference point for rolling out of the turn.

Stopping—Efficiency   *See* Reverse Thrust.

Stopping Distance—Performance   In general the performance gained by using reverse thrust is not applied to the takeoff emergency stopping distance (EMDR) or landing stopping distance, although a 10 percent safety factor is commonly applied to landing distance in the event of an inoperative thrust reverser.

The performance gained by using an antiskid system is applied to both the takeoff distance and the landing stopping distance. If the antiskid system is inoperative, then takeoff from a wet runway is normally prohibited and the landing calculation has a large safety factor for the landing distance required, usually about 50 percent.

Maximum braking performance is applied to both the takeoff emergency stopping distance (EMDR) and the landing distance required.

Stopway (Runway)   The stopway is the length of an unprepared surface at the end of the runway in the direction of the takeoff, which is capable of supporting an aircraft, if the aircraft has to be stopped during a takeoff run.

A stopway may sometimes extend beyond the clearway if the length of the clearway is limited because of an obstruction within 75 m of the runway/stopway centerline. However, this obstruction does not limit the stopway, which only needs to be as wide as the runway.

Stress   The main reasons why a person suffers stress are

1. Personal relationships

2. Bereavement

3. Financial and time commitments

4. Health problems, including fatigue

5. Work matters, such as immediate situation (workload) on the flight deck, general operating/company environment, and relationship with colleagues and management

You should deal with stress by getting help for the issue that is causing the stress, before it starts to affect your performance; e.g., for marriage problems get marriage counseling. Stress reduces the attention, motivation, and performance of pilots in their duties, especially an emergency task, which is obviously a danger to the safe operation of the aircraft. Therefore it is important to manage stress levels to remain as stress-free as possible.

The environmental stresses on the flight deck are

1. *Heat.*   Comfort zone is around 20°C. Discomfort and therefore stress are promoted by temperatures >30°C and <15°C.

2. *Humidity.*   Comfort humidity is between 40 and 60 percent.

3. *Noise.*   However, with a low-arousal personal condition, i.e., lack of sleep, noise may actually improve performance.

4. Vibration.

5. Heavy workload on the flight deck.

Structural Weight—of an Aircraft   The structural weight of an aircraft is limited because the main force generated to balance the aircraft's gross weight is the lift force, and if the lift cannot equal the aircraft's weight, then the aircraft cannot maintain level flight. Therefore the aircraft weight is directly restricted by the lift capabilities of the aircraft.

NB: The lift force generated is limited by the size (design) of the wing, the attainable airspeed (airspeed is limited by the power available from the engine/propeller), and the air density.

Sublimation   Sublimation is the process of turning water vapor immediately into ice when the dew point/actual temperature is less than 0°C.

NB: The dew point is now called the *frost point.*

The usual result of sublimation is *hoar frost,* which is in fact regarded as a type of icing.

Supercharger   A supercharger simply increases the air delivered to the engine cylinder above its normally aspirated capacity by compressing the intake air, which in turn requires more fuel to be delivered to the carburetor to maintain the correct mixture ratio, which in turn produces a greater power output (horsepower). Therefore a supercharged engine is capable of producing a greater power output than a normally aspirated engine of the same cylinder size.

Superchargers are used to artificially raise the engine manifold pressure to compensate for low atmospheric pressure to either

- Increase engine power output for takeoff and the initial climb (ground-boosted engine) and/or

- Maintain MSL engine power at high altitudes (altitude-boosted engine)

Turbo/superchargers are needed to artificially raise the engine manifold air pressure because the engine's power output basically depends on the weight of mixture (*charge*) which can be burned in the cylinders in a given time, and the *charge* depends on the temperature and pressure of the atmospheric air. On a normally aspirated engine, the pressure in the engine manifold is slightly less than atmospheric pressure because of intake duct losses, and the manifold pressure drops with an increase of altitude because of reduced air pressure with altitude; therefore power output decreases with altitude. Therefore turbo/supercharger is used to

- Increase sea-level engine power output (ground-boosted engine charger)
- Maintain MSL engine power at higher altitudes (altitude-boosted engine charger)

Turbo/superchargers increase the pressure of the air delivered to the engine manifold by using a centrifugal impeller, which causes the air to be accelerated (creating mainly kinetic energy and some pressure energy). The air is then passed onto a diffuser, which consists of divergent ducts, which decreases the velocity and therefore increases the air pressure. In this manner, considerable compression ratios are attainable, but the degree to which the air pressure is increased is dependent on physical constraints, which limit the size of the impeller. This limits the compression ratio and consequently the power output of the engine and/or the maximum operating altitude of the aircraft.

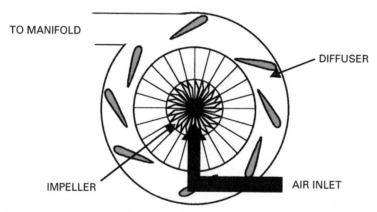

Figure 183 Supercharger impeller/diffuser.

Ground-boosted turbo/superchargers increase ground pressure above the standard MSL 14.7 psi. This causes additional stress on the engine, and so the engine's main components have to be reinforced.

Altitude-boosted turbo/superchargers maintain approximately MSL pressure at higher altitudes. This type can be fitted to normally aspirated engines without any significant engine design changes.

The main difference between a supercharger and a turbocharger lies in how their impellers are driven. A supercharger is driven internally, by gears from the crankshaft directly to the impeller/diffuser assembly (either a single or dual assembly), and is fitted downstream of the throttle valve. Because the engine drives the supercharger via gears off the crankshaft, the engine loses output power. Therefore rated power/maximum continuous power is less than the engine power required to drive the supercharger. It is commonly used on heavy/medium propeller aircraft. A supercharger's purpose is to compress the fuel/air mixture.

An auto boost control (ABC) is incorporated with the supercharger, to keep the boost pressure/MAP constant during a climb or a descent. The supercharger is designed to maintain a given pressure at differing altitudes. At low altitudes where the air is denser, the supercharger produces too much pressure. Therefore to avoid high combustion pressures, the ABC restricts the delivery of the pressure. As the aircraft climbs, via its throttle butterfly valve the ABC opens progressively to deliver a greater air pressure and thereby maintains a constant boost pressure. This relieves the pilot of the responsibility of constantly varying the position of the throttle levers during a climb or descent.

Supercooled Water Droplets (SWDs)   Supercooled water droplets are liquid water drops sized between 40 and 2000 μm that exist in a nonfrozen liquid form, in the atmosphere, at temperatures down to as much as −45°C, well below the normal freezing point of water (0°C). This is known as being *supercooled*.

Such supercooled water drops will only partially freeze on contact with a colder, subzero surface such as the skin of an aircraft, or propeller blade, etc. Usually SWDs will progressively turn into ice as they are washed back along the colder aircraft surface. This is so because the SWDs release latent heat (into the remaining water) as they start to freeze, which reduces the rate of freezing of the remaining water/liquid, allowing it to stay longer in a liquid state, which in turn gets spread backward, by the airflow, along the aircraft's surface. As the remaining liquid/water comes into direct contact with another part of the subzero surface, it in turn freezes progressively. This can cause long trails of ice to build up on and from the leading edges, which the deicing systems of the aircraft need to be able to remove. For every degree of supercooling, one-eightieth of the water drop will change to ice on impact.

The temperature of SWDs determines the severity of the icing experienced. Namely, the closer the SWDs are to 0°C, the less of the water drops is frozen on impact. Therefore more of the water drops remain in liquid form that is released to flow backward on the airframe. SWDs between 0 and 23°C produce a clear ice sheet over the aerofoil.

Figure 184   Clear ice on an aerofoil.

Colder SWDs, i.e., closer to −45°C, freeze more of the water drops on impact. Therefore it releases less water to flow backward on the airframe; i.e., SWDs between −23 and −45°C produce rime ice mainly at the leading edge.

Figure 185   Rime ice on an aerofoil.

Obviously the larger the supercooled water drops, the more severe the icing.

NB: SWDs do not exist below approximately −45°C, where any moisture is already an ice crystal in the atmosphere, and do not stick to the aircraft's airframe. Therefore airframe icing is only possible between 0° and −45°C.

Superstall　A *superstall* may also be referred to as a *deep stall* or a *locked-in stall* condition which, as the name suggests, is a stall from which the aircraft is unable to recover. It is associated with rear-engine, high T tail swept-wing aircraft, which because of their design tend to suffer from an increasing nose-up pitch attitude at the stall with an ineffective recovery pitching capability.

The BEA Trident crash in 1972 at Slough, England, is probably the most famous and tragic outcome of a superstall.

A superstall has two distinct characteristics:

1. The nose-up pitching tendency at the stall is due to the following:
   a. Near the stall speed, the normal rooftop pressure distributions over the wing chord line change to an increasing leading-edge peaky pattern because of the enormous suction developed by the nose profile. At the stall this peak will collapse.
   b. A simple "virgin" swept or tapered wing will stall at the wing tip first (if the wing has not been designed with any inboard stall properties) mainly due to the greater loading experienced, leading to a higher angle of incidence and causing the wing tip to stall. Because of the wing sweep, the center of pressure (cp) moves inboard to a point where the cp is forward of the cg (center of gravity), therefore creating an increasing pitch-up tendency.
   c. The forward fuselage creates lift, which usually continues to increase with incidence until well past the stall. This destabilizing effect has a significant contribution to the nose-up pitching tendency of the aircraft.

      However, these phenomena themselves are not exclusive to high T tail, rear-engine aircraft and alone do not create a superstall. For a superstall to occur, the aircraft will have to be incapable of recovering from the pitch-up tendency at the stall.
2. An ineffective tailplane makes the aircraft incapable of recovering from the stall condition, which is due to the tailplane being ineffective because the wing wake, which has now become low-energy disturbed/turbulent air, passes aft and immerses the high-set tail when the aircraft stalls. This greatly reduces the tailplane's effectiveness and thus loses its pitching capability in the stall, which it requires to recover the aircraft. This is so because a control surface, especially the elevator, requires clean laminar stable airflow (high-energy airflow) to be aerodynamically effective.

See Fig. 186 at the top of the next page.

The superstall is followed by a rapidly increasing descent path due to greatly reduced lift and increased drag, which compounds the rate of increase of incidence, leading the aircraft to extreme angles of incidence and a deep stall with an ineffective pitching control from the tailplane.

The recommended recovery is to command a full nose-down attitude, and hopefully the high T tail elevator will receive clean air which will enable it to reestablish the pitching capability of the aircraft before ground contact. However, as this is very much a known problem, *stick pushers* are designed as protection on high T tail aircraft to auto engage after the *stick shaker* and before the aircraft achieves the high angle of incidence that results in a super/deep stall.

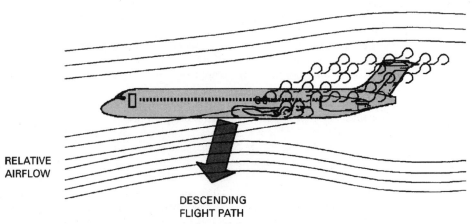

Figure 186   Superstall.

Surface Turbulence   *See* Turbulence—Surface.

Surface Wind   At the surface, the wind weakens in strength (speed) and backs in direction in the northern hemisphere (veers in direction in the southern hemisphere).

The wind speed reduces near the surface, compared to the free air geostrophic/gradient wind at 2000 ft, due to the friction forces between the moving air and the ground. The wind direction also changes, compared to the 2000-ft geostrophic/gradient wind. In the northern hemisphere the surface wind backs in direction, because as the wind speed reduces, the geostrophic force is reduced proportionally. Therefore it can no longer balance the pressure gradient force, and this results in the wind backing in direction near the surface in the northern hemisphere (veers in direction in the southern hemisphere). Therefore at the surface the wind flows inward to a low-pressure area and outward from a high-pressure area, rather than flowing parallel to the curved isobars, for both northern and southern hemispheres.

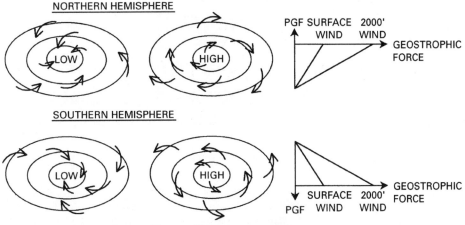

Figure 187   Surface wind around low- and high-pressure systems.

NB: Over the sea the change of speed and direction of the surface wind is less marked because there is less friction.

The diurnal variation (time of day) affects the degree to which the underlying trend of the surface wind is altered, i.e., to weaken in strength and back in direction (northern hemisphere) compared to the free air gradient wind (at approximately 2000 ft).

By day, the heating of the earth's surface and consequently the heating of the air in contact with it cause a vertical motion in the lower layers of the atmosphere convective currents. This results in a turbulent layer near the surface, which rises up to mix with the higher free stream of air. Mixing of the air in this fashion reduces the effect of friction at the surface by spreading it over a greater depth, and consequently the wind at the surface is only slightly changed compared to the free air gradient wind at altitude. That is, the *day surface wind* (1) loses less of its strength/speed and therefore backs only slightly compared to the free air gradient wind or (2) is a stronger wind that has veered compared to the night surface wind.

By night, the friction layer near the surface is shallower, due to the lack of heating. The gradient wind will continue to blow at altitude, but its effects will not be mixed with the surface airflow to the extent experienced during the day. Therefore the night surface wind will drop in strength, and the Coriolis effect will weaken, resulting in a marked directional difference between the gradient wind and the night surface wind.

That is, the *night surface wind* will drop in strength (speed) and back in direction significantly compared to the free air gradient wind, and to a slightly lesser extent compared to the daytime surface wind.

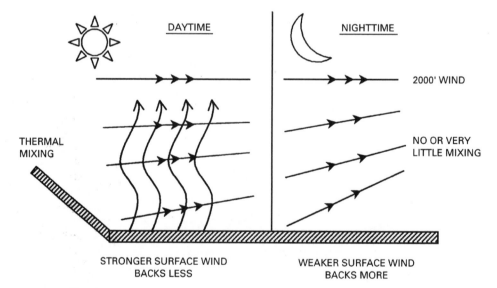

Figure 188 The diurnal variation of wind.

Hence, the surface wind by day will resemble the gradient wind more closely than the surface wind at night. Therefore the degree to which the surface wind changes direction and speed, compared to the free air gradient 2000-ft wind, is a function of

- Diurnal variation, i.e., nighttime or daytime
- Surface variation, i.e., land or sea

The following table gives a good approximation of the variation of the surface wind compared to the free air gradient wind (at approximately 2000 ft) above the surface layer of turbulence, considering both diurnal and surface variations, in the northern hemisphere.

	Day		Night	
	Direction	Speed	Direction	Speed
Land	−20°	×0.5	−40°	×0.3
Sea	−15°	×0.7	−15°	×0.7

NB: In the southern hemisphere the surface wind veers in direction compared to the free air gradient wind.

The surface wind is important to pilots because it relates directly, in terms of both its speed and direction, to the effect it has on the aircraft during takeoff and landings.

Its direction is normally measured in relation to magnetic north, so that it is relative to the runway direction, which is also aligned to magnetic north. However, it is measured at an approximate height of 10 m above the ground, i.e., where windsocks, etc., are placed; and therefore a slight discrepancy between the reported and actual surface wind may be present. However, any discrepancy is usually insignificant.

Surge—Engine A *surge* is the reversal of airflow through the engine, where the high-pressure air in the combustion chamber is expelled forward through the compressors, with a loud bang and a resultant loss of engine thrust. A surge is caused when

- All the compressor stages have stalled, e.g., Bunt negative g maneuver.
- An excessive fuel flow creates a high pressure in the rear of the engine. The engine will then demand a pressure rise from the compressors to maintain its equilibrium; but when the pressure rise demanded is greater than the compressor blades can sustain, a surge occurs, creating an instantaneous breakdown of the flow through the machine.

A surge is indicated by

- Total loss of thrust
- Large increase in TGT

The required action in response to an engine surge is to

- Close the throttles smoothly and slowly.
- Adjust the aircraft's attitude, to unstall the engines, which leads to the surge.
- Slowly and smoothly reopen the throttles.

Surveillance Radar Approach (SRA) *See* SRA (Surveillance Radar Approach).

Swept Wing The primary reason for the swept-wing design is to increase the wing's M_{Crit} speed, allowing the aircraft to achieve the high Mach cruise speeds its powerful jet engines are capable of delivering. The swept-wing design achieves this because it is sensitive to the (airflow) airspeed vector normal to the leading edge for a given aircraft Mach number. A swept wing makes the velocity vector normal (perpendicular) (*AC*) to the leading edge a shorter distance than the chordwise resultant (*AB*). As the wing is responsive only to the velocity vector normal to the leading edge, the effective chordwise velocity is reduced (in effect the wing is persuaded to believe that it is flying more slowly than it actually is). This means that the airspeed can be increased before the effective chordwise component becomes sonic, and thus the critical number is raised.

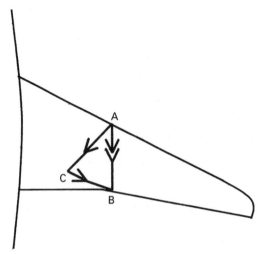

Figure 189 Airflow velocity vectors over a swept wing. *AC* = velocity vector normal to the leading edge, *AB* = effective chordwise velocity.

However, the designer also had to provide a satisfactory degree of lift from the swept wing that satisfied our main lift concern, that of adequate airfield performance. To accomplish this, the designer examined and developed the following features of the swept wing:

1. *Clean wing design areas*

    Of the seven main design areas of a jet aircraft wing, only the sweep of the wing does not positively contribute to the lift capabilities of the wing. However, the other six areas of the wing's design (as follows) do contribute to the lift capabilities of the wing. (However, remember that they also contribute to the high Mach speed capability of the aircraft, and they also have structural restraints. Both can limit the degree of lift design in an individual area.)

    a. Area.   The area of the wing is based on the basic lift formula

$$\text{Wing area} = \tfrac{1}{2}\rho V^2 S \,(\text{CL})$$

    A smaller than optimum area will spoil airfield performance as lift is directly proportional to wing area.

    b. Aspect ratio.   The induced drag of an aircraft is inversely proportional to the aspect ratio. The higher the aspect ratio, the less the induced drag, the better the lift/drag ratio, and therefore the greater the lift. However, an extremely high aspect ratio becomes limited by structural weight and by a very demanding cantilever ratio (i.e., structural depth at the root sufficient to support a long semispan), and too low an aspect ratio produces a high induced drag.

Therefore a compromised aspect ratio is designed to achieve maximum lift, for minimum induced drag without compromising structural requirements.

c. *Taper ratio.*   The taper ratio is the ratio of the root chord to the tip chord. The chosen ratio provides the optimum lift ratio across the span of the wing. A typical taper ratio is approximately 2.5:1.

A correct taper ratio will employ elliptical loading (each section of a wing produces the correct proportion of the total lift, e.g., the greatest amount of lift is produced at the root and minimum lift is produced at the tip).

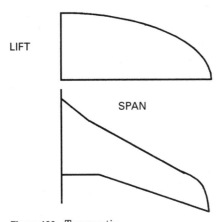

Figure 190   Taper ratio.

If a taper ratio from the optimum were employed, then this would result in an upset of the lift pattern across the span. If the ratio were too small, say, 1:1, then this would incur a penalty in structural weight of the aircraft. If the ratio were too large, say, 4:1, then this would incur a high local coefficient of lift at the wing tips, which would lead to a tendency for the tips to stall.

d. *Section.*   There are basically two types of section, which determine the lift pattern over the wing chord.

(1) *Rooftop* produces a significant proportion of lift distributed along the length of the wing chord, and it controls the local airflow, ensuring that it does not go supersonic and therefore remains shock-free. Because it is more predictable it is the normal section type employed by the designer.

(2) *Peaky* produces lift concentrated at the wing leading edge and allows for higher cruise Mach speeds before the onset of significant drag, but it is more penalizing at low airspeeds. It is less forgiving on errors and more dependent on correct pitch attitudes.

See Fig. 191 at the top of the next page.

e. *Chamber and twist.*   The amount of chamber on a wing is a compromise between its lift and speed requirements. A minimum amount of chamber is commonly used which trades lift for higher speeds. However, the chamber design does ensure the optimum lift distribution across the span, while twist/washout produces a lower angle of incidence at the wing tips, which

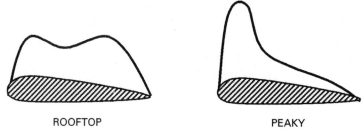

ROOFTOP   PEAKY

Figure 191 Section pressure distribution.

helps to prevent the tips from stalling and therefore ensures the lift capability at the wing tip. A sophistication of twist and chamber is now aimed at balancing conditions for the lowest cruise drag against the clean stall qualities.

f. *Thickness/chord ratio.* This is a compromise between the best aerodynamic qualities and the structural requirements. Structurally, the thicker the wing, the easier is the job to build and maintain the wing and to house gear, fuel tanks, etc. Also the thicker the wing, the greater the difference between the upper and lower surfaces and therefore the greater the lift. However, the aerodynamic argument is for thin wings with a small wetted area and therefore low profile drag, which allows higher cruise speeds or less sweep for the same cruise speed.

The aerodynamic argument for higher speeds is normally dominant in the design of the wing's thickness/chord ratio. Therefore the wing is designed as thin as possible that just allows for the structural requirements to be met while optimizing its aerodynamic high-speed qualities, which means that the lift qualities of the wing's thickness/chord ratio are sacrificed.

In the end, the final design of the clean swept wing is a compromise between meeting the conflicting requirements of speed, structural needs, and lift. Simply, the clean swept-wing design has traded lift for higher cruise speeds, which results in an insufficient coefficient of lift at low speeds for the required field performance levels.

Therefore the designer still has to further optimize the lift design on a swept wing to meet the field performance lift parameters, which she or he achieves by adding high-lift devices to the clean wing.

2. *High-lift devices*—trailing-edge flaps, leading-edge flaps, and slats and slots—are added to the wing to increase the wing's chamber, chord line, and area to produce a greater coefficient of lift, while slots are used to introduce new clean air to prevent/delay the separation of the boundary layer and so suppress the stall and therefore increase the overall lift.

Again the degree of sophistication of these high-lift devices is determined by the airfield performance required to be added to the clean wing.

Finally the designer has a wing with a CL sufficient to fulfill its field performance requirements that doesn't reduce its speed or structural abilities.

The advantages to a jet aircraft from the swept wing are as follows:

1. *High Mach cruise speeds.* The swept wing is designed to enable the aircraft to maximize the high Mach speeds that its jet engines can produce.

The straight-wing aircraft experiences sonic disturbed airflow, resulting in a loss of lift at relatively low speeds. Therefore a different wing design was required for the aircraft to be able to achieve higher cruise speeds. The swept-wing design delays the airflow over the wing from going supersonic, and as such it allows the aircraft to maximize the jet engine's potential for higher Mach cruise speeds.

Additionally the swept wing is designed with a minimal chamber and thickness, thereby reducing profile drag, which further increases the wing's ability for higher speeds.

2. *Stability in turbulence.* Ironically a disadvantage of the swept wing is its poor lift qualities, which is an advantage in that it is more stable in turbulence compared with a straight-wing aircraft. This is so because the swept wing produces less lift and therefore is less responsive to updraughts, which allows for a smoother, more stable ride in gusty conditions.

Because the swept wing is designed for high cruise speeds, it suffers from the following disadvantages as a consequence:

1. Poor lift qualities are experienced because the swept-back design has the effect of reducing the lift capabilities of the wing. Therefore the swept-wing aircraft requires a higher angle of attack and/or speed to maintain a given lift value, which is especially evident on the approach. The higher angles of attack employed result in poorer visibility for the pilot. (See Fig. 189, airflow velocity vectors over a swept wing.)

The free stream resultant AB is felt by the wing as the velocity vector AC normal (perpendicular) to the leading edge and the washed-out spanwise vector CB. This is so because the wing is only responsive to velocity vectors normal (perpendicular) to the leading edge. Because AC is less than $AB,$ the apparent airspeed is lower than the real airspeed and there is a corresponding reduction in lift. Therefore sweep reduces the wing's basic lift capabilities.

2. Higher stall speeds are a consequence of the poor lift qualities of a swept wing. A higher angle of attack is required to maintain a given amount of lift, which creates a higher induced drag value and results in the airflow over the wing stalling at a higher speed.

3. Speed instability is the second consequence of poor lift at lower speeds for the swept-wing aircraft, and it is due to the manner in which lift and drag vary with speed in the lower ranges. Around VIMD on the rather flat jet drag curve, the aircraft does not produce any noticeable changes in flying qualities, except the rather vague lack of speed stability. Speed is unstable below VIMD because the aircraft is now sliding up the back end of the jet drag curve, where power required increases with reducing speed. This means that despite the higher CL associated with lower speeds, the drag penalty increases faster than the lift; therefore the lift/drag ratio degrades, and the net result is a tendency to progressively lose speed. So speed is unstable because of the drag penalties particular to the swept wing.

4. A wing tip stalling tendency is particular to a swept-wing aircraft mainly because of the high local CL loading it experiences. Uncorrected (in the design), this effect would make the aircraft longitudinally unstable, which is a major disadvantage.

A simple swept and/or tapered wing will stall at the wing tip first, if it is not induced/controlled to stall at another wing section first by the designer.

This is true because the outer wing section experiences a higher aerodynamic loading due to the wing taper, which causes a greater angle of incidence to be experienced to a degree where the airflow stalls at the wing tips. The boundary layer spanwise airflow, also a result of sweep, further contributes to the airflow stalling at the wing tips. A stall at the wing tip causes a loss of lift outboard and therefore aft (due to the wing sweep) which moves the center of pressure inboard and therefore forward; this produces a pitch-up tendency, which continues as the wing stalls progressively farther inboard.

A wing tip stall is resolved in the wing design with the following better aerodynamic stalling characteristics:

1. The greater chamber at the tip increases airflow speed over the surface, which delays the stall.

2. Washout or twist creates a lower angle of incidence at the wing tips, which delays the effect of the outboard wing loading that causes the stall. Because spanwise airflow aggravates the outboard stall, fences are used to prevent this airflow to the wing tips. Also leading-edge breaker strips on the inboard (root) section of the wing are used to induce the breakup of air over the root section of the wing before the tips, thereby resulting in an inboard wing stall, which produces a safer, more stable nose-down aircraft stall.

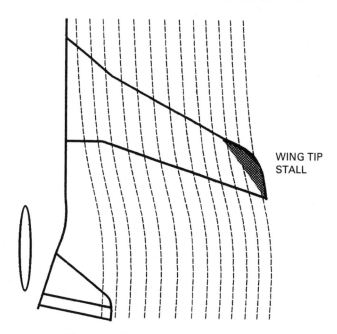

WING TIP
STALL

Figure 192 Wing tip stall.

T

Tab Surface   A tab surface is a small hinged surface found on a primary flight control surface. Usually it forms part of the trailing edge on a manual control surface.

A tab can be used to reduce/balance the opposing hinge moment (air load force) on the associated control surface, with the following systems.

1. *Trimming tab.*   A trimming tab is used to balance the main forces (hinge moment and pilot input) on its associated control surface, to maintain its deflected position. For example, an aircraft requiring a pull force to maintain its required attitude deflects the elevator upward. To relieve the pilot of this control force, a trimming tab is provided, which is moved in the opposite direction (downward) to the elevator and produces an upload that balances the hinge moment (air load force) and thereby maintains the elevator in its required position. Trimming tabs are commonly found on stabilizer/elevator, rudder, and ailerons and are controlled by trim wheels or switches on the flight deck.

2. *Control balance.*   A control balance tab is required to assist in moving a control surface when the hinge moment (air load force), which is trying to return the surface to its neutral position, is too great for the pilot to handle unaided. This is so because the hinge moment (air load force) experienced on the control surface increases with an increase in the lift force which itself increases with speed (hinge moment = lift force × arm) to a point (speed) whereby the hinge moment (air load force) felt by the pilot through the control column or rudder pedals is too great for the pilot alone to manually move the control surfaces. Therefore the control surface requires this hinge moment (air load force) to be reduced to enable the pilot to operate the controls. This can be achieved in a number of aerodynamic ways, such as control balance tabs, hinge/horn balance, and mass balance devices.

NB: Such balancing also helps to prevent a reversal of the control surface, e.g., aileron reversal.

Control tabs are commonly found on elevators, rudders, and ailerons.

3. *Servo tab.*   The purpose of a servo tab is to operate the control surface. The movement of the servo tab is brought about directly by the control column or rudder pedals (with no direct connection of the pilot's controls to the primary control surfaces). This movement of the tab will then produce a movement of the primary control surface in the opposite direction. For example, a pull on the control column produces a downward movement of the servo tab which causes an upward movement of the elevator, thereby pitching the aircraft up in response to the command.

The servo tab is used to reduce the loads experienced by the pilot, which it does because the servo tab's surface is smaller and the smaller the size of the surface, the less the lift force and the less the hinge moment (air load force) experienced.

The servo tab is usually found on aircraft where a normal balance tab is insufficient to balance the control surface hinge moment (air load force) to an acceptable manageable level for the pilot.

TAF Terminal aerodrome forecasts (TAFs) are coded routine weather forecasts for an aerodrome. It is a weather forecast given by a qualified meteorological forecaster based at the aerodrome.

NB: Cloud base in a TAF is given AAL (above aerodrome level).

TAFs are usually issued for a 9-h period and updated every 3 h. However, they may be issued for up to 24 h with updates every 6 h, but their accuracy is not as high.

Example of TAF: EGCC / 15 0600 / 0716 / 210/15 / 5000 / SHRA / BKN 025 / TEMPO / 1116 / 3000 / SHRA / PROB30 / 1416 / TSRA / BKN 006 CB   This can be decoded as follows:

EGCC	Location identifier (Manchester, United Kingdom)
15 0600	Time the report was taken (15th day 06.00 h UTC)
0716	Validity time of forecast (0700 to 1600 h UTC)
210/15	Wind direction and speed (210° true at 15 kn)
5000	Horizontal visibility (5000 m)
SHRA	Weather (rain showers)
BKN 025	Cloud (broken at 2500 ft)
TEMPO	Variation
1116	Validity time of Tempo (1100 to 1600 h UTC)
3000	Horizontal visibility (3000 m)
SHRA	Weather (rain showers)
PROB30	Probability
1416	Validity time of Prob (1400 to 1600 h UTC)
TSRA	Weather (thunderstorm rain showers)
BKN 006 CB	Cloud (broken at 600 ft with cumulonimbus clouds)

Tailwind Component   *See* Wind Component.

Takeoff—on a Contaminated (Wet) Runway   The following operational points need to be used, or at least considered, for a takeoff on a contaminated runway:

1. Make the correct performance calculations for a contaminated runway.

2. Use engine ignition (manual or override) to lessen the possibility of ingestion causing an engine malfunction (flameout) on the ground.

3. Use the highest takeoff power available to keep the distance on the ground to a minimum; e.g., on some aircraft types TOGA may be required, while on other types a reduction of the flex temperature may be used.

4. Use a wet $V1$ speed, where possible.

NB: When distance is available, a progressive reduction in $V1$ to a wet $V1$ speed reduces the risk associated with an aborted takeoff (overrun) without unduly compromising the continued takeoff.

5. Where alternative flap settings are available, use a higher angle setting to keep distance on the ground to a minimum.

6. Don't forget the importance of clearing snow, frost, or ice from the aerofoil surfaces before attempting a takeoff.

The following additional operational points need to be followed, or at least considered, for a takeoff on a slush-contaminated runway:

7. Make the correct performance calculations for a slush-contaminated runway.

8. Remember during planning that small temperature changes can make rapid and significant alterations to the runway surface conditions. Therefore either have some distance margin in hand or have available a recent runway survey.

9. When performance data are quoted for standard and chimed tires, make sure you use the appropriate data for the tires fitted.

NB: Chimed nosewheel tires deflect the nosewheel spray pattern downward, thus reducing the drag due to slush impinging on the airframe and reducing the amount of slush ingestion by the engines.

10. Wherever possible, avoid operations in crosswind conditions, because of the difficulties involved in maintaining directional control.

NB: On takeoff, the rudder control usually becomes effective around 70 to 80 kn. Therefore directional control has to be maintained with just a touch of steering and some delicate use of differential braking below 70 to 80 kn. If the operation looks as if it's getting out of hand, keep the aircraft straight at all costs and accept that running off the end of the runway, straight, at low speed, is a lesser hazard than sliding off sideways at high speed.

11. On some aircraft early nosewheel raising is permitted during the takeoff roll in slush. This technique is desirable to reduce ingestion as well as reduce take-off distance and damage to the airframe.

NB: The earliest speed at which the nosewheel can be raised is a function of the cg position. The farther aft the cg, the lower the speed. This technique should not be employed unless it is specifically permitted in the flight manual.

12. After takeoff, beware of the possibility of the effects of the slush accumulation on the control surfaces, either directly or after freezing.

Takeoff—Crosswind   In a crosswind, an aircraft will weathercock into wind because of the large keel surface behind the wheels.

Planning   Ensure that the crosswind is within the limits of the aircraft.

On the runway   Use the rudder to keep the aircraft straight during the takeoff roll. Providing that the aircraft's crosswind limit is not exceeded, this directional control will be possible. A crosswind from the left will require right rudder to counteract the weathercock effect. Less rudder is required as the speed increases due to the greater airflow over the surface, which increases the rudder's effectiveness. Hold the nosewheel firmly on the ground until liftoff assists in directional control. Counteract a crosswind lifting the into wind wing with aileron; i.e., move the control column into the wind, to raise the aileron on the into-wind wing to keep the wings level. Less aileron is required as the speed increases due to the greater airflow over the surface, which increases the aileron's effectiveness. A crosswind from the left requires the control column to be moved to the left. Hence the controls are crossed in a crosswind takeoff, i.e., right rudder and control column to the left.

Takeoff   Adopt a smooth but positive rotation to obtain the normal climb attitude. Simultaneously center the rudder and aileron controls.

Climb-out   Once well clear of the ground, turn into the wind sufficiently to counteract the drift, and climb out normally on the extended runway centerline. Any remaining "crossed" controls should be removed once airborne.

Takeoff Distance Available (TODA)   Takeoff distance available (for all engine opera-tions) is the length of the usable runway available plus the length of the clearway available, in which the aircraft initiates a transition to climbing flight and attains a screen height at a speed not less than the takeoff safety speed (TOSS), or $V2$.

$$\text{TODA} = \text{usable runway} + \text{clearway}$$

TODA is not to exceed 1.5 times TORA (takeoff run available).

The TODA/R and TOSS/V2 are the vital performance figures for a pilot during the takeoff.

Takeoff Distance Required (TODR)   The takeoff distance required (for all engine operations) is the measured distance required to accelerate to the rotation speed V_R and thereafter effect a transition to a climbing flight and attain a screen height at a speed not less than the takeoff safety speed (TOSS) or $V2$. The whole distance is factored by a safety margin, usually 15 percent.

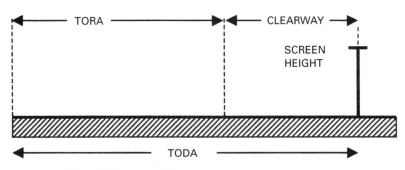

Figure 193   Takeoff distance available.

Takeoff Flight Path—Options   The takeoff flight path is based upon the following technique and options.

The flight path begins at reference zero, where the jet aircraft has attained a height of 35 ft (in dry conditions) and $V2$ after the failure of its critical engine at $V1$. Landing gear retraction is completed at the end of the first segment, and the climb is continued at $V2$ with takeoff flaps and takeoff thrust on the operating engine(s) until the second- to third-segment level-off transition takes place at one of these:

1. A minimum gross height of 400 ft
2. The maximum standard gross height, usually 1000 ft

After option 1 or 2, the aircraft is accelerated in level flight, flaps are retracted, and acceleration is continued to the final segment where the climb is reassumed on maximum continuous thrust.

3. The maximum height, which can be reached using one of these:
    a. Takeoff thrust for its maximum period after the brakes are released, usually 5 min
    b. Maximum continuous thrust to an unrestricted height

Option 3 is used to clear distant obstacles, i.e., in the normal third segment, using an extended $V2$ technique to gain a greater climb gradient.

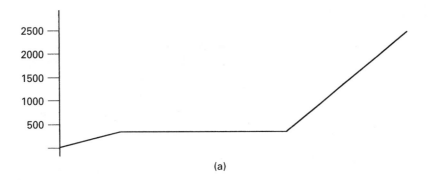

(a)

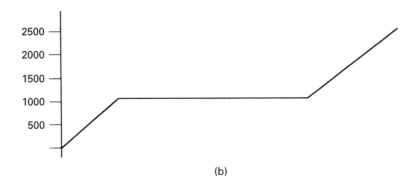

(b)

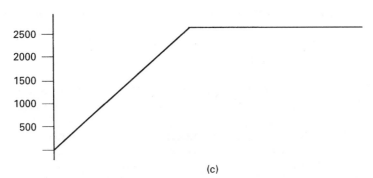

(c)

Figure 194 Takeoff flight path options. (*a*) Second- to third-segment level-off at 400 ft gross height. (*b*) Second- to third-segment level-off at maximum standard gross height, usually 1000 ft. (*c*) Second segment continued to either takeoff thrust limit or maximum continuous thrust unrestricted height.

Takeoff Performance   An aircraft's takeoff performance is subject to many variable conditions, including these:

Aircraft weight   An increased aircraft weight results in a greater takeoff distance required (TODR) and a reduced net takeoff climb gradient. An increased aircraft weight has the following variable effects that increase the TODR and NTOFP:

1. On the ground, the greater aircraft weight results in a slower acceleration, for a given thrust setting; hence a greater takeoff run distance is required.

2. The greater aircraft weight on the wheels during the takeoff run increases the friction forces with the runway surface, for a given thrust setting; hence a greater takeoff run distance is required.

3. The greater aircraft weight has a greater momentum that increases its EMDA stopping distance required in a RTO. Therefore on field length-limiting runways, the takeoff run available has to be reduced to provide adequate stopping distance.

4. The stalling speed is increased as the aircraft weight becomes greater. This is so because the liftoff speed V_R is indirectly related (via V2) to the stalling speed V_S; the V_R speed is therefore proportionally increased with increased aircraft weight, which results in an increased takeoff run required for a given thrust setting.

5. After liftoff, the increased weight degrades the aircraft's climb performance, i.e., ROC, which results in the TODR over the ground being increased, and so the net takeoff climb gradient is reduced.

Aircraft flap setting   The effect of flaps on the takeoff performance, that is, TOR/TOD, and climb performance varies according to aircraft type, especially between swept-wing (jet) and straight-wing (turboprop) aircraft, and further with the degree of flap deployed on individual aircraft.

Swept-wing aircraft require a low flap setting, i.e., takeoff flap, to improve the CL during the takeoff. This has two positive effects. First, a low flap setting reduces the aircraft's takeoff run required (TORR), because the higher CL lowers the stalling speed VS which in turn reduces the V2 and VR speeds and results in the aircraft reaching its liftoff speed from a shorter ground takeoff run. Second, a low flap setting reduces the aircraft's TODR because the increased CL benefits outweigh the increased airborne drag and thereby improve the climb performance of the aircraft so that the aircraft reaches the screen height over a shorter distance.

The maximum takeoff flap may be used to reduce the ground takeoff run required when the field length is limiting or the runway surface is poor. However, airborne climb performance may be compromised due to an increase in drag, which reduces the lift/drag ratio and results in a reduced rate of climb (climb gradient) performance.

The use of flap settings outside its takeoff range would increase aerodynamic drag during the ground run, causing a slower acceleration that results in an increased TORR. That, once airborne, would significantly degrade the climb gradient performance because of the poor lift/drag ratio that results in an unacceptable increase in TODR.

Straight-wing (turboprop) aircraft usually require a low flap setting to improve the CL during takeoff. This has the effect of reducing the aircraft's takeoff run required because the higher CL lowers its stalling speed V_S, which in turn reduces its V_R speed and results in the aircraft reaching its liftoff speed V_R from a shorter ground run. However, the TODR to the screen height may not be reduced significantly, if at all, because, as well as increasing lift, flap deployment increases drag, thus reducing the lift, drag ratio and resulting in a lower rate of climb.

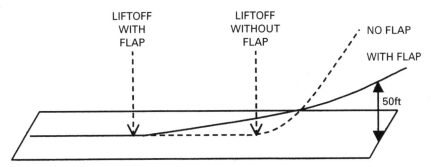

Figure 195 Effect of flaps on the takeoff ground run.

For this reason only a small takeoff flap setting is used for the takeoff to maintain an adequate airborne climb performance. However, large takeoff flap settings may be used to reduce the ground takeoff run as much as possible, so long as the climb performance to the screen height is not compromised, when the field length (runway) is limiting or the runway surface is poor.

Conversely, no flap may often be used when takeoff distance is limiting. For large flap settings above the takeoff range, say >20°, the increased aerodynamic drag during the takeoff ground run causes a slower acceleration that results in an unacceptable increase in the takeoff run required. That, once airborne, significantly reduces the lift/drag ratio that degrades the climb gradient performance to an unacceptable level.

The principles of flap deployment on the takeoff performance for both swept- and straight-wing aircraft are similar, but the effects are much more acute for swept-wing aircraft. Therefore the use of takeoff flap deployment is much more rigid on swept-wing aircraft, whereas the straight-wing (turboprop) aircraft has greater flexibility and variation between different aircraft types.

Aerodrome pressure altitude. A high-pressure altitude (common to high-elevation aerodromes) decreases an aircraft's performance which results in an increased takeoff distance required. The higher the pressure altitude of the aerodrome, the lower the air pressure and the lower the air density. A low air density reduces an aircraft's engine and aerodynamic (CL) performance and results in an increase in takeoff run and distance required for a given aircraft weight.

Ergo, *high* means a decrease in performance that results in greater takeoff distance required or a lower takeoff weight.

Air density/density altitude (temperature and pressure altitude) An increase in density altitude (which represents a decrease in air density) increases the takeoff distance required. Density altitude is a function of pressure altitude and temperature; and the higher the density altitude, the lower the air density. That is, high-pressure altitude and/or temperature decreases air density.

The low air density associated with a high-density altitude decreases the aircraft's engine and aerodynamic (CL) performance, resulting in an increased takeoff run/distance and reduced climb performance for a given aircraft weight due to the following effects:

1. The engine (including the propeller) will not produce as much power/thrust, because the ambient air density is less and therefore the mass of air entering the engine is reduced, causing a decrease in engine thrust produced. Therefore the ground acceleration and airborne climb performance will be poorer as the density altitude increases. Hence a longer TOR/D is required.

2. The aerodynamic performance of the aircraft is also reduced as air density decreases, because true airspeed (TAS) increases, which requires a longer takeoff distance. This is true because to produce the required aerodynamic lift force [remembering the lift formula $L = CL \frac{1}{2} \rho V^2 (span)$] the increase in TAS compensates for the decrease in air density.

Ergo, *hot* and *high* means a decrease in performance that results in either a greater TODR or a lower TOW.

Takeoff performance tables and graphs allow for variations in the pressure altitude and the ambient temperature of the aerodrome, i.e., the density altitude.

Humidity. High humidity decreases air density, which decreases an aircraft's aerodynamic (CL) and engine performance and results in an increased TOR/D required for a given aircraft weight. This is so because when humidity is high, lighter water molecules replace some of the air's heavier molecules, which results in a lowering of the overall air density. And the lower the air density, the lower the aircraft's aerodynamic and engine performance, resulting in an increased TOR/D required for a given aircraft weight.

Ergo, *hot, high,* and *humid* means a decrease in performance that results in either a greater TODR or a lower TOW.

Wind Wind has a profound effect on the takeoff performance of an aircraft. An aircraft may experience a headwind, tailwind, or crosswind.

A headwind reduces the takeoff distance required for a given aircraft weight, or it permits a higher TOW for the TOR/D available. This is so because the field length-limited TOW will be favorably influenced by a headwind component; i.e., the kinetic energy is expressed in terms of ground speed corresponding to a given V2 speed. This will result in an earlier takeoff for the same weight or enable an increase of weight for a given runway. Or in simpler terms, an aircraft requires a certain speed relative to the air it is flying through to generate lift. With a headwind an aircraft that is stopped at the end of the runway and facing into a headwind, say of 30 kn, is already 30 kn closer to its liftoff speed compared to the nil-wind situation. The aircraft therefore reaches its liftoff speed at a lower ground speed, and so less ground run is required. In the air the angle of climb relative to the ground is also increased by the headwind, giving a better obstacle clearance performance. Furthermore the wind usually increases in strength as you move farther away from the ground, due to the friction forces between the ground and the wind causing the wind to slow down near the surface; i.e., friction forces diminish as you move away from the surface. Therefore when climbing into an increasing headwind, it tends to increase your airspeed (the speed of the aircraft relative to the air), and also this increases the aircraft's gradient of climb (angle) over obstacles along the flight path.

Ergo, the greater the headwind, the better the aircraft performance.

A tailwind increases the takeoff distance required for a given aircraft weight, or it requires a lower TOW for the TOR/D available. This is so because the tailwind worsens the aircraft's performance, as it has to accelerate to a ground speed that matches the wind speed before it has an effective speed of zero, relative to the air that it is moving in; and then it has to accelerate to its liftoff speed. The effect of this is to lengthen the ground run and to flatten the climb performance. Furthermore, a tailwind would increase in speed with altitude, reducing your airspeed further, and hence the aircraft's pitch attitude has to be lower to retain the desired IAS. Therefore the climb angle (gradient) is further degraded as you climb out.

Ergo, the greater the tailwind, the worse the aircraft performance.

It is for this reason that a takeoff with a tailwind shows very poor airmanship.

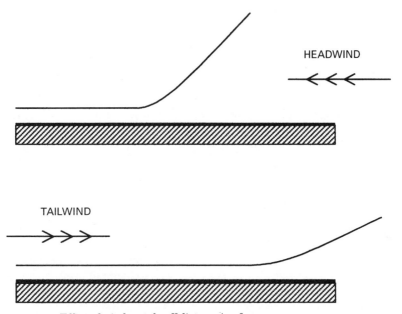

Figure 196   Effect of wind on takeoff distance/performance.

Not more than 50 percent of the reported headwind or not less than 150 percent of the reported tailwind should be used to calculate the takeoff performance.

These adjustments provide a safety margin to the reported wind, which covers acceptable fluctuations of the actual wind experienced, i.e., the actual headwind not being as strong as reported or the actual tailwind being stronger than reported. This ensures that the aircraft can achieve its takeoff run, distance, and climb performance or landing distance requirements.

NB: Most performance charts and tables already factor in these adjustments.

It is extremely bad airmanship to attempt a takeoff with a tailwind.

An aircraft must not take off in a crosswind that exceeds its certified maximum crosswind limitation for the aircraft type, to safeguard directly the directional and lateral control and indirectly the takeoff run performance of the aircraft.

If the certified crosswind limitation of the aircraft is exceeded, it has the following effects:

1. *Directional control is reduced.*

The effect of the crosswind on the aircraft's keel surface causes the aircraft to weathercock into the wind. This can be balanced by the aerodynamic lift force of the rudder. However, above the certified maximum crosswind limit the rudder has an insufficient force to balance the weathercock effect and maintain an adequate VMCG directional control ability to an unacceptable level.

2. *Lateral control is reduced.*

A further effect of the crosswind is that it tries to lift the into-wind wing, which has to be held down with aileron. However, above the certified maximum crosswind limit, the aileron will have insufficient force to balance the lift force on the into-wind wing; therefore the aircraft's lateral control is reduced.

3. *Drag is increased.*

Above the maximum certified crosswind limit, some aircraft types may be able to maintain directional and/or lateral control with maximum deflected rudder and/or aileron control surfaces, but suffer from an increase in drag on these surfaces. This results in an unacceptable decrease in acceleration during the takeoff run that is detrimental to the aircraft's takeoff performance.

Runway length, slope, and surface (including wet or icy conditions)   The length of the available runway is one of the performance limitations that restrict the maximum weight of the aircraft. The greater the runway length, the greater the acceleration the aircraft can gain and the higher the liftoff speed VR the aircraft can obtain. And because the VR speed is related to the aircraft weight, it can be seen that the longer the runway available, the greater the possible aircraft takeoff weight.

NB: Other limitations, such as the maximum permissible structural certificate of airworthiness (C of A) weight of the aircraft, may be more limiting than the runway length.

A hard and dry runway surface allows good acceleration on the ground and therefore reduces the takeoff run required for a given aircraft weight, or allows a higher aircraft weight for a given runway length. On other surfaces, i.e., grass or wet contaminated hard surfaces, the acceleration is retarded on the ground; therefore the takeoff run and distance required are increased.

NB: Takeoff performance graphs and tables (i.e., field length) normally allow for such variations in surface conditions.

A downward slope allows the aircraft to accelerate faster; therefore, the takeoff run and distance required for a given aircraft weight are reduced; or a higher maximum takeoff weight is possible for a given runway length, compared to a level runway.

An upward slope hinders the aircraft's acceleration; therefore, the takeoff run and distance required for a given aircraft weight are increased; or the aircraft maximum takeoff weight is reduced for a given runway length, compared to a level runway.

NB: Takeoff performance graphs and tables (i.e., field length) normally allow for such variations in runway slope.

The various takeoff performance charts and tables are presented logically and allow you to enter known conditions such as temperature and pressure altitude and to proceed through the chart or table to end up with an answer for TODR, given a set aircraft weight. Or you can find the highest allowable aircraft weight given a set TODA for the ambient conditions.

Takeoff Performance—Limitations   The normal performance limitation for an aircraft's takeoff profile relates to the aircraft's weight. That is, the aircraft must weigh less than the most restrictive weight to meet all the various performance criteria during the takeoff profile. This weight is known as its *maximum takeoff weight* MTOW.

The normal takeoff performance criteria are as follows:

1. Determine the most restrictive field length (TORA) available from
   a. All-engine runway length (TOR/D)
   b. Emergency distance available (EMDA)
   c. One-engine inoperative runway length (TOR/D)
2. Determine the most restrictive MTOW from
   a. The most limiting field length (TOR/D) available (as determined from point 1)
   b. Weight, altitude, and temperature (WAT) limit
   c. Tire speed limit
   d. Net takeoff flight path
      (1) Climb gradient required for obstacle clearance with one engine inoperative
      (2) Different types of climb gradient (profiles) employed are
          Normal
          Increased V2
          Extended V2
          Climbing turns
   e. Aircraft certified (ca) maximum structural weight
3. Determine the performance speeds, given the calculated most limiting aircraft weight.
   a. $V1$, V_R, and V2.

NB: Performance V speeds are a function of the stall speed V_S, which itself is a function of the aircraft weight. Ergo performance V speeds are weight-determined.

Check that

$$V1 > VMCG \quad \text{and} \quad V_R \geq V1$$

   b. VMBE. Check that $V1 <$ VMBE. If $V1 >$ VMBE, then the aircraft's gross weight has to be reduced until $V1 <$ VMBE.

This gives the maximum takeoff weight for normal operation conditions, which allows you to determine the takeoff EPR given the aircraft weight and V speeds.

Takeoff Roll (40 to 100 kn)   A 40- to 100-kn call during the takeoff roll is used to check the requirements that need to be established by the called speed. These requirements include the following:

1. Directional control surface (vertical tailplane) starts to become effective with all engines operating.
2. Takeoff EPR should be set by this check speed so that the pilot is not chasing engine needles for a prolonged period during the takeoff roll.

3. Cross-check the airspeed indicator gauges to ensure their accuracy and reliability.

In addition, type-specific requirements might need to be established by the take-off roll check speed.

Takeoff Run Available (TORA)   Takeoff run available (for all engine operations) is the usable length of the runway available that is suitable for the ground run of an aircraft taking off. In most cases this corresponds to the physical length of the runway.

Takeoff Run Required (TORR)   Takeoff run required (for all engine operations) is the measured run (length) required to the unstick speed V_R plus one-third of the airborne distance between the unstick and the screen height. The whole distance is then factored by a safety margin, usually 15 percent.

TAT   The TAT is the total air temperature indicated on the air temperature instrument, which is a product of the SAT and the adiabatic compression (RAM) rise in temperature experienced on the temperature probe.

NB: Therefore TAT is a higher temperature than OAT whenever there is an airflow into the temperature probe, which is sometimes referred to as a *heating error* when you need to calculate the actual OAT. This heating error can remain after a flight, due to the residual heat left in the probe, which is why you sometimes have a difference between OAT and TAT shortly after a flight.

Taxiway—Lights   The two types of taxiway lighting systems are

1. One line of green taxiway centerline lights

2. Two lines of blue taxiway edge lights

NB: At certain points on the taxiway, there may be red stop bars, to indicate the position of a hold, etc., for instance, before entering an active runway.

TCAS   The Traffic (Alert) Collision Avoidance System provides traffic information and maneuver advice between aircraft if their flight paths are conflicting. TCAS uses the aircraft's SSR transponders and is completely independent of any ground-based radar units. It is rapidly becoming a mandatory requirement around the world and is already established in U.S. airspace.

TCAS I is an early developed system that provides traffic information only.

TCAS II is a later developed system that provides additional maneuver advice, but in the main it is restricted to vertical separation.

TCAS IV is a new TCAS under development (1998), which will give resolution advisories (RAs) in the horizontal as well as the vertical plane. However, further development of TCAS IV is likely to be canceled in preference to ADS-B.

An aircraft's TCAS will interrogate the SSR transponders of nearby aircraft to plot their positions and relative velocities. Directional finding aerials obtain the relative bearings of other aircraft, and distance is calculated by using the time delay between the transmitted and received signals. With this information the TCAS computer can determine the track and closing speeds of other aircraft fitted with transponders; and where it determines a collision is possible, it provides visual and aural warnings as well as command actions on how to avoid the collision. This is all done with vertical avoidance commands only. There is as yet no turn command given.

The warnings and advice (advised actions) from the system have different levels:

Initially   A traffic advisory (TA) warning is generated for other traffic that "may" become a threat. "No maneuver is advised or should be taken."

Collision threat   A resolution advisory (RA) warning is generated when a threat is considered to be on a collision course.

Advice on a maneuver in the pitching plane, i.e., rate of climb or descent to avoid the collision, is generated which can be increased or decreased as the threat increases or reduces until a "clear of conflict" advice is given. Therefore you only respond to an RA, which should be done promptly and smoothly and should take precedence over ATC clearance to avoid immediate danger.

RA use should be restricted in the following circumstances:

- In dense traffic area (limited to TA use)

- Descent recommendations inhibited below 1000 ft

- All RAs inhibited below 500 ft

    NB: All TAs also are restricted below 400 ft.

TCAS can cope with mode A, C, and S transponders. However, when both aircraft are equipped with TCAS II and mode S, the advice on how to avoid a collision will be coordinated by the mode S data link between the two aircraft.

The pilot's order of priority when given a wind shear, GPWS, and TCAS warning at the same time is

First,   wind shear

Second,   GPWS

Third,   TCAS

Teardrop/Base Turn   The teardrop/base turn is used to reverse the direction by more than 180° onto an inbound final-approach track. It consists of the following:

- A specified outbound track, from a fix to a set distance or time

    NB: Different outbound tracks exist for different categories of aircraft based on their speed, to allow sufficient space for the inbound turn. That is, faster aircraft require a larger turning radius.

- Followed by a turn to intercept the inbound track of an instrument approach

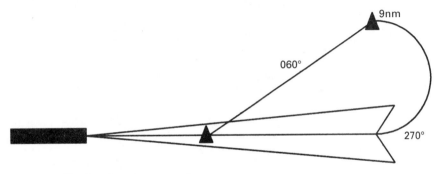

Figure 197 Teardrop/base turn procedures.

Temperature   Temperature is a measure of molecule agitation in a substance, which is represented as the hotness of a body. Therefore temperature can be thought of as a measure of the hotness of a body. As a body of matter absorbs heat energy, its molecules become more agitated. This agitation and motion are measured as a temperature. The molecular agitation in a body is theoretically zero at the absolute zero temperature $-273°C$ (or 0 K).

It is important to clearly distinguish between heat and temperature. Heat is a form of energy which, once absorbed by a substance, agitates its molecules. Temperature is a measure of the degree of molecular agitation in the substance, which is represented as the hotness of the body. Therefore measuring the temperature of a body is not the same as measuring the heat energy within a body, because the amount of heat energy corresponding to a given temperature depends upon the capacity of the substance to hold the heat energy. This is known as *specific heat capacity*. However, in a general sense the temperature of a body can give you a rough indication of the heat energy.

The temperature at the earth's surface, and the lower atmosphere where most of the weather is found, is dependent upon two factors:

1. How much heat energy in the form of shortwave energy (solar radiation) reaches the earth's surface. This is dependent upon the following factors:

    a. *Latitude.*   The more directly the sun's rays hit the earth's surface, the greater the quantity of heat energy transferred and so the greater the temperature. The sun is directly overhead the tropics, and this is where the intensity of insulation is greatest. As we move away from the equator, the radiation hits the surface obliquely and the energy is spread over a wider area, reducing the heating effect. Hence this contributes to the warmer climates at the tropics and the colder climates at the earth's poles.

    b. *Season.*   The earth's axis of rotation is tilted with reference to its orbit. In December the north pole is tilted farthest from the sun, and in June it is tilted at its closest point toward the sun. This tilting creates the same effect as a change of latitude and is the primary cause of the different seasons, and associated temperature changes, throughout the year.

    c. *Time.*   The time of day determines the amount of the sun's heating on the earth's surface. When the sun is highest in the sky, at midday, the sun's heating is at its maximum. However, due to the time delay associated with the cumulative heating effects of reradiating this energy into the lower atmosphere, the highest temperature of the day is not experienced until later in the afternoon, at approximately 1500 h during the summer months.

2. The energy absorption (and retention) capacity of the surface.

    a. *Absorption.*   The type of surface determines how much heat energy is absorbed, i.e., its reflective quality. At its simplest level a white surface, say, snow or ice, reflects a large amount of radiation, approximately 80 percent. Hence this also contributes to the colder climates at the earth's pole. Grass, water, and built-up areas absorb approximately 50 percent of the solar radiation. Dark areas, i.e., dense forests such as rainforests in the tropics, will absorb approximately 90 to 95 percent of the solar radiation, which contributes to the warmer climate at the tropics.

    b. *SHC.*   Specific heat capacity of the surface.

These reasons explain why it is hot in the tropics and cold at the poles.

Temperature Deviation Temperature deviation is the measurement of the actual temperature against the international standard atmosphere (ISA) temperature for the corresponding altitude. Temperature deviation is expressed as ISA ±6°.

In aviation it is common practice to use ISA deviations as a means of describing temperature. Remember, ISA temperature is +15°C at sea level and decreases at an environmental lapse rate of 2°C per 1000 ft up to an altitude of 36,000 ft, where it stays at a constant −57°C.

> **Example** If you are at 33,000 ft with an OAT of −45°C, what is the temperature deviation from IAS?

ISA temperature is +15°C at MSL with a lapse rate of 2°C per 1000 ft. So at 33,000 ft

$$\text{ISA temperature} = +15 + [33 \times (-2)] = -51°\text{C}$$

$$\text{Actual temp.} - \text{ISA temp.} = -45°\text{C} - 51°\text{C} = +6°\text{C}$$

$$\text{Temp. deviation} = \text{IAS} + 6°$$

Temperature (Air)—Flight Level An air temperature that differs from the ISA temperature will result in a different actual flight level or height above the ground, than the pressure level read by the altimeter.

A higher than ISA air temperature makes the air less dense and lighter in weight, causing the density altitude to differ from the pressure altitude. This results in the actual flight level being higher than the pressure level read by the altimeter. However, a lower than ISA air temperature makes the air denser and heavier, causing the density altitude to differ from the pressure altitude. This results in the actual flight level being lower than the pressure level read by the altimeter. Therefore when flying from a high to low (temperature), beware below, because your actual flight level (and therefore ground clearance) is lower than indicated by your altimeter. Or in other words, your altimeter *overreads*. This "high to low" mnemonic applies equally to pressure values.

Temperature Inversion/Layer A temperature inversion occurs when the air closest to the ground, or even the ground itself, is cooler than the air above it. In other words, the air temperature increases with height (rather than the usual decrease). A temperature inversion layer is the height/altitude where the air temperature changes from the temperature increasing with height (i.e., temperature inversion) to a normal decrease of temperature with height state.

See Fig. 198 at the top of the next page.

When a temperature inversion occurs, it acts as a blanket, stopping vertical movement/currents; i.e., air that starts to rise meets an inversion layer and so stops rising.

A temperature inversion is of interest because it is the opposite of the normal pattern of temperature distribution in the atmosphere. Therefore it is important to remember that the normal temperature distribution in the atmosphere is such that *temperature decreases with height*. In the international standard atmosphere (ISA) (which is purely a theoretical model of the atmosphere used as a measuring stick), the temperature is assumed to decrease by 2°C for each 1000 ft of height gained in a stationary air mass. However, sometimes conditions exist that lead to a temperature inversion:

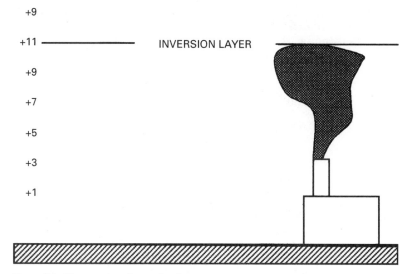

+9

+11 ———————————— INVERSION LAYER

+9

+7

+5

+3

+1

Figure 198 Temperature/inversion layer.

1. On clear nights, when the earth loses a great deal of heat by terrestrial radiation, it cools down the air in contact with it, i.e., the lower layers, by conduction. The cooler air will tend to sink and not mix with air at the higher levels, leading to a temperature inversion where the air at ground level is colder than the air at altitude. Such a temperature inversion layer may exist only for 20 ft to a few hundred feet, but it is important to recognize because it can have by-products such as ground fog or wind shear.

2. Frontal systems, when a cold air mass undercuts a warm air mass, produce a temperature inversion of warmer air above the colder air. Such a temperature inversion can be found at various altitudes, due to the nature of its origin. But again it is important to recognize because it may also have its own by-products such as deteriorating weather conditions, maybe even icing at lower altitudes, and of course wind shear.

NB: Wind shear is possible with any change of temperature.

Temperature (Air) Measured   *See* Air Temperature—Measured.

Tempo   Tempo relates to a temporary variation to the general forecasted weather lasting less than 1 h, or if recurring lasting in total less than one-half the trend or TAF period it is included within.

NB: A tempo can relate to improvements as well as deterioration in wind, visibility, weather, or clouds.

Once the tempo weather events have finished, the original prevailing weather reasserts itself.

Thermal Efficiency   This is the energy which can be converted to useful workload, expressed as a fraction of the total energy available in the fuel.

Thermal Equator   The thermal equator is the position of the maximum thermal temperature around the earth's surface.

NB: It should not be confused with the global equator, which is the greatest parallel of latitude distance around the earth's surface.

The position of the thermal equator moves according to the sun's heating, which varies widely with season; i.e., in the summer the thermal equator moves toward the pole (in both the northern and southern hemispheres), due to the greater heating experienced at higher latitudes. In addition the seasonal temperature range over the land is greater than that over the sea. Therefore the thermal equator varies considerably compared to the global equator over the land, but remains close to the global equator over the sea.

Thermal Wind   Thermal winds (of quite different direction and strength to the low/medium level geostrophic and gradient winds) are generated by a difference in temperature (thermal gradient) between two columns of air, over large areas and at great upper heights.

The direction of a thermal wind is parallel to the isotherms; i.e., the thermal wind direction is such that if you stand with your back to the thermal component, in the northern hemisphere, the low temperature will be to your left. The strength or speed of the thermal wind is directly proportional to the temperature gradient, i.e., the spacing of the isotherms between the two columns of air.

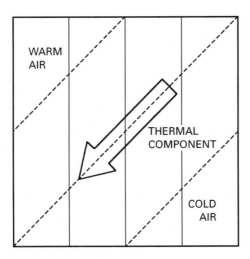

Figure 199   Thermal wind component.

Thermal winds take effect as height increases, where the mean temperature difference (temperature gradient) between the two columns of air is greatest. This is due to either a different initial surface temperature and/or a different change in temperature with height (lapse rate), for the two columns of air.

NB: Thermal wind is temperature gradient times height or depth of the column of air, in 1000 ft.

The temperature gradient is the mean temperature difference divided by the distance, in 100 nm, between the two columns of air.

Thermodynamics   Thermodynamics is the study of heat/pressure energy, or the behavior of gases (including air) and vapors under variations of temperature and pressure.

First law of thermodynamics   *Heat and mechanical energy are mutually convertible, and the rate of exchange is constant and can be measured.* For example, when fuel is burned in a piston engine, the heat energy in the fuel is converted to mechanical energy by the action of the pistons and crankshaft.

Second law of thermodynamics   *Heat flows from a hot substance to a cold substance unaided.* For example, heat is exchanged in an aero engine as cold air flows past the cylinders, removing some of the combustion heat. *But heat will not flow from a cold substance to a hot substance without the expenditure of energy from an external source.* Therefore fuel must be burned inside the engine cylinder to raise the temperature to increase the pressure.

Thrust Reversers   Thrust reversers on a jet/gas turbine engine reverse the airflow forward, thereby creating a breaking action. There are two types of thrust reversers:

1. Blockers or bucket design
2. Reverse flow through cascade vane, etc.

See also Reverse Thrust.

Thunderstorm   Thunderstorms are associated with cumulonimbus clouds, with several thunderstorm "cells" possible within a single cloud.

Four conditions are required for a cumulonimbus cloud to develop:

1. Moisture content in the air must be high.
2. A trigger lifting action (or catalyst) causes a parcel of air to start rising.

NB: The four main lifting actions are convection, turbulence, frontal, and orographic.

3. There is adiabatic cooling of the rising air.
4. There is a highly unstable atmosphere, so that once the air starts to rise, it will continue rising. Effectively the ELR must be greater than the SALR, for over 10,000 ft.

Next we have to examine the life cycle of the cumulonimbus cloud and its associated thunderstorm. This life cycle can be divided into three phases: the developing stage, the mature stage, and the decaying stage.

Developing stage   During the development of the cumulonimbus cloud, updrafts move air aloft, allowing condensation to take place throughout the ascent of the convective currents. There may be several cells within the one cumulonimbus cloud, each with its own convective current of air moving at different speeds, e.g., up to 3000 to 4000 ft/min. The various currents of air represent turbulence and possible wind shear hazards to aircraft flying through this stage of the storm.

Mature stage   During this stage, water drops start to fall through the cloud, drawing air down with them. Although it is dependent on the shape of the storm and the prevailing wind gradient, this downdraft is often in the middle of the cloud/storm, surrounded on all sides by strong continuing updrafts, which are providing further fuel for the storm.

NB: During this stage downdrafts can reach 3000 ft/min, and updrafts can reach 5000 to 6000 ft/min. When the downdraft breaks out of the base of the cumulonimbus cloud, it is colder than the surrounding air as it has only been warmed at the SALR during its descent within the cloud. As it is cooler and therefore denser, the downdraft will continue down to the surface, where it can be felt as the *first gust.* As the air reaches the surface, it spreads out to create the highly dangerous pattern of winds that cause wind shear. Particularly intense microburst downdrafts can be created if the precipitation that accompanies a downdraft is reevaporated in the drier air below the cloud base (Virga), which cools the descending air still further and intensifies the downdraft into a microburst.

NB: Evaporated precipitation intensifies a downdraft because as the liquid changes to a vapor, it absorbs latent heat from the air, thereby cooling the descending air, which makes it denser and heavier.

The mature phase of a cumulonimbus cloud is also the most hazardous stage of its thunderstorms. In short, the dangers include:

- Torrential rain
- Hail
- Severe turbulence
- Severe icing
- Wind shear and microbursts
- Lightning

Decaying stage This is the final stage of the cumulonimbus cloud. It starts with the end of the thunderstorm, which is marked by the end of continuous rain and the start of sporadic showers, sometimes as Virga due to a temperature inversion beneath the cloud base, which can still cause a marked wind shear. At the higher levels it may take on the familiar anvil shape, as upper winds spread out under the tropopause. An anvil can have marked down vertical currents beneath it, which cause a strong wind shear that should also be avoided. During the decay stage, the vertical currents weaken, and the air within the cloud subsides more than it ascends, causing the cloud to collapse.

See Fig. 200 at the top of the next page.

Thunderstorms can produce the following hazards to all aircraft types:

1. Severe wind shear, which can cause
   a. Handling problems
   b. Flight path deviations, especially vertically
   c. Loss of airspeed
   d. Possible structural damage
2. Severe turbulence, which can cause
   a. Possible loss of control
   b. Possible structural damage
3. Severe icing, especially clear ice formed from SWDs striking a surface with a subzero temperature
4. Airframe structural damage from hail
5. Reduced visibility
6. Lightning strikes, which can possibly cause damage to the electrical system
7. Radio communication and navigation interference from static electricity present in the thunderstorm

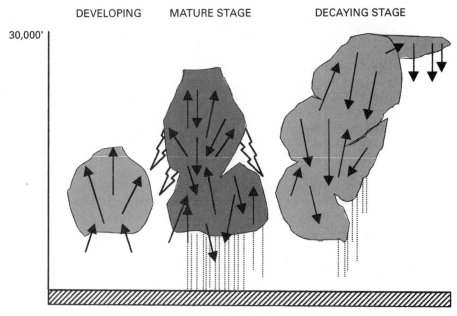

Figure 200 Cumulonimbus clouds and thunderstorm formation.

These hazards exist inside, under, and, for some distance, around the thunderstorm's associated cumulonimbus cloud. Therefore cumulonimbus clouds (and thunderstorms) should be avoided by a minimum of 10 nm and in severe conditions (i.e., the mature stage of cumulonimbus cloud) by at least 20 nm.

Tire Creep   Creep is the tendency of the tire to rotate slowly (creep) around the wheel hub as a result of a millisecond landing friction on the tire before wheel spin occurs, usually because of a low tire pressure. This creep, if excessive over a period of time, will cause the tire to tear out the inflation valve and burst on touchdown.

When the tire is initially installed, a mark, usually a red dot, is placed on the tire sidewall and a second dot is placed adjacent on the wheel hub. From then on any displacement of the red dots from each other indicates tire creep, and this can be easily seen during a preflight inspection. The following can counteract tire creep:

1. Maintaining the correct tire pressure ensures instantaneous wheel spin on touchdown, thereby preventing tire creep.
2. Knurled flange wheels form a cog wheel effect between the tire rubber and the wheel hub.
3. Tapered bead seat of the wheel is tapered so that the flange area is of a greater diameter than at the center of the rim. When the tire is inflated, the side pressure forces the bead outward to grip the rim.

Tire Speed Limits   Tire speed limit restricts the aircraft's MTOW, given the ambient aerodrome pressure altitude, temperature, and wind conditions, so that V_R

(liftoff speed) is less than the maximum rated ground speed limit for the tires. This protects against the tires blowing out during the takeoff roll.

NB: Aircraft tires are built to have a structural integrity up to a specified ground speed, and a maximum specified vertical speed and lateral velocity impact on landing. Any speeds experienced by the aircraft above the tire's rated speed can result in the tire failing, or a blowout, as a result of excessive friction causing thermal expansion in the tire air pressure.

If the takeoff weight determined for the maximum tire speed limit is lower than that determined from all the other performance criteria, then this becomes the overall limiting takeoff (or landing) weight.

Tire—Temperatures   Prior to the takeoff run, the following factors affect the tire temperature:

- OAT (reflects the tire ambient temperature)
- Aircraft weight
- Taxi time
- Amount of braking (which generates heat in the brakes that is transferred to the tires)

During the takeoff run:

- TOR time (i.e., determined by head/tailwind component, runway slope, pressure altitude)

All generate a temperature rise in the tires.

It is important to monitor tire temperature because an increase can lead to an increase in tire pressure, which causes the tire to expand, possibly to its blowout limit.

TODA   *See* Takeoff Distance Available (TODA).

TODR   *See* Takeoff Distance Required (TODR).

TOGA Switches   The TOGA (takeoff go-around) switches (located on the aft edge of each throttle lever on the B737), or the gate on the airbus aircraft, provide a means of engaging the auto throttle and the flight director in the takeoff or go-around mode.

TOGA selection depends on the aircraft type. But two common systems are as follows:

1. Move the thrust levers into a forward TOGA gate, as on the Airbus 320.

2. Press the TOGA switches on the thrust levers, as on the Boeing 737.

These will give TOGA thrust and flight director pitch commands.

TORA   *See* Takeoff Run Available (TORA).

Torque   Torque is a force causing rotation. In particular, it is the force created within an engine, which causes rotation of its rotating parts, e.g., the crankshaft. It is a measure of the load experience, expressed in pounds per inch or feet. A quantity of torque, or twisting moment, is involved in the measurement of the engine brake horsepower (BHP).

$$\text{Torque} = \text{force} \times \text{distance} \qquad \text{(at right angles to force)}$$

TORR *See* Takeoff Run Required (TORR).

Touchdown Point The touchdown aiming point is 1000 ft along the runway, given a 3° glide path. It is normally distinguished by marker board markings on the runway surface.

Trade Winds Trade winds are steady and predictable surface winds, which rarely exceed 15 kn at the surface, but can extend up to 10,000 ft. They blow from the subtropical highs into the equatorial low (ITCZ), i.e., blowing from the northeast in the northern hemisphere and from the southeast in the southern hemisphere.

As the ITCZ moves throughout the year, the geographic places that experience the trade winds and associated weather change with the season. This is very important in recognizing what the weather is likely to be, for instance, in Bombay in July.

Traffic Collision Avoidance System (TCAS) *See* TCAS.

Transformer A transformer is an electric device without any moving parts that utilizes magnetic induction between two windings, known as the primary (input) and secondary (output) coils, to increase or decrease the alternating current.

NB: When the secondary coil has more turns than the primary, the voltage is increased or *stepped up*. When the secondary coil has fewer turns than the primary, then the voltage is decreased, or *stepped down*.

Transformers are used in ac electrical systems, to step up or down the voltage, to meet the requirement of the load upon it.

NB: Often transformers are paired with rectifiers, namely as a transformer rectifier unit (TRU), to convert the current to dc as well as the voltage level.

Transponder The information carried on the airborne transponder reply pulses (secondary radar) is determined by the operating functions selected on the transponder unit by the pilot. The transponder facilities/functions are as follows:

1. *Code selection.* Knobs are provided for the pilot to select the aircraft's assigned identification transponder code. This four-number code is displayed in digital form on the transponder.

   The transponder should be selected to STDBY while the code is being changed to avoid inadvertently transmitting an emergency code or another aircraft's code.

2. *Transmission function switch.*
   a. STDBY. Usually the transponder is warmed up in the standby position prior to takeoff. From standby the transponder is ready for immediate use.
   b. ON. This transmits the selected aircraft identification code in mode A—alpha, at the normal power level. (Mode A is the aircraft's identification mode only.)
   c. LOW SENS. This means low sensitivity. It transmits the selected code at a lower power level than that in the ON position. This may be requested by the

radar controller to avoid overstrong blips on the screen from aircraft close to the ground antenna.

    d. ALT. This means altitude. This selection, known as mode C—Charlie, transmits the aircraft's altitude in addition to the aircraft's normal selection code. If mode C is not installed, the transponder still transmits in mode A, i.e., the aircraft's identification code only.

    e. TST. This selection causes the transponder to generate a self-interrogation signal to provide a check of its operation. If the transponder is operating correctly, its reply monitor light will illuminate.

3. *Reply/monitor light.* The reply/monitor light on the transponder will flash to indicate that the transponder is replying to an interrogation pulse from a ground station. The reply monitor light will glow steady for transmission of an ident pulse and to indicate that the transponder is functioning correctly under a test (TST) function.

4. *Ident switch.* This selection is made if ATC requests a pilot to "Squawk ident." The pilot would press his ident button/switch, which creates a special pulse to be transmitted with the aircraft's transponder reply. This causes a special symbol to appear for a few seconds on the radar screen around the aircraft's return, thus providing positive identification for the radar controller.

NB: The ident button should be pressed only upon request from the radar controller.

5. *System switch.* This switch selects transponder 1 or 2, when the aircraft is fitted with two transponders.

6. *Fault light.* This light illuminates when the selected transponder fails.

Transponder Codes

0000	Transponder mode C malfunction; i.e., a 200-ft or greater difference between the aircraft and ATC's radar height readings; if you cannot separately turn off mode C
7700	Emergency
7600	Two-way communications failure
7500	Unlawful interference, hijacking, or unlawful interception

NB: Some local authorities may advise that 7700 be selected in the event of an unlawful airborne interception.

7000	VFR conspicuity code (1200 in the United States)
2000	IFR conspicuity code
0033	Aircraft engaged in parachuting operations

Transport Wander Transport wander is a form of apparent wander of an uncorrected gyro.

If a DI gyro is aligned to true north at one place on the earth and then the aircraft is moved to another east/west position on the globe, the gyro axis will be out of alignment. This is known as *transport wander* on uncorrected gyros.

Transport wander + apparent wander = total apparent wander

Flights north/south produce no transport wander but will produce apparent wander as the latitude changes.

In the northern hemisphere, flights to the east will increase total apparent wander, and flights to the west will decrease total apparent wander. That is, flying west at the same speed as the earth's rotation, you would maintain your position in space and therefore suffer *no* total apparent wander. That is, transport wander +ve, and apparent wander −ve.

The formula used to correct for transport wander is

$$\text{Transport wander} = \frac{\text{E/W ground speed} \times \text{tangent latitude}}{60}$$

NB: Transport wander in an easterly direction will have a different sign (−ve in the northern hemisphere and +ve in the southern hemisphere) from transport wander in a westerly direction (+ve in the northern hemisphere and −ve in the southern hemisphere) when applied to total wander (drift) calculations.

Transverse Mercator Chart   A transverse Mercator chart is a cylindrical projection which uses a central meridian as its great circle of tangency (as opposed to the equator for a normal Mercator chart).

Scale and convergency are correct at the central meridian and can be assumed to be correct within 350 nm of the central meridian, the chart's usable limits. Therefore within its usable limits the transverse Mercator chart has the properties of a *Lambert chart* and is considered to be *orthomorphic,* with great circle tracks as straight lines and rhumb lines as curves.

The advantage of a transverse Mercator chart is that it is ideal for long north/south routes, based on a particular meridian as the central meridian.

Trend   A weather *trend* is usually attached to an aerodrome weather report, i.e., metar, and is commonly referred to as a *landing forecast*. The trend is a forecast of any significant weather changes expected in the next 2 h after the time of the report, and it is described using the normal coded weather format and abbreviations. If no significant change is expected, the observation (report) will be followed by NOSIG (no significant change) as the trend.

Because a trend forecast period is much shorter than a normal aerodrome forecast (i.e., a TAF), it should be much more accurate.

NB: Cloud bases in a trend are given AAL (above aerodrome level).

A qualified forecaster can only give a trend, while a report can be given by just an observer.

Trip Fuel—Climb Departure   The best rate-of-climb V_y departure uses the least trip fuel. This is true because it ensures the aircraft reaches its optimum cruise altitude as quickly as possible, and so the aircraft spends a greater part of its flight time at its optimum altitude than with any other climb profile.

NB: The optimum en route altitude has the best aerodynamic and engine performance qualities. Being at this optimum cruise altitude as long as possible means that the best fuel economy and SFC are obtained for the largest percentage of the flight (trip), and therefore the least trip fuel is used.

A reduced-power climb also uses more trip fuel. This is so because using a reduced-power climb means the aircraft has a slower rate of ascent and so it takes longer to reach its cruise altitude; consequently it spends less time at its optimum cruise altitude, and therefore it uses more trip fuel.

A derated takeoff also uses more trip fuel. Similar to using a reduced-power climb, the aircraft has a slower initial rate of ascent and so takes longer to reach its transition to its en route climb profile and then its cruise altitude. Consequently it spends less time at its optimum cruise altitude, and therefore it uses more trip fuel.

NB: Remember, at its optimum cruise altitude the aircraft experiences the best aerodynamic and engine performance, which results in the best fuel economy and SFC.

Tropical Revolving Storm (TRS)   Tropical revolving storms are deep, intense depressions or *lows* found in equatorial regions around the ITCZ. They are known as cyclones in the Indian and Pacific oceans, hurricanes in the Caribbean and Americas, and typhoons in the China Sea area.

Tropical revolving storms start at the edge of the ITCZ as it retreats in late summer or earlier autumn. The storms are enormous heat engines, and they take their energy from water vapor off the warm sea, which is condensed aloft, releasing latent heat. Sea temperatures of at least 27°C are needed to form a TRS, which is a temperature that the sea achieves only after prolonged heating, such as after a long, hot summer. Also for this reason a TRS does not form over cold seas and dies out when it passes over land, or moves to a colder sea region. However, the TRS does not form at the equator because the Coriolis effect is zero at the equator; so it forms at approximately 5° to 20° latitude, where the sea temperature is high enough and where the Coriolis effect is present.

In a TRS, air converges in the lower levels, flows into the depression, and then rises. This warm, moist air forms large cumulus and cumulonimbus clouds. Freshening winds start to become circulatory, and the normal semidiurnal pressure variations give way to a continuous fall in pressure, deepening the low. Circulating winds become stronger to gale force, and the TRS tracks away from the ITCZ at approximately 15 kn, mainly in a westerly direction and then gradually turning east. Winds can exceed 100 kn, with thunderstorm activity, which becomes increasingly frequent as the center of the storm approaches. The *eye* of the TRS, however, is often only about 10 nm across with light winds and broken clouds. Once the eye has passed, very strong winds from the opposite direction will occur.

On crossing land TRSs can cause immense damage.

Trough   A trough is a V-shaped extension of a low-pressure system. Air flows into a trough (convergence) and rises. If the air is unstable, weather similar to that found in a depression or a cold front will occur, such as cumuliform clouds, cumulonimbus clouds, and thunderstorm activity.

True Airspeed (TAS)   The TAS is simply the actual speed of an aircraft through the air mass in which it is flying. TAS is therefore dependent upon the air density, properties of the air mass in which the aircraft is flying. The higher the air density, the greater the resistance to the motion of the aircraft and therefore the lower the true airspeed. Conversely, the opposite is true: the lower the air density, the less the resistance to the motion of the aircraft and therefore the higher the true airspeed.

Air density varies for two main reasons: temperature (warm air is less dense than cold air) and pressure. The higher the altitude, the lower the air density; and the lower the altitude, the greater the air density. Therefore at a constant IAS/RAS, the TAS will increase with an increase in altitude or temperature because of the reduction in air density. This can be expressed by using the formula

$$\text{Dynamic pressure (IAS)} = \tfrac{1}{2}\,\rho V(\text{TAS})^2$$

To maintain a constant IAS as ρ (density) decreases (due to either altitude and/or temperature), you have to increase V (TAS). Remember, IAS is only equal to TAS at ISA MSL conditions, because ASI is calibrated to these conditions; that is, 1225 g/m^3. Any deviation from these conditions, i.e., a higher pressure altitude or temperature, requires the application of a density error correction to the IAS to find the TAS.

With a constant TAS, the CAS (IAS) will reduce with altitude, but the Mach number increases with an increase in altitude.

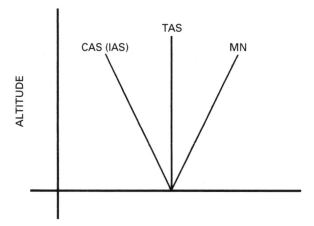

Figure 201   Constant TAS versus MN and CAS (IAS).

If you climb at a constant CAS (IAS), your TAS will increase. This is so because air density decreases with altitude, and this can be expressed as

$$\text{CAS (IAS)} = \tfrac{1}{2}\rho V^2$$

where CAS (IAS) is constant and ρ (density) decreases with an increase in altitude. Therefore V (TAS) will increase.

If you fly at a constant CAS (IAS) into a warmer area, your CAS (IAS) remains constant, because CAS (IAS) is unaffected by temperature while the TAS will increase because of the decrease in air density associated with warmer temperatures.

True Direction   True direction is measured with reference to true north.

Turbocharger   The main difference between a supercharger and a turbocharger lies in how their impellers are driven. A turbocharger has externally driven turbines, powered by engine exhaust gases. Because a turbocharger employs exhaust gases to drive the turbine (the turbine is connected to the impeller by a rotor shaft) the engine does not suffer any loss of power output. The workload of the turbine is governed by the amount of exhaust gas directed by the *waste gate* to pass over the turbine. A turbocharger's purpose is to compress the *air only*. The turbocharger is not necessarily an integral part of the engine and therefore may be mounted on the engine or on the fireproof bulkhead and away from combustible fuel lines in the engine bay.

Turboprop Aircraft—Short Regional Operations   Turboprop aircraft are in general better suited for short regional operations because short sector routes normally have a restricted cruise altitude, which the turboprop aircraft is normally better suited to, for the following reasons:

1. A jet engine is suited to high altitudes (30,000 ft and above) which an aircraft is not capable of reaching during a short sector route.

2. The turboprop engine is designed to operate at its most efficient at a medium altitude, which is associated with short regional operations.

3. Many short regional operations operate out of restrictive airfields, and the turboprop aircraft's high-lift straight wing is capable of meeting the field length, climb, and descent gradients of these restrictive airfields.

4. Short-range sectors are usually flown more frequently and with a smaller passenger demand per trip. Therefore a turboprop aircraft, which typically has a smaller passenger capacity than a jet aircraft, is a better economical design for this type of short sector demand, i.e., loads of 20 to 60 passengers.

However, note that this trend is changing, with 30- to 70-seat jet airliners now being built that are efficient at medium altitudes over short regional routes and as such are replacing the turboprop fleets operated by regional airlines.

Turbulence   Turbulence is the eddy motions in the atmosphere, which vary with time, from place to place, and in magnitude. Some form of turbulence is always present in the atmosphere, with severe or even moderate turbulence at best being uncomfortable and at worst capable of overstressing some airframe types.

Turbulence is mainly considered to be vertical gusts. The two main forms of vertical turbulence are

1. Convection turbulence, caused by solar radiation heating the ground and producing rising thermal currents

2. Obstruction and orographic (terrain-generated turbulence)

Such vertical gusts will increase the aircraft's angle of attack, causing an increase in the lift generated at a particular airspeed and therefore an increased load factor on the airframe. The angle of attack can even be increased beyond its critical *stall*, causing the wing to stall at a speed well above the published *g* stalling speed, due to the increased load factor experienced. It is therefore important not to exceed the aircraft's critical angle (load factor) when flying in turbulence. The main way of achieving this is to fly at the *turbulence penetration speed* VRA/MRA.

The other main forms of turbulence (considered to be horizontal) are

3. Jetstreams

4. Wake turbulence

In addition, other (horizontal) forms of less dynamic turbulence are frontal, temperature, and any general wind direction and/or speed changes.

Always try to avoid areas of severe weather and turbulence.

- Do not take off or land in areas of *severe* weather and turbulence.
- In flight, avoid predicted areas of turbulence by
  Flight planning adjustments if possible, namely, changing routing
  Using the weather radar, monitoring weather reporting stations, in-flight reports on ATC frequencies, and visual observations

However, if you are committed to areas of moderate/severe weather in flight, adopt the following technique:

- Ignition on/autopilot on.
- Passenger signs on (seat belts and no smoking). In thunderstorms, turn the flight deck lighting full on, night or day. Stow all loose equipment, and strap yourself in tight, using the full harness.
- Hold the attitude (climb, cruise, or descent).

  NB: Sacrifice altitude to maintain attitude.

  However, be fairly gentle with the use of the elevator to avoid overstressing the airframe structure through large changes in angle of attack and lift produced. Also for this reason do not try to climb over a storm from a straight and level attitude. Check height for maneuverability and terrain clearance. Use aileron to maintain lateral control.

- Auto thrust off, to protect against engine flame-out.
- Select the rough airspeed, which may require reducing power. Allow the airspeed to fluctuate more than usual, but be prepared to correct for any large speed diversions by adjusting the power.

Further,

- Continue to apply avoidance techniques.
  Monitor the weather radar for the best passage through the severe weather.
  Use the deicing systems as required, and ensure that the pitot heat is on.
- Be prepared for other disturbances.
  High levels of static on some radios may necessitate that they are turned off.
  Aircraft electrical systems and compasses may be affected temporarily by strong local electric fields. Therefore check all flight instruments and supplies.
  You may get sparks or small electric shocks if the aircraft is struck by lightning.
    Or the whole aircraft may be surrounded by St. Elmo's fire.
  You may hear loud battering noises if you fly into hail.

Turbulence—Surface   Surface turbulence is caused by the surface wind being blown over and around surface obstacles, such as hills, trees, and buildings. This causes the surface wind to form turbulent eddies, the size of which is dependent upon the size of both the obstruction and the wind speed.

Turbulence—Wake   *See* Wake Turbulence.

Turn(s)   A rate 1 turn is a 3° per second or 180° per minute turn. A rate 2 turn is a 6° per second or 360° per minute turn. A rate 3 turn is a 9° per second or 540° per minute turn. The radius of a turn can be calculated with the formula

$$\text{Radius of turn} = \frac{\text{TAS}}{\text{rate of turn} \times 60 \times 3.14}$$

The angle of bank required to achieve a rate 1 turn is calculated from

$$\text{Angle of bank} = \frac{\text{TAS}}{10} + 7°$$

The distance of the circumference traveled in a turn can be calculated as follows:

$$\text{Circumference (nm)} = \frac{\text{TAS}}{\text{rate of turn} \times 30}$$

Turn Coordinator   The turn coordinator is an advanced development of the earlier turn indicator. It is similar to the turn indicator instrument, except its single gimbal is raised at the front by 30° so that the instrument is sensitive to both roll and yaw, and it begins to indicate a turn as soon as the roll in begins. However, the turn coordinator only indicates rate 1 turns accurately, and it should not be confused with an artificial horizon as it displays no attitude information. A warning "No Attitude Information" is often written on the instrument face.

Turn and Slip Indicator Instrument   The turn and slip indicator is in effect two instruments combined as a single unit. One measures *turn,* and the other measures *slip* or *skid.*

NB: *Turn* is the movement about the aircraft's yaw axis (the aircraft's vertical) which results in a change of direction. *Slip* is a lateral force into the turn. *Skid* is a lateral force out of a turn.

The turn and slip instrument consists of the following:

1. There is a rate gyro.
2. The gyro rotates about a horizontal axis.
3. There is one gimbal, which is pivoted about the aircraft's fore and aft axis, that measures the aircraft's yaw when the precessed force in this plane of freedom is sensed.
4. There are two planes of freedom.
   a. The gyro's spin axis
   b. Yaw axis of the gimbals
5. The gyro's axis is aligned to the aircraft's lateral axis.

The *turn instrument* may be either electrically or air-driven, and its direction of rotation has the top of the gyro moving away from the pilot. When the aircraft is turned, the change of direction produces a sideways turning force which is precessed to act on the bottom of the gyro that tilts the gyro/rotor away from the aircraft's vertical. As a result, the gimbal attached to the rotor moves a needle that indicates the yawing force as a rate of turn on the instrument face. The tilt of the gyro mass, which makes the gyro precess, stretches a spring until the rate of the gyro's tilt and the spring stretch are equal, which indicates a constant-angle-of-bank turn.

If there is insufficient or too much bank applied, the turn will not be balanced and a yawing force will be felt by the occupants as either a skid or a slip. The slip indicator measures this, which is usually a ball in a curved liquid-filled tube. In straight and level balanced flight, the force of gravity keeps the ball in the central position, and therefore no slip or skid is indicated.

In a balanced turn, the combination of gravity opposed by the centrifugal force experienced in the turn acts as a balanced force through the earth's vertical, and therefore no slip or skid is indicated. However, if the turn is not balanced, because of either insufficient or too much bank, then the resultant force will displace the ball from the aircraft's vertical.

An unbalanced slip, with the ball displaced to the side of the turn, represents either too much aileron bank or too little rudder.

Unbalanced skid, with the ball displaced to the opposite side of the turn, represents either too little aileron bank or too much rudder.

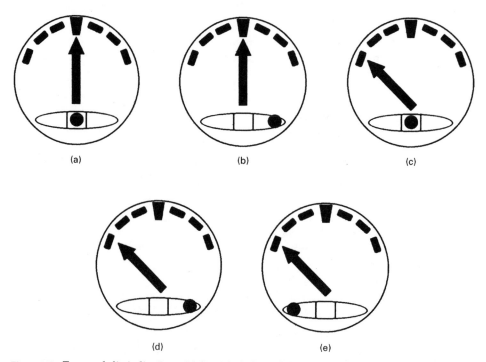

(a)   (b)   (c)

(d)   (e)

Figure 202 Turn and slip indications. (*a*) Straight balanced flight. (*b*) Straight flight, right wing low. (*c*) Balanced left turn, correct bank. (*d*) Unbalanced left turn, skidding because of insufficient bank or too much rudder. (*e*) Unbalanced left turn, slipping because of too much bank or too little rudder.

NB: If the ball is to the left, use the left rudder; and if the ball is to the right, use the right rudder. Hence we have the old adage to "kick or squeeze the ball back into the center."

Turn and Slip Indicator Instrument—Errors   The turn indicator gyro suffers from the following:

1. System failure of the gyro, due to an air or electrical drive failure. Air-driven instruments will underread if its operating speed is reduced as a result of a low air supply to the instrument rotor. A complete air supply failure will result in the turn indicator reading no movement. Indications of this are

   a. Suction gauge reading below normal
   b. Possible warning flags on instrument face
   Electrically driven instruments will underread if the operating speed is reduced as a result of a power reduction to the instrument. A complete power failure will result in the turn indicator reading no movement. An indication of this is a "Power Failure" warning flag on the instrument face.

2. Looping error. This is an inherent design error in the instrument that in any yaw condition the gyro will tilt or has tilted. If the aircraft is then pitched, the rate of this pitch under these circumstances can affect the gyro in the same manner as a yaw rate and precesses the gyro to indicate an increase in the rate turn.

   NB: The slip indicator is not subject to any errors as long as the capsule is not damaged.

3. Real wander of the gyro's spin axis.

Unusual Attitude   There are two fundamental unusual attitudes:

1. Nose high with a decreasing airspeed (stall)
2. Nose low with an increasing airspeed (spiral dive or spin)

NB: Unusual attitudes generally result from some form of mishandling by the pilot.

Upper Winds   Upper winds are determined by their thermal gradient. A difference in temperature between two columns of air will cause a pressure difference at height, even if both columns of air had the same sea-level pressure. This is true because a warmer temperature makes the air less dense and lighter in weight, which allows the column of air to rise with the effect of a greater actual distance (height) for a 1-mbar change of pressure. This results in an altitude/flight level, say, 18,000 ft, of one column of air having a higher pressure level, say, 500 mbar, than a second column of air that has a colder temperature, say, at 18,000 ft/490 mbar. Therefore a pressure difference is born that creates a wind (parallel to the isobars) at altitude, which is different from the wind experienced at sea level, even if no wind was present at sea level. The temperature difference (gradient) between these two columns of air at height becomes significant in that the thermal component, which runs parallel to the isotherms, has an effect on the upper winds. The vector sum of the isotherm thermal wind component and the surface and upper isobar pressure-driven geostrophic (or gradient) wind produces the direction and speed of the upper wind.

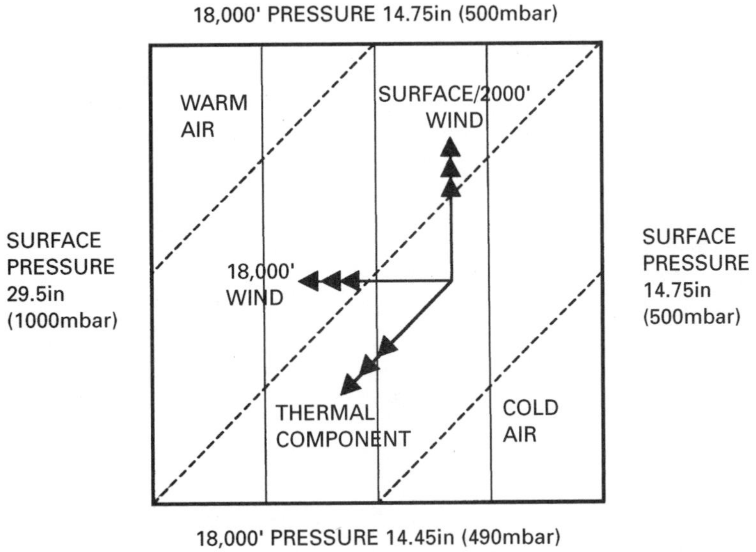

18,000' PRESSURE 14.75in (500mbar)

WARM AIR

SURFACE/2000' WIND

SURFACE PRESSURE 29.5in (1000mbar)

18,000' WIND

SURFACE PRESSURE 14.75in (500mbar)

THERMAL COMPONENT

COLD AIR

18,000' PRESSURE 14.45in (490mbar)

Figure 203   Upper wind, vector sum of the wind components.

In the northern hemisphere the thermal gradient is generally north/south (north/cold and south/warm), and therefore the upper winds are generally westerly (i.e., from the west), with the highest wind speed where the thermal gradient is greatest, e.g., jetstreams.

NB: However, there is a light easterly wind over the thermal equator that can extend all around the globe at certain times of the year and can reach speeds of up to 70 kn.

V

Variable Incidence Tailplane An all-moving tailplane is normally called a *variable incidence tailplane* when it does not have an elevator surface. Therefore the variable incidence tailplane provides pitch maneuverability, by the control column, and longitudinal balancing, by the trim system.

Variable Load This is the weight of the crew, crew baggage, and removable units, i.e., catering loads, etc.

$$\text{Variable load} = \text{APS} - \text{basic weight}$$

Variable-Pitch Propeller Variable-pitch propellers are a development of the fixed-pitch propeller that has a variable and controllable blade angle between its coarse and fine positions. This is used to adjust the blade angle of attack to its optimum setting in order to maintain the propeller efficiency and aircraft thrust over a wide range of aircraft speeds and differing operating conditions.

The propeller produces thrust by accelerating a mass of air rearward; therefore the propellers have to accelerate a constant mass of air rearward to maintain a constant thrust. So at high forward speeds (high altitudes) where the blade angle of attack is reduced (increase in forward airspeed causes a decrease in blade angle of attack), a coarse blade/pitch setting is required to maintain a constant blade angle of attack. This produces a constant air mass displacement and therefore a constant thrust and propeller efficiency. At low (altitude) speeds, the blade angle of attack is not reduced, and a fine blade/pitch setting maintains the constant angle of attack, which produces a constant air mass displacement and a constant thrust and propeller efficiency. Therefore it is seen as desirable to employ a propeller that can progressively increase its blade angle with an increasing forward airspeed, so that it can operate at low blade angles (fine pitch) for low airspeeds and a higher blade angle (coarse pitch) for higher airspeeds, thus providing a near-constant positive blade angle of attack at all times and a constant thrust and propeller efficiency over a range of speeds and conditions.

See Fig. 204 at the top of the next page.

The density of the air will affect the amount of power absorbed by the propeller. Should the density of the air increase (due to a decrease in temperature), the greater weight of air on the propeller will absorb more engine power, and so the rpm will decrease. Therefore an increase in engine power is required to maintain a certain rpm. Conversely, a decrease in the air density (due to an increase in temperature) will require less engine power to maintain a specific rpm. However, since the engine power output is directly affected by the air density itself because of the weight of charge induced into the cylinders (the greater the weight, the greater the power output), it is usually more dominant than the effects of the air density changes on the propeller. In conditions of increased air density, the increased engine power more than compensates for the increased propeller demands; thus the power required to maintain a certain rpm with an increase in air density conditions would in fact be less. The opposite will apply in the event of air density decreasing.

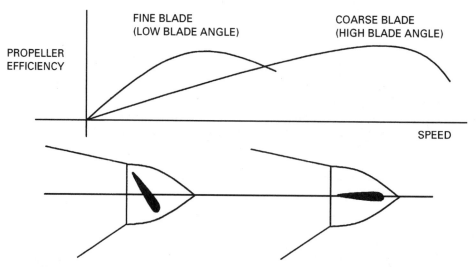

Figure 204   Propeller blade angle/pitch settings and range.

To summarize, variable-pitch propellers maximize the propeller efficiency through a large speed range by maintaining a constant blade angle of attack, which thereby produces a constant thrust value.

Variable-Thrust Takeoff   A variable (reduced/flexible) thrust takeoff uses the takeoff thrust (EPR/N1) required for the *aircraft's* actual takeoff weight, which is a reduced/derated thrust value from the MTOW thrust value that meets the aircraft's takeoff and climb performance requirements, with one engine inoperative.

The full takeoff thrust is calculated against an aircraft's performance-limited (field length, WAT, tire, or net flight path, obstacle clearance climb profile) maximum permissible takeoff weight, and not its actual takeoff weight. The reduced (flexible) takeoff thrust is the correct thrust setting for the aircraft's actual takeoff weight that achieves the aircraft's takeoff and climb, one engine inoperative, performance requirements.

A variable takeoff thrust is calculated by using the *assumed/flexible temperature* performance technique. Using this reduced takeoff EPR means the aircraft is now operating at or near a performance-limiting condition, whereby following an engine failure the whole takeoff would still be good enough in terms of performance. In fact, the lower-weight aircraft using a derated thrust will mirror the takeoff run/profile of a MTOW aircraft using a full-thrust setting.

NB: A function of a variable (assumed/flexible) takeoff is as follows: A higher assumed/flexible temperature relates to a lower air density delivered to the engine. The fuel flow is reduced correspondingly to maintain the correct air/fuel mixture. This produces a lower theoretical thrust from the engines, and practically this is even lower, as the actual air density is denser and therefore the mix is air-dominant and therefore less explosive. This lower thrust therefore makes the aircraft's momentum to be more dominant in reaching its $V1$ and V_R speeds. This greater dependence on momentum requires a longer runway (TOR/D) for a given weight; or it relates to the same TOR performance (i.e., rotation point) for a lower-weight aircraft using a derated takeoff thrust against a MTOW aircraft using a maximum thrust setting.

A variable/flexible thrust takeoff can be employed on most occasions when the aircraft's takeoff weight is lower than the performance-limited MTOW for the ambient air temperature. Its use is optional, but the higher EPR required for the MTOW must always be available whenever a reduced-thrust takeoff is being conducted. It is recommended that full takeoff power be restored in the event of an engine failure above $V1$, or whenever deemed necessary. The use of MTOW EPR for an aircraft with a lower actual weight will obviously provide the aircraft with a better all-round takeoff performance. It should be clearly understood that a variable/reduced takeoff is a reduced thrust in relation to the thrust setting required for the maximum permissible takeoff weight. It is in fact the correct thrust takeoff setting for the aircraft's actual weight that meets its performance requirements.

There are two main reasons for using a variable/reduced takeoff.

1. To protect engine life and to improve engine reliability. Variable/derated/reduced thrust takeoffs reduce the stress and attrition of the engine during the takeoff period when the highest levels of loads are placed on the engine.

2. To reduce the noise generated by the aircraft. (Noise suppression of this type is normally used for takeoff and occasionally on approaches over noise abatement areas, and for nighttime flying noise restriction.)

When a variable/reduced thrust is used for takeoff, the risk per flight is decreased because of the following main reasons. (1) The assumed/flexible temperature method of reducing thrust to match the takeoff weight does so at a constant thrust/weight ratio. This makes the actual takeoff distance from the reduced-thrust setting to be less than that at full thrust and full weight by approximately 1 percent for every 3°C that the actual temperature is below the assumed temperature. (2) The acceleration stop distance is further improved by the increased effectiveness of full reverse thrust at the lower temperature. (3) The continued takeoff after engine failure is protected by the ability to restore full power on the operative engine.

A variable/reduced (flexible) thrust takeoff can only be used when the full takeoff thrust is not required to meet the various performance requirements on the takeoff and initial climb-out.

Therefore the limitations of using a variable/reduced takeoff are as follows:

1. Not MTOW limited by
   a. Takeoff field length
   b. Takeoff WAT curve (engine-out climb gradient at takeoff thrust)
   c. Net takeoff flight path (engine-out obstacle clearance)
2. Maximum OAT limitation
   Where the proposed takeoff weight is such that none of the above considerations are limiting, then the takeoff thrust may be reduced until one of the considerations listed below becomes limiting.

   NB: This is based upon one engine inoperative.

3. A variable/derated/reduced thrust takeoff is limited by the following:
   - The variable/derated/reduced thrust must not be reduced by more than a set amount (engine-specific), say, 25 percent below the MTOW full-rated takeoff thrust.

   NB: This limitation is normally restricted by a maximum flexible or assumed temperature constraint; namely, flexible temp $< T_{max}$.

- The variable/derated/reduced temperature cannot be lower than TREF (flat rating cutoff temperature which guarantees a constant rate of thrust at a fixed temperature, which equates to the climb thrust). Flexible temperature > TREF.
- The variable/derated/reduced thrust cannot be lower than the maximum continuous thrust used for the final take off flight path.
- The actual OAT, i.e., flexible temperature > OAT.

4. Variable/derated/reduced takeoffs should not be used on the following occasions:
   a. On icy or very slippery runways
   b. On contaminated runways, i.e., precipitation-covered
   c. When reverse thrust is inoperative
   d. When the antiskid system is inoperative
   e. When an increased $V2$ procedure is used in order to improve an obstacle-limited takeoff weight

A MTOW aircraft can still use a variable/reduced takeoff technique, providing the TOR/D is not limiting. This is true because you can trade momentum gained from a longer TOR/D to achieve the $V1$ and V_R speeds at the performance-limiting conditions for a lower thrust setting.

Variation   *See* Magnetic Variation.

VASI Lights   The visual approach slope indicator (VASI) lights are a system used to provide slope guidance during the visual stage of the approach. They will assist the pilot to maintain a stable descent path down to the runway surface for the flare and landing.

Two-bar VASI   A typical two-bar VASI system has two wing bar single lights, one above the other, which are positioned at the side of the runway, usually at 300 m from the approach threshold. A duplicate set is positioned on the opposite side of the runway.

It is sometimes known as the red/white system, because the colors seen by the pilots tell them if they are right on slope, too high, or too low, as follows:

All bars white	High on the slope
Near bars white and far bars red	Right on slope
All bars red	Low on the slope

See Fig. 205 at the top of the next page.

Three-bar VASI   This system has an extra wing bar single light, farther down the runway, to assist the pilots of larger aircraft, such as the B747. This is done because the guidance given by VASIs depends on the position of the pilot's eyes. Because the wheels of a large aircraft are much farther beneath the pilot's eyes, it is essential that his or her eyes follow a parallel but higher slope to ensure adequate clearance over the runway threshold. Pilots of large aircraft should use the second and third wing bar lights and ignore the first, while pilots of smaller aircraft should use the two nearer wing bar lights and ignore the third farthest one. The red/white guidance system is the same as the two-bar VASI system.

See Fig. 206 at the bottom of the next page.

V_a Speed   The V_a (maneuvering) speed is the airspeed at which maximum elevator deflection causes the stall to occur at the airframe's load factor limit. The V_a for maximum aircraft weight is specified in the flight manual.

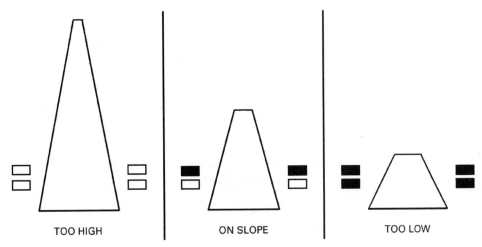

Figure 205 Two-bar VASI guidance.

VAT/VREF Speed Velocity at threshold/velocity reference speed is the target approach threshold speed above the fence height for a specified flap setting which ensures that the landing field length is constantly achieved.

VAT/VRE = 1.3 V_S in landing configuration

VAT0 = target threshold, all-engine operation, speed with maximum flap setting

VAT1 = target threshold, 1 engine (critical) inoperative, speed

VAT2 = target threshold, 2 engine inoperative, speed

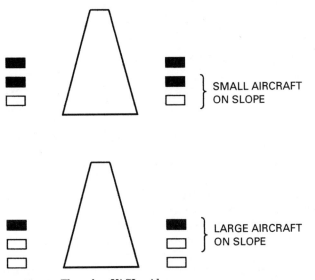

Figure 206 Three-bar VASI guidance.

Adjustments to VAT/VREF are made for headwind values; that is, 50 percent of the headwind and the full gust value is added to the VAT/VREF speed up to a maximum permitted predetermined limit, say, 20 kn.

Above a maximum VAT speed at the threshold, the risk of exceeding the landing field length is unacceptably high, and so a go-around should be initiated.

VDF/MDF Speed   The VDF/MDF (IAS velocity/Mach) is the maximum flight diving speed for a jet aircraft. It is established as the highest demonstrated speed during flight certification trials.

Venturi   A venturi is a practical application of Bernoulli's theorem. Sometimes it is called a convergent/divergent duct.

A venturi tube has an inlet, which narrows to a throat, forming a converging duct, resulting in

- Velocity increasing
- Pressure (static) decreasing
- Temperature decreasing

It is followed by an outlet section, which is relatively longer with an increasing diameter forming a diverging duct, resulting in

- Velocity decreasing
- Pressure (static) increasing
- Temperature increasing

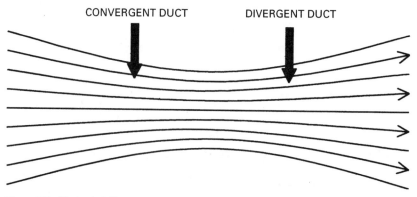

Figure 207   Venturi airflow.

For a flow of air to remain streamlined, the mass flow through a venturi must remain constant. To do this and still pass through the reduced cross section of the venturi throat, the speed of flow through the throat must be increased. In accordance with Bernoulli's theorem, this brings about an accompanying drop in pressure and temperature. As the venturi becomes a divergent duct, the speed reduces and thus the pressure and temperature increase.

Vertical Speed Instrument (VSI)   The vertical speed instrument measures the rate of change of static pressure and displays this as a rate of climb or descent (expressed as feet per minute, fpm) on the VSI face.

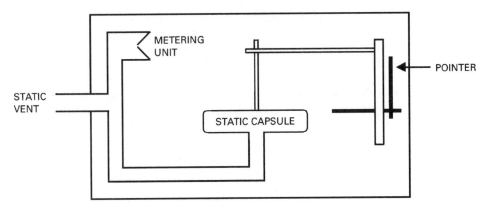

Figure 208 Vertical speed indicator.

The capsule is fed with static pressure and reacts immediately to any change in static pressure, whereas the static pressure feed into the case is restricted, or slowed, via a metering unit, thereby creating a differential static pressure between the capsule and the case. As long as the aircraft continues to climb or descend, it will translate this as a rate-of-climb or rate-of-descent measurement on the instrument dial face.

The VSI uses a logarithmic scale display, which has a greater sensitivity at small ROC/ROD values. Zero is usually at the nine o'clock position with an ROC scale above and an ROD scale below.

Vertical Speed Instrument (VSI)—Errors The vertical speed instrument suffers from the following errors:

1. *Instrument time lag error.*   The instrument suffers from a time delay before it starts to display a ROC/D.

2. *Pressure error.*   This is caused only by variations of the static and pitot probe relative position with respect to the airflow over the aircraft's surface. This can lead to sensed pressure readings that are not representative of the free atmosphere when the airflow pattern is disturbed whenever the airspeed changes. (This is also known as *position error.*)

3. *Maneuver error.*   Maneuvers that disturb the airflow pattern lead to incorrect sensed pressure readings and are unpredictable.

A static line blockage causes the static pressure in the VSI capsule and case via the metering unit to remain a constant value. Therefore the VSI display will read zero at all times regardless of any actual change in the aircraft's ROC or ROD.

The blockage is normally caused in flight by ice formation or small insects covering the static port. The actions for a blocked static line causing an unreliable VSI reading would be to

1. Ensure the pitot static probe anti-icing heating (pitot heat) is *on,* if applicable.

2. Use an alternative—static source or air data computer, if applicable.

3. Use a limited flight panel, i.e., altimeter if available.

4. Fly correct attitude and power settings.

For climbs and descents calculate the ROC/ROD from either the altimeter or approximations given known aircraft attitude and power performance.

NB: If the blockage has been caused by ice formation and you have an unserviceable or no pitot heat system, then you should avoid icing conditions.

VHF Communications   Very high-frequency (VHF) radio transmissions are used for short-range communications. It uses line-of-sight propagation paths and allows reception and transmissions at any point within its area of coverage, namely, from the ground station to its maximum range.

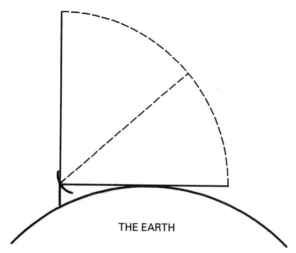

THE EARTH

Figure 209   Line-of-sight VHF propagation path.

Military and civil agencies use different frequency bands. Military agencies use the upper VHF and lower UHF bands, while civil agencies use 1520 VHF frequencies in the 118 to 137 MHz band at 12.5-kHz intervals. This range usually gives good reception and only slight interference from static tied to atmospheric attenuation.

NB: Atmospheric attenuation is the loss of signal strength due to the radio wave's energy being absorbed by the electrons in dust and water droplets in the atmosphere. This becomes significant above the UHF band and reduces the transmission range to an unacceptable range.

VHF radio transmissions have the ICAO designator A3E.

A   double sideband (amplitude-modulated and vertically polarized)

3   analog information

E   voice

The maximum theoretical coverage of a VHF transmission can be calculated using the *line-of-sight formula*

$$\text{Coverage} = 1.25 \sqrt{H_1 \, (\text{ft AMSL})} + 1.25 \sqrt{H_2 \, (\text{ft AMSL})}$$

where H_1 is the height of the transmitter and H_2 is the height of the receiver.

For example, for an aircraft at 35,000 ft,

$$\text{Coverage} = 1.25 \sqrt{0} + 1.25 \sqrt{35{,}000} = 234 \text{ nm}$$

The following factors affect the range of a VHF communication:

1. *Transmitter power:* The greater the power, the greater the range.
2. *Frequency:* The higher the frequency, the slightly greater the signal's vertical path and therefore the shorter its range.
3. *Height of transmitter and receiver:* At low altitudes an aircraft can be shielded by the curvature of the earth from ground VHF line-of-sight signals. Therefore the higher the aircraft, the higher it is above the curvature of the earth, and the greater its ability to intercept a VHF line-of-sight signal from a distant ground-based transmitter, hence the greater the range. Also, the higher the ground-based transmitting station, the greater its range because the curvature of the earth is effectively reduced.
4. *Obstructions:* Because it is a reflective signal, any obstructions will reflect the signal and therefore stop its usage past this point, hence its line-of-sight properties.
5. *Fading:* Occasionally, line-of-sight signals and ground-reflected signals, which are slightly out of phase, can interfere to produce the effect of signal fading.

VHF Directional Finding (VDF)   VHF directional finding (VDF) uses communication radio waves and does not require any additional instrumentation in the cockpit. It can be requested from ATC at aerodromes equipped with radio aerials that can sense the direction of the VHF-com radio signals from an aircraft whenever the pilot transmits. This is known as automatic *direction finding* (DF) in the VHF band. This system allows the ATC to provide the pilot with aural bearing information on request, as a QDM, QDR, QTE, or QUJ.

A VDF is commonly used to track to or from a station, or as a DF letdown procedure.

NB: It is the pilot's responsibility to allow for drift, etc., to any bearing passed from the DF ground station.

Bearings are classified by their quality,

Class A   Bearings are accurate to ±2°
Class B   Bearings are accurate to ±5°
Class C   Bearings are accurate to ±10°
Class D   Accuracy is outside ±10°

Ground DF stations can decline to give a bearing if the accuracy is poor; however, the normal ADF bearings passed to the pilot are class B.

Virga Rain   Virga is rain that falls from the base of a cloud but evaporates at a lower level in drier, warmer air, before it reaches the ground. This is a sign of a temperature inversion, which in turn is an indication of possible wind shear.

Virga is not really a form of precipitation because it does not reach the ground; however, it is generally considered to be, and it is important for pilots to recognize it because it can affect an aircraft's flight path.

Visibility—Atmospheric/Meteorological   Meteorological visibility is defined as the greatest *horizontal* distance at which a specified object can be seen in daylight conditions. It is therefore a measure of how *transparent* the atmosphere is to the human eye.

Visibility is reduced whenever particles are present in the atmosphere because they absorb the light. The atmospheric particles that reduce visibility include the following:

1. Small particles of
   a. Pollution/smoke, causing haze and smog
   b. Solid particles, e.g., dust or oil, causing haze
   c. Water droplets or ice, causing fog or cloud
2. Larger particles of
   a. Sand, dust, volcanic ash or sea spray, with strong winds to hold them in suspension
   b. Precipitation, especially heavy rain, snow, and hail

Poor visibility is usually associated with stable air, where the moisture and/or atmospheric contamination is kept in situ especially at low levels. Unstable air produces a convection current that carries moisture and contamination aloft and thereby produces good visibility at low levels.

Visibility may be further reduced by the position of the sun. Visibility may be much greater flying "down sun," where the pilot can see the sunlit side of objects, than when flying into the sun. As well as reducing the visibility, flying into the sun may cause a glare. If landing is into the sun, consideration should be given to altering the time of arrival.

NB: Remember that the onset of darkness is earlier on the ground than at altitude, and even though visibility at altitude may be good, flying low in the circuit area and landing at a darkening field may cause problems.

The most important visibility to the pilot is from the aircraft to the ground, i.e., the slant or oblique visibility, especially during, and in the direction of, takeoff and landing. This may be very different from the horizontal visibility distance (normal meteorological measured visibility direction). For example, low-level stratus, smog, or fog may severely reduce the slant visibility (especially on an approach) when the vertical visibility might be unlimited.

Visibility is reported in the following two ways:

1. *General visibility,* from overall meteorological reports. In general meteorological reports, the visibility passed is always the *least* distance visible from the point of observation in all directions. Visibility is reported in kilometers, or in meters in very poor conditions. A forecast/report that contains the term *7000 0.5 HZ* means 7000-m visibility in haze (0.5 is not significant to the pilot). If the visibility is 10 km or greater (i.e., in excess of 9999 m), then it would be written as 9999; and a visibility less than 100 m will be written as 0000.
2. *Runway visual range* (RVR) for instrument landings.

VMBE Speed The VMBE (maximum brake energy speed) is the maximum speed on the ground from which a stop can be accomplished within the energy capabilities of the brakes.

If the $V1$ speed exceeds the VMBE speed, then the aircraft's takeoff weight has to be reduced until the $V1$ speed is less than or equal to the VMBE speed, to ensure that the aircraft does not exceed its brake energy limit. Hence VMBE can limit $V1$, and thereby MTOW, especially on downward-sloping runways with a tailwind.

An aircraft will have a set weight reduction for each knot of speed.

NB: Both V_R and $V2$ need to be redetermined for the lower aircraft weight.

VMCA Speed   The VMCA is the minimum control speed in the air for a multiengine aircraft in the takeoff and climb-out configuration at and above which it is possible to maintain directional control (heading) of the aircraft around its normal/vertical axis by use of the rudder (i.e., the turning moment produced by the vertical tailplane with maximum rudder deflection is sufficient to balance the yawing moment of the aircraft nose) within defined limits after the failure of an off-center engine (asymmetric power).

The heading of an aircraft is determined by the direction of the nose of the aircraft, which is pivoted about the normal/vertical axis at the center-of-gravity point. With an off-center engine failure, and assuming a constant power and configuration setting, the aircraft will yaw about the cg point to the dead engine, due to the asymmetric thrust properties. This yaw changes the aircraft's direction. The magnitude of the yaw is a function of the asymmetric thrust times aircraft nose to cg arm, or

$$\text{Yawing moment} = \text{asymmetric thrust} \times \text{nose to cg arm}$$

Whenever the aircraft is committed on the takeoff run, i.e., past $V1$, or in the air, this yawing moment has to be balanced to maintain directional control of the aircraft. The directional control is provided by the aircraft's vertical tailplane and its rudder control surface, which is used to produce a turning moment to oppose the yawing moment of the aircraft. The vertical tailplane produces a turning moment that is based on the design of the surface, i.e., size; assumes a maximum rudder deflection; and is a product of the vertical tailplane to cg arm and the weight on the tailplane, where the weight is the air load force experienced.

$$\text{Turning moment} = \text{rudder to cg arm} \times \text{weight (air load force)}$$

Because the rudder to cg arm is a constant during a particular takeoff run or phase of flight, the vertical tailplane turning moment is influenced/determined by the weight on the vertical tailplane. The weight, as previously stated, is the force experienced, which is a product of:

- Angle of deflection—we always assume a maximum rudder deflection

- Air density (has a minimal influence)

- Airspeed

Therefore the speed determines the aerodynamic force/weight over the vertical tailplane. The greater the speed, the greater the rudder turning moment and the greater the directional control for a given atmospheric condition, i.e., air density. Therefore weight is a product of airspeed, and the formula can be expressed as

$$\text{Turning moment} = \text{rudder to cg arm} \times \text{speed (VMCA)}$$

In essence, the magnitude of the two opposing moments—the yawing moment of the aircraft's nose and the turning moment of the vertical tailplane—produces a seesaw effect. When the aircraft's yawing moment is dominant, the aircraft cannot maintain directional control, which results in the aircraft yawing off heading. When the vertical tailplane turning moment is dominant, the aircraft can maintain directional control which results in the aircraft being able to maintain heading. Which moment is dominant depends on the airspeed; therefore VMCA relates to the minimum control speed on the ground or in the air, at which and above the vertical tailplane turning moment is dominant over the yawing moment, and therefore directional control of the aircraft is guaranteed.

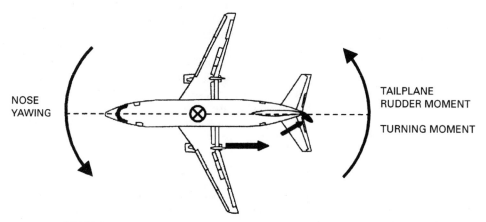

Figure 210   VMCG/A—yawing and turning moments.

When an aircraft is loaded, the position of the cg is determined, and the cg position affects the size of both the yawing and turning moments. Therefore the VMCA speeds vary with the cg position.

The turning moment acts around the cg. If the cg is in the aft position, the vertical tailplane (rudder) moment arm will be shorter, and therefore the vertical tailplane turning moment is less for a given airspeed. Thus the aircraft requires a higher minimum control speed (VMCA) with an aft cg position.

Turning moment = rudder to cg arm × speed (VMCA)

Conversely, the opposite is true for a forward cg position. A forward cg will have a longer arm, so the vertical tailplane turning moment is greater for a given speed and thus the aircraft can have a lower VMCA.

To conclude, remember that VMCA relates to asymmetric power configurations and guarantees directional control of the aircraft (not climb capabilities) whenever the aircraft's airspeed is equal to or greater than VMCA.

VMCA must be less than $V2$ (normally $V2$ is equal to or greater than 1.1 × VMCA).

NB: $V2$ relates to the directional control and a minimum climb performance of the aircraft, in the event of an engine failure.

VMCG Speed   The VMCG is the minimum control speed on the ground for a multiengine aircraft at a constant power setting and configuration, at and above which it is possible to maintain directional control (heading) of the aircraft around its normal/vertical axis by use of the rudder (i.e., the turning moment produced by the vertical tailplane with maximum rudder deflection is sufficient to balance the yawing moment of the aircraft nose), to maintain runway heading after the failure of an off-center engine (asymmetric power).

The heading of an aircraft is determined by the direction of the nose of the aircraft, which is pivoted about the normal/vertical axis at the cg point. With an off-center engine failure and assuming a constant power and configuration setting, the aircraft will yaw about the cg point to the dead engine, due to the asymmetric thrust properties. This yaw changes the aircraft's direction. The magnitude of the yaw is a function of the asymmetric thrust × aircraft nose to cg arm.

$$\text{Yawing moment} = \text{asymmetric thrust} \times \text{nose to cg arm}$$

Whenever the aircraft is committed on the takeoff run, i.e., past $V1$, or in the air, this yawing moment has to be balanced to maintain directional control of the aircraft. The directional control is provided by the aircraft's vertical tailplane and its rudder control surface, which is used to produce a turning moment to oppose the yawing moment of the aircraft. The vertical tailplane produces a turning moment that is based on the design of the surface, i.e., size; assumes a maximum rudder deflection; and is a product of the vertical tailplane to cg arm and the weight on the tailplane, where the weight is the air load force experienced.

$$\text{Turning moment} = \text{rudder to cg arm} \times \text{weight (air load force)}$$

Because the rudder to cg arm is a constant during a particular takeoff run or phase of flight, the vertical tailplane turning moment is influenced/determined by the weight on the vertical tailplane. The weight, as previously stated, is the force experienced, which is a product of

- Angle of deflection—we always assume a maximum rudder deflection
- Air density (has a minimal influence)
- Airspeed

Therefore the speed determines the aerodynamic force/weight over the vertical tailplane. The greater the speed, the greater the rudder turning moment and the greater the directional control for a given atmospheric condition, i.e., air density. Therefore weight is a product of airspeed, and the formula can be expressed as

$$\text{Turning moment} = \text{rudder to cg arm} \times \text{speed (VMCG)}$$

In essence, the magnitude of the two opposing moments, i.e., the yawing moment of the aircraft's nose and the turning moment of the vertical tailplane, produces a seesaw effect. When the aircraft's yawing moment is dominant, the aircraft cannot maintain directional control, which results in the aircraft yawing off heading. When the vertical tailplane turning moment is dominant, the aircraft can maintain directional control that results in the aircraft being able to maintain heading. Which moment is dominant depends on the airspeed; therefore VMCG relates to the minimum control speed on the ground or in the air, at which and above the vertical tailplane turning moment is dominant over the yawing moment and therefore directional control of the aircraft is guaranteed. (See Fig. 210.)

When an aircraft is loaded, the position of the cg is determined, and the cg position effects the size of both the yawing and turning moments. Therefore the VMCG speeds vary with the cg position.

The turning moment acts around the cg. If the cg is in the aft position, the vertical tailplane (rudder) moment arm will be shorter, and therefore the vertical tailplane turning moment is less for a given airspeed. Thus the aircraft requires a higher minimum control speed (VMCG) with an aft cg position.

$$\text{Turning moment} = \text{rudder to cg arm} \times \text{speed (VMCG)}$$

Conversely the opposite is true for a forward cg position. A forward cg will have a longer arm, and so the vertical tailplane turning moment is greater for a given speed. Therefore the aircraft can have a lower VMCG.

If VMCG is limiting for the weight of the aircraft, you have to *reduce takeoff thrust*. The vertical tailplane (rudder) turning moment is used to oppose/balance the asymmetric thrust yawing moment to maintain directional control. Therefore by reducing thrust, any off-center engine loss during the takeoff run has a reduced asymmetric thrust imbalance, which thereby reduces the yawing moment experienced and requires a reduced tailplane turning moment to maintain directional control. And because the magnitude of the tailplane turning moment is a product of airspeed, a lower (VMCG) ground speed maintains the aircraft's directional control. Ergo, reduced takeoff thrust gives rise to a lower VMCG.

To conclude, remember that VMCG relates to asymmetric power configurations and guarantees directional control of the aircraft (not climb capabilities) whenever the aircraft's airspeed is equal to or greater than VMCG.

During the ground TOR, if you lose an off-center engine below VMCG, you will not have directional control of the aircraft. Therefore you have to remove the asymmetric thrust by closing the engines to regain directional control, and obviously this results in an aborted takeoff. To allow for this action VMCG must always be less than or equal to $V1$, thus ensuring that you retain directional control at and above $V1$.

The VMCG has to be equal to or less than $V1$. This ensures that the aircraft can maintain directional control with an off-center engine failure at or above $V1$, when the aircraft is committed to the takeoff and directional control of the aircraft is essential for the safe operation of the aircraft.

VMCL Speed   The VMCL is the minimum control speed in the air for a multi-engine aircraft in the approach and landing configuration at and above which it is possible to maintain directional control of the aircraft around the normal/vertical axis by use of the rudder after the failure of an off-center engine, within defined limits while applying variations of power, i.e., idle to maximum thrust on the live engine(s).

VMO/MMO Speed   The VMO/MMO (IAS velocity/Mach) is the maximum operating speed permitted for all operations. It is normally associated with jet aircraft.

VMU Speed   This is the minimum demonstrated unstick speed at which it is possible to get airborne on all engines and to climb out without hazard.

VNAV   *See* LNAV and VNAV.

VNE Speed   The VNE (IAS velocity) is the never-exceed velocity. It is associated with propeller-driven aircraft and is a higher speed than the VNO speed, which can be used when operationally desired, but must never be exceeded.

VNO Speed   The VNO (IAS velocity) is the normal operating speed permitted for all normal operations. Normally it is associated with propeller aircraft.

Volcanic Ash   The hazard associated with flying in a region of volcanic ash is flameout of the engines. This is due to the engine being starved of air, because

1. Ash builds up on fan and compressor blades, which upsets the airflow through the engine.
2. High volcanic ash content in a parcel of air also deprives the combustion chamber of the air it requires for a controlled explosion.

If you ever find yourself in a region of volcanic ash, do the following:

1. On first observation or contact with volcanic ash, make an immediate 180° turn away from region of volcanic ash.
2. If inadvertently you encounter a region of volcanic ash, select all engine high-bleed systems on, e.g., air conditioning, engine and airframe deicing or anti-icing systems, etc.

NB: This helps to reduce the high volume of ash from the engine, thereby helping to maintain a higher percentage of air delivery to the combustion chambers to protect against a flameout.

3. Make sure the APU is on, to assist engine relight if necessary.

In addition, you should apply any further procedures that your flight manual may recommend.

Volmet   Volmet is a continuous broadcast on a VHF/HF setting including

1. Actual weather report
2. Landing forecast
3. A forecast trend for the 2 h following
4. A sigmet (significant weather, if any)

of several selected aerodromes, that produces met reports within a given region.

Volmets are updated each hour and each half-hour, in line with the new issued metars at the various aerodromes.

NB: Although volmet carries aerodrome met reports, it is commonly used for route and/or destination weather briefings, either preflight or during flight.

Volt   A volt (abbreviated *V*) is a unit of electric force/pressure. Volts are a measure of

- Electromotive force (emf), i.e., the electrical pressure available from a source of electric energy used to produce electron flow (current).
- Potential differences (pd) in the emf level, between any two circuit points, that create an electrical pressure, which can produce electron flow from the less positive to the more positive circuit point.

NB:

$$1 \text{ kilovolt (1 kV)} = 1000 \text{ V}$$

$$1 \text{ millivolt (1 mV)} = 1/1000 \text{ V (0.001 V)}$$

Voltmeter (Gauge)   A voltmeter indicates the amount of volts produced by an electrical source, e.g., generator, or the potential difference in a system between any two points, as in a transformer.

Volumetric Efficiency   Volumetric efficiency is the ratio of the weight of gas induced into the cylinder on the induction stroke to the weight of gas which would fill the cylinder at normal temperatures and pressures, expressed as a percentage.

Volumetric efficiency =

$$\frac{\text{weight of gas induced}}{\text{weight of gas that would fill the cylinder at normal temperature and pressure}}$$

A method of improving the engine's performance is to increase its volumetric efficiency, e.g., by increasing the valve overlap.

VOR (VHF Omni Range)   A VOR (VHF omni range) is a short-range, sophisticated and accurate VHF navigational radio aid which outwardly generates specific track/position lines.

A VOR ground transmitter radiates line-of-sight signals in all directions. However, unlike an NDB, the signal in any particular direction differs slightly from its neighbor. These individual direction signals can be thought of as tracks or position lines, radiating out from the VOR ground station. By convention, 360 different and separate tracks away from the VOR are used, each with its position related to magnetic north, that is, 000° to 359°. These VOR tracks or position lines are called *radials*. A radial is the magnetic bearing outbound from a VOR (QDR).

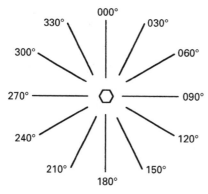

Figure 211   VOR signals—track/position lines.

The principle of how a VOR ground station works, i.e., with 360 separate radial signals, is achieved because it transmits 360 different VHF radio signals, with both

1. An FM *reference phase,* which is omnidirectional (i.e., the same in all directions)

2. An AM *variable phase* which rotates uniformly at a rate of 1800 rpm

with its phase varying at a constant rate throughout the 360°. Thereby a unique signal is created for all 360 radials. The receiving aircraft determines each individual radial by examining the phase difference between the FM reference signal and the AM variable signal, and displays it as a corresponding radial. For example, no phase difference represents a QDR of 000°, and a phase difference of 280° represents a QDR of 280°.

A VOR cockpit display will indicate the aircraft's orientation to a VOR radial and consequently the QDR/QDM relative to the selected VOR station. However, the VOR cockpit display is not heading-sensitive, which means the display will not change as a result of the airplane changing heading, as long as its position to the radial remains unchanged.

A VOR can be used for the following reasons:

1. Orientation/fixing position. This is accomplished by establishing a position line along which the aircraft is known to be, by
   a. A single VOR radial/DME couple
   b. Two VOR cross-cutting radials
   c. A single VOR radial and either an NDB or VDF position line, which produce a cross-cut position
2. Intercepting and tracking en route either along a radial or crossing known radials from an off-track VOR, or as a VOR letdown procedure, with or without paired DME information. VOR indications can be displayed on any of the following cockpit instruments: an OBI, RMI, or HSI.

The VOR transmits in the 108- to 118-MHz VHF band.

NB: The 108- to 112-MHz band is shared with ILS stations, so VORs are only allocated frequencies at *even* 100-kHz spacing, or 108.2 kHz. Between 112 and 118 MHz, this band is used exclusively for VOR stations, and therefore the frequency spacing is reduced to 50 kHz.

A VOR is often associated/paired with a DME (distance-measuring equipment) so that both range and bearing information are provided.

There are several different types of VORs:

Standard VOR is used for en route navigation, usually to define the centerline of an airway.

Terminal VOR is a low-power beacon used as part of an airfield approach.

Broadcast VOR is normally a terminal VOR, with a voice broadcast giving the airfield weather and ATIS information. The voice information is transmitted on the carrier wave.

Doppler VOR is a second-generation beacon that uses a complex aerial array to reduce the problems of site error. A Doppler VOR has an AM *reference signal* and an FM *variable signal* that rotate anticlockwise. This is opposite to a standard VOR.

NB: In an aircraft it is impossible to tell the difference between a Doppler and a standard VOR signal.

The errors experienced by a VOR are as follows:

1. Equipment errors are both ground-based and internal equipment errors. The aircraft's reception equipment has internal errors, which must be accurate to within ±3°. The ground station transmitting equipment has internal errors, which must be accurate to within ±2°. This constitutes the ICAO bearing accuracy requirement of ±5° on 95 percent of occasions.

The total error of a standard airway VOR determines the distance between two VORs, so it maintains the airway's centerline within an airway's maximum permissible width, i.e., 5 nm with a maximum error of 5°.

See Fig. 212 at the top of the next page.

A VOR has a published designated operational coverage (DOC) which gives a protected range, sector, and altitude for a VOR signal, where freedom from terrain and manmade obstruction that may reflect a signal (site error) is guaranteed.

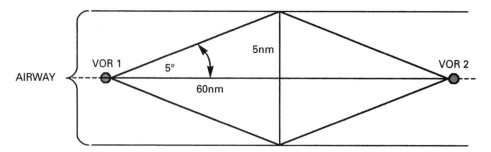

Figure 212   VOR spacing along an airway. 1:60 rule: distance along track = [(distance off track)/angle] × 60.

2. Site error is the effect of the signal being reflected off objects near to the beacon that causes tremendous confusion in the reading.

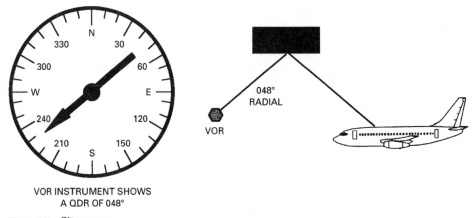

VOR INSTRUMENT SHOWS
A QDR OF 048°

Figure 213   Site error.

Because of this VORs are carefully positioned so that these reflections are minimized and therefore do not worsen a VOR's bearing accuracy.

3. Propagation error with its scalloping effect is the buildup of errors, particularly at the limit of a beacon's range due to terrain and artificial obstructions. Atmospheric ducting can lead to synchronous reception (duct propagation) even within a VOR's designated operational coverage. This is so because ducting is unpredictable and no allowance can be made for it. However, confirming a station ident and a commonsense reading is always good airmanship.

Therefore because propagation errors are unpredictable, they cannot be legislated for and so do not contribute to a VOR's calculated total error.

The *cone of confusion* is the area overhead a VOR transmitter where it is not possible to obtain any accurate bearing information. As such, this is sometimes considered an error, because the VOR fails to provide any bearing information. However, some might say that its lack of information is an indication of the aircraft being directly overhead the beacon, and therefore it cannot be considered an error.

Because of its VHF line-of-sight transmissions, a VOR is free from sky wave interference (night effect), coastal refraction, and quadrantal error and experiences only slight atmospheric interference, i.e., static.

An RMI cockpit display will indicate the aircraft's orientation to a selected VOR station, i.e., the QDM or QDR. And an OBI or HIS cockpit display will indicate the aircraft's orientation to a selected VOR radial.

The following technique is an example of how to intercept and maintain a VOR radial.

Intercepting a specified radial

1. Tune and identify the required VOR. The RMI needle will now indicate the QDM to the beacon.

2. Select the radial track required on an OBI or HSI.

3. Turn the aircraft to intercept the required track. Using an RMI, intercept the required track in the same manner as you would intercept a track to and from an NDB/ADF. That is, if the RMI needle head is right of the required QDM track, then fly an intercept heading farther to the right of the needle.

   Using an OBI or HSI, center the deviation bar of the selected radial, by flying a heading toward the deflected deviation bar.

   NB: The OBI and HSI both display tracking guidance relative to the selected radial. If the aircraft is on the selected radial, the VOR needle or CDI (course deviation indicator) will be *centered*. If the aircraft is not on the selected course/track, then the CDI will be to the left or right. That is, if the selected radial is to the left of the aircraft, then the deviation bar will be to the left.

   However, the OBI is a track-up display; i.e., the selected track accommodates the top of the dial position, regardless of the aircraft's heading. The HSI always acts as a command instrument, because the head-up display of the remote indication compass ensures its correct orientation at all times; i.e., the aircraft's heading is always displayed at the top of the instrument.

4. When steady on an intercept heading, check that the RMI needle is moving toward the required QDM.

5. Using the appropriate amount of lead time, turn onto a heading to capture the required QDM track.

Maintaining a specified track Lay off drift to maintain the QDM.

NB: The drift allowance is the angle between the aircraft heading and the VOR radial.

VOR Range The VOR uses VHF (radio signal) line-of-sight propagation paths. Its theoretical maximum range, based on line-of-sight propagation, is

$$\text{Range} = 1.25 \ \sqrt{H_1 \ (\text{ft AMSL})} + 1.25 \ \sqrt{H_2 \ (\text{ft AMSL})}$$

where H_1 is the height of the transmitter and H_2 is the height of the receiver.

NB: However, its range is also a function of the power output of the transmitter.

Vortex Generators/Fences These reduce spanwise airflow and thereby reduce its effects. One of the effects of spanwise airflow over the wing is the reduced effectiveness of the ailerons, due to the diagonal airflow over the control surfaces.

Vortex generators are located on the upper surface of a wing to create a slightly disturbed airflow perpendicular to the leading edge of the wing, which helps to maximize the effectiveness of the control surfaces, especially the ailerons.

Fences also help to maximize the effectiveness of the control surfaces in a similar yet cruder manner. However, they are normally used to reduce the reverse spanwise airflow on the upper wing surface from reaching the wing tips. This reduces the airflow, which contributes to a wing tip stall.

Vortex generators can also be used in other areas of the aircraft where a disturbed airflow is required (disturbed air tends to be denser and of a slower velocity) such as inlets to some types of APUs.

Vortices   Vortices at the wing tips are created by spanwise airflow over the upper and lower surfaces of a wing/aerofoil that meets at the wing tips as turbulence and therefore induces drag, especially on a swept wing.

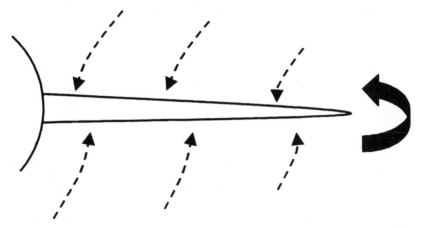

Figure 214   Spanwise airflow/vortices on the wing.

Spanwise airflow is created because a wing producing lift has a lower static pressure on the upper surface than on the lower surface. At the wing tip, however, there can be no pressure difference, and the pressure is equalized by air flowing around the wing tip from the higher pressure on the lower surface to the lower pressure on the upper surface. There is therefore a spanwise pressure gradient, i.e., pressure changing along the wing span.

This pressure gradient will cause the air flowing over the upper surface to incline inward toward the root where the pressure is lowest, and the air flowing over the lower surface to incline outward toward the tip, away from the higher pressure at the root. When the two spanwise airflows meet, at the trailing edge, they are moving at an angle to each other, and in joining they form a sheet of vortices behind the trailing edge. These vortices eventually combine to form a single large vortex behind each wing, just inboard of the wing tip.

See Fig. 215 at the top of the next page.

For practical purposes, these vortices in the wake of an aircraft may be regarded as being made up of two counterrotating cylindrical air masses trailing aft from the aircraft. Typically the two vortices are separated by about three-quarters of the air-

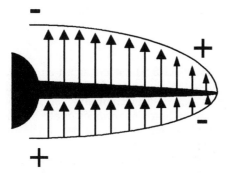

Figure 215 Wing span pressure gradients.

craft's wing span, and in still air they tend to drift slowly downward, usually leveling off not more than 1000 ft below the aircraft's flight path.

The largest vortices are created when the wing pressure differential is greatest, when a high angle of attack is employed to produce the required lift force at low speeds.

The effects of wing tip vortices are as follows:

1. Vortices create aircraft drag (induced because the vortices induce a downward velocity in the airflow over the wing, causing a change in the direction of the lift force, so that it has an induced drag component, creating a loss of energy).
2. Vortices create turbulence, which may affect the safety of other aircraft within approximately 1000 ft below and behind the aircraft.
3. Downwash affects the direction of the relative airflow over the tailplane, which affects the longitudinal stability of the aircraft.

VRA/MRA Speed   The VRA/MRA is an airspeed for rough air conditions, or *turbulence penetration speed*. The rough-air speed is recommended for flight in turbulence that is based on the aircraft's VB speed (design speed for maximum gust intensity). It provides speed protection against the two possibilities, which stem from the effect of a disturbance in rough-air conditions. Namely, the VRA/MRA speed is high enough to allow an adequate margin between the aircraft stall speed and also low enough to protect against structural damage from a high-speed gust disturbance.

VREF Speed   *See* VAT/VREF Speed.

V_R Speed   The rotation speed (V_R) is the speed at which the pilot initiates rotation during the takeoff to achieve V2 at the screen height, even with an engine failure.

- Speed V_R cannot be less than 1.05VMCA or 1.1VMU or 1.05VMU.
- Speed V_R is either greater than or equal to V1, but never less than V1.

V_S Speed   The stall speed V_S is the speed at which the airflow over the wings will stall. The stall speed varies with aircraft weight and configuration.

The stall speed is the *reference speed* for the other performance speeds, such as V2 and V_{REF}.

V_x *See* Maximum Angle (V_x) Climb Profile.

V_y *See* Maximum Rate (V_y) Climb Profile.

V1 Speed   Speed V1 is the *decision speed* in the event of an engine failure during the takeoff roll, at which it is possible to continue the takeoff and achieve the screen height within the normal takeoff distance available, or to bring the aircraft to a full stop within the emergency distance available (accelerate stop distance). The takeoff must be abandoned with an engine failure below V1, and the takeoff must be continued with an engine failure above V1.
   NB: If the takeoff weight is limited by TODA, TORA, or EMDA, the V1 speed relates to a single point along the runway where the pilot will have the decision to continue or abort the takeoff in the event of an engine failure.
   Speed V1 cannot be less than VMCG and cannot be greater than V_R or VMBE.
   If the field length is limiting, then the greater the aircraft weight, the lower the V1 speed. This means the lower V1 speed provides a greater stopping distance while ensuring that V1 remains greater than VMCG and VMU. And if the field length is not limiting, then the greater the aircraft weight, the higher the V1 speed, providing V1 remains less than the VMBE speed and the field length emergency stopping distance is not compromised.

V1 Speed—Range   When the planned takeoff weight (TOW) is not field-length-limited (i.e., not limited by TORR/TODR/EMDR), there may be a range of V1 speeds, e.g., between 125 and 135 kn, for which the pilot has a stop/go option in the event of an engine failure. The minimum V1 speed in the range is still restricted by the VMCG speed, and the maximum V1 is still restricted by the VMBE.
   A range allows the V1 speeds to be raised or lowered to achieve a different departure profile or climb technique, i.e., an increased TOW or an increased V2 climb for obstacle clearance. However, normally a single V1 speed is chosen within the range prior to commencing the takeoff run, so that if an engine failure does occur, the decision has effectively already been made. This removes any delayed response and indecision that an analysis of a V1 range at the time of the engine failure might precipitate, which improves operational safety.

V2 Speed   The V2 speed is the takeoff safety speed achieved by the screen height, in the event of an engine failure that maintains adequate directional control and climb performance properties of the aircraft. Speed V2 cannot be less than $V_S \times$ 1.20 or VMCA $\times$ 1.10.
   NB: Speed V2 is also known as the *takeoff safety speed* (TOSS).

V3 Speed   The V3 speed is the all-engine operating takeoff climb speed the aircraft will achieve at the screen height.

V4 Speed   The V4 speed is the all-engine operating takeoff climb speed the aircraft will achieve by 400 ft, and it is used as the lowest height where acceleration to flap retraction speed is initiated.

Wake Turbulence Wake turbulence is the phenomenon of disturbed airflow, i.e., wing tip vortices created behind an aircraft's wing, as the aircraft moves forward.

NB: Wake turbulence can be a serious hazard to lighter aircraft following heavier aircraft. Wake vortex turbulence is present behind every aircraft, including helicopters, when in forward flight. It is extremely hazardous to a lighter aircraft with a smaller wing span during takeoff, initial climb, final approach (including the circuit), and landing phases of flight, following a heavier aircraft with a larger wing span.

NB: Wake turbulence can also be encountered in the cruise, but is not really a hazard, just an uncomfortable disturbance.

The characteristics of the wake vortex turbulence generated by an aircraft in flight are determined initially by the aircraft's gross weight, wing span, airspeed, configuration, and attitude. That is, heavy aircraft with a large wing span create the greatest wake turbulence (wing tip vortices) because they create the greatest pressure differential between the upper and lower surfaces of the wing. Wake turbulence (wing tip vortices) is increased for *a particular aircraft type* at low airspeeds because the aircraft has to employ a greater angle of attack to generate the required lift, which also creates a greater pressure differential between the upper and lower surfaces of the wing. The greatest wake turbulence (wing tip vortices) for *a particular aircraft type* is created at high configuration and attitudes, because they create a greater pressure differential between the upper and lower surfaces of the wing.

Subsequently these characteristics are altered by interactions between the vortices and the ambient atmosphere; and eventually, after a time varying according to the circumstances from a few seconds to a few minutes after the passage of an aircraft, the vortices and therefore the wake turbulence become undetectable.

The wake turbulence created from trailing wing tip vortices can be strong enough to upset a following aircraft, especially lighter aircraft if it flies into the turbulence. To avoid wake turbulence accidents and incidents, air traffic control delays the operation of lighter aircraft behind heavier aircraft for up to 5 min, to allow the vortices that create the wake turbulence to dissipate.

NB: Aircraft of the same size or heavier aircraft following lighter aircraft do not suffer from dangerous levels of wake turbulence, and therefore they are not usually spaced per ICAO separation minima.

The main aim of wake turbulence avoidance is to avoid passing through it at all. There are two methods of avoiding wake turbulence:

1. *Separation minima (time).* This is a time and distance technique employed by ATC to allow the wake turbulence of preceding aircraft to dissipate. This technique is typically used during final approach and takeoff phases of flight, where the flight path track cannot be altered.

NB: Wake turbulence can be a serious hazard to lighter aircraft following heavier aircraft.

The separation minima stated cannot entirely remove the possibility of a wake turbulence encounter. The objectives of the minima are to *reduce* the probability of a vortex wake encounter to an acceptable low level and to minimize the magnitude of the upset if an encounter does occur. Therefore even when employing a time and distance separation, you must still be prepared to alter your flight path if required.

NB: Care should be taken when following any substantially heavier aircraft, especially in conditions of light winds, as the majority of serious incidents occur close to the ground when the winds are light.

2. *Alteration to the flight path.* This is the ultimate action to be taken to avoid wake turbulence, and this is the pilot's responsibility. Therefore you must understand and be able to visualize the formation and movement of the invisible wing tip vortices from preceding aircraft, to be able to avoid these danger areas, especially heavier aircraft at low speed and high angles of attack at low altitudes, which produce the greatest wake turbulence.

Wake vortex generation begins when the nosewheel lifts off the ground on takeoff and continues until the nosewheel touches down on landing. The vortices will tend to lose height slowly, up to 1000 ft, and will drift with the wind, which is especially important in crosswind conditions, as you can encounter wake turbulence when passing downwind of a preceding aircraft.

NB: A *crosswind* will cause the vortices to drift downwind of the preceding aircraft. A *headwind* or *tailwind* will carry the vortices in the direction of the wind. A *nil, light,* or *variable* wind condition will make the vortices just "hang around." This condition can be very dangerous, and delaying your takeoff or approach or even changing runway is worth considering.

Takeoff When you are departing behind a larger aircraft on the same runway, commence your takeoff run at the end of the runway (an intersection departure, less than the full length of the runway, may bring your flight path closer to the preceding aircraft's wake turbulence). Note the preceding aircraft's rotation point, and rotate your aircraft before it. Climb above and stay above the preceding aircraft's flight path until you turn clear of its wake.

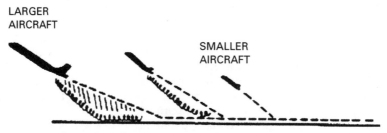

LARGER
AIRCRAFT

SMALLER
AIRCRAFT

Figure 216 Departing behind a larger aircraft on the same runway.

When you are departing on a crossing runway behind a larger aircraft on the other runway, note the preceding aircraft's rotation point. If it occurs before the runway intersection point, either delay your takeoff or plan to be airborne well past the intersection point (allowing for wind, which could drift the vortices past the intersection point along your runway). Further, avoid headings which will cross behind and below the preceding aircraft.

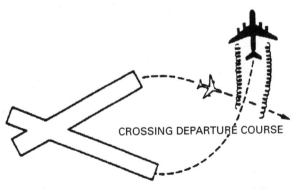

Figure 217 Departing behind a larger aircraft on a different runway.

When you are departing behind a larger aircraft landing on the same runway and are taking off after a heavy aircraft has landed, plan to become airborne well past the point where the preceding aircraft's nosewheel touched down.

Approach and landing When you are landing behind a larger aircraft on the same runway, stay at or above the preceding aircraft's final-approach path; note its nosewheel touchdown point, and land beyond it.

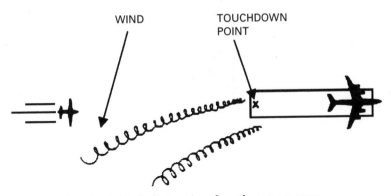

Figure 218 Landing behind a larger aircraft on the same runway.

When you are landing behind a larger aircraft on a parallel runway which is closer than 760 m, consider possible drift toward your runway. If this is possible, then stay at or above the preceding aircraft's final-approach path, note its nosewheel touchdown point, and land farther down your runway.

See Fig. 219 at the top of the next page.

When you are landing behind a larger aircraft on a crossing runway, if the preceding aircraft touches down after the intersection, cross above its flight path and land past the intersection, allowing for any vortices' drift.

See Fig. 220 in the middle of the next page.

When you are landing behind a departing heavier aircraft on the same runway, note the departing aircraft's rotation point and allowing for drift, land well before it.

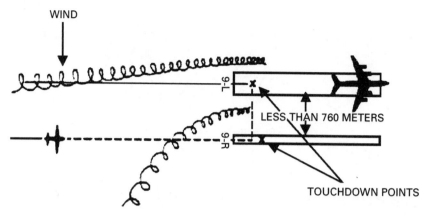

Figure 219 Landing behind a larger aircraft on a parallel runway closer than 760 m.

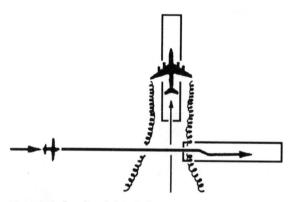

Figure 220 Landing behind a larger aircraft on a crossing runway.

NB: The normal touchdown zone will probably ensure this.

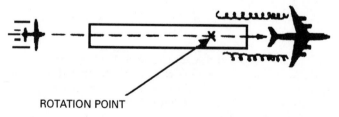

Figure 221 Landing behind a departing heavier aircraft on the same runway.

When you are landing behind a departing heavier aircraft on a crossing runway, note the departing aircraft's rotation point. If it is beyond the runway intersection, you can continue the approach and land normally.

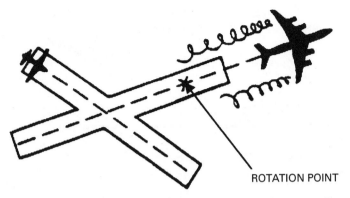

ROTATION POINT

Figure 222   Landing behind a departing heavier aircraft on a crossing runway.

If the preceding aircraft's rotation point is prior to the intersection, avoid flight below its flight path and, if necessary, abandon the approach unless a landing is ensured well before you reach the intersection.

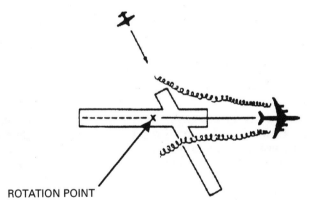

ROTATION POINT

Figure 223   Landing behind a departing heavier aircraft on a crossing runway.

Missed approach   After a preceding heavier aircraft has executed a low missed approach or a touch-and-go landing, its turbulent wake will be a hazard for up to 2 min along the runway for any following aircraft, particularly in light wind conditions. Therefore you should consider changing your flight path in these circumstances.

See Fig. 224 at the top of the next page.

Circuit   Remember, the greatest risk of wake turbulent conditions exists at low level and low speeds. Avoid flying below and behind larger aircraft in the circuit. Fly a few hundred feet above them or at least 1000 ft below them or upwind of them.

En route   In VMC conditions, avoid flight below and behind a larger aircraft's flight path. If a large aircraft is observed above on the same track, consider adjusting your position laterally, preferably upwind.

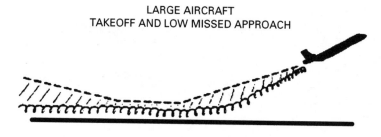

LARGE AIRCRAFT
TAKEOFF AND LOW MISSED APPROACH

TAKEOFF AND LANDING HAZARD

Figure 224 Takeoff or landing behind a missed-approach aircraft on the same runway.

NB: At high altitude, en route wake turbulence is considered a disturbance, not an upset.

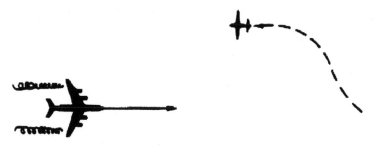

Figure 225 En route in VMC conditions.

Wander The wander of a gyro is any movement of the gyro's spin axis away from its fixed direction. If this occurs, it gives rise to inaccurate instrument readings.

The rigidity of a gyro system is responsible for maintaining the direction of the spin axis fixed in space. A gyro wander can be either *drift* or *topple*. Gyro drift occurs when the spin axis turns in the horizontal plane, and gyro topple occurs when the spin axis tilts in the vertical plane.

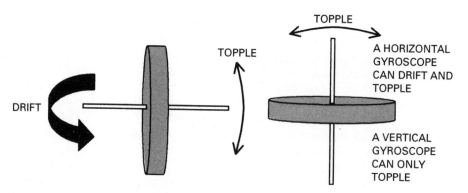

DRIFT

TOPPLE

TOPPLE

A HORIZONTAL
GYROSCOPE
CAN DRIFT AND
TOPPLE

A VERTICAL
GYROSCOPE
CAN ONLY
TOPPLE

Figure 226 Drift and topple.

See also Real Wander.

Warm Front   A warm front is the boundary produced between two air masses (i.e., warm air behind cold air) where the warmer, less dense air mass rises up and replaces at altitude (slides over) the colder air mass at the surface. A front is shown as a line on a weather chart, which represents the front's *surface* position. However, the frontal air at altitude is actually well ahead of its depicted position on a weather chart.

NB: A *warm* front is represented by half circles along the front line, which are "pointed" in the front's direction of movement.

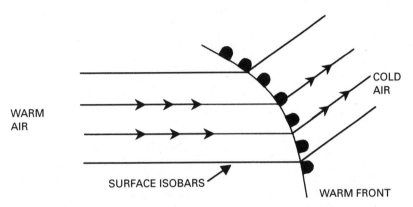

Figure 227   Warm front shown on a weather chart.

In fact, the slope of the warm front is typically 1:150, which is much flatter than that of a cold front. Therefore the top of a warm front, which is usually characterized by cirrus cloud, could be up to 600 nm ahead of the surface front, and altostratus rain-bearing cloud could be up to 200 nm ahead of it. A warm front is at the leading edge of the warm (air mass) sector and ahead of the cold front in a depression (low-pressure system).

The speed of a warm front is two-thirds of the geostrophic wind component. This can be determined by measuring the gap between the isobars along the front, comparing this distance against a geostrophic wind scale, and then calculating two-thirds of this value. The direction of travel for a warm front is perpendicular to its surface position, provided the geostrophic component is greater than 10 kn.

See Fig. 228 at the top of the next page.

However, if the geostrophic wind component is less than 10 kn, then this rule is unreliable and the front will tend to move from high pressure to low.

The speed of the warm front is also slower than that of a cold front because of the tendency of warm air (in the warm front) to rise up and slide over the cold, more resistant, denser air (mass) in front of it. Therefore the force (geostrophic wind component) moving a warm front is used to move the warm air upward as well as horizontally forward. Hence the speed of a warm front is only two-thirds of the geostrophic wind component. However, the denser, cold air in the cold front meets very little resistance from the warm, less dense sector air in front of it, and therefore its driving force is used solely to move it horizontally forward, at roughly the full geostrophic wind component. The passage of a warm front has the following general characteristics and associated weather:

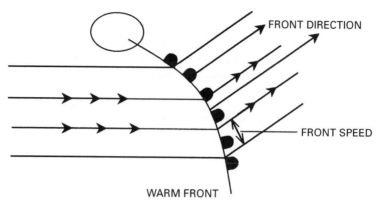

Figure 228   Speed and direction of a warm front.

As the warm front approaches:

1. A lowering of the cloud base is experienced. This is so because the shallower slope of the warm front gets closer to the ground as the front approaches. Therefore as the front gets nearer, the vertical displacement of the warm air above the slope, which experiences adiabatic cooling and therefore cloud, occurs at a lower and lower height. This is represented by cirrus clouds, giving way to cirrostratus clouds, giving way to altostratus clouds with possible virga rain, giving way to nimbostratus clouds with increased rainfall.

2. Poor visibility is experienced with the passage of the warm front. The visibility is reduced due to the increase of low-level clouds and the more consistent rainfall, and possibly rain ice.

3. The atmospheric pressure will usually fall as a warm front approaches.

As the warm front passes:

4. A rise in air temperature is experienced. This is due to the arrival of the warm sector air mass.

5. There is a low-level cloud base or fog. This is due to the warm sector air mass (possibly tropical maritime) that is now warm and moist giving rise to low-level stratus/cumulostratus cloud cover with a turbulent lower layer or fog due to the evaporation of the warm air.

6. Wind veers in direction in the northern hemisphere and backs in the southern hemisphere.

7. The atmospheric pressure will usually stop falling and may even rise.

8. Good visibility is experienced. In general, the visibility will be good, especially above the low-level cloud, although obviously in the cloud or fog the visibility will be poor.

See Fig. 229 at the top of the next page.

Washout   Washout on a wing is a decrease in the angle of incidence from the wing root to the tip. This compensates for the early stall due to the higher levels of loading experienced at the wing tips.

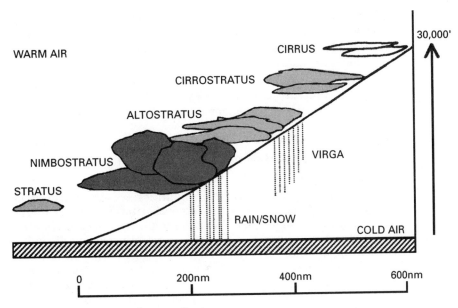

Figure 229 Cross section of a warm front.

Water Equivalent Depth *See* WED (Water Equivalent Depth).

WAT (Weight, Altitude, and Temperature) Limits The weight, altitude, and tempera-ture (WAT) conditions limit the aircraft's MTOW (or MLW) so that it meets the required second-segment (and missed-approach) climb gradient performance with one engine inoperative, given the ambient aerodrome conditions of pressure alti-tude and temperature (density).

The minimum height an aircraft must be able to achieve at its WAT limiting (MTOW) condition should always be the circuit height, say, 1500 ft.

Watt A watt (abbreviated as W) is a unit of electric power (expressed by the sym-bol *P*). The electric power (watt) is a product of the volts and amperes being used in the electric circuit.

$$\text{Watts} = \text{volts} \times \text{amperes}$$

NB:

1 kilowatt (1 kW) = 1000 W

1 horsepower (1 hp) = 746 W   modern engines are measured by horsepower

Wattmeter (Gauge) A wattmeter measures and indicates the amount of watts (elec-tric power) being consumed.

Weather These main properties are the basis of all weather phenomena:

1. Solar radiation and its interaction with the atmosphere. For example, solar radi-ation is the primary heat source for the earth, and as such it is an important

fundamental property for the interaction of other properties (pressure, density, temperature, etc.) and the movement of the earth's weather systems.

2. Atmospheric pressure, temperature, density, and the relationships between them, e.g., creation of air masses and pressure system.

3. The earth's surface and its interaction with the atmosphere. For example, water adds the property of moisture to the earth's weather system, which obviously is a highly important ingredient.

Weather Minima   Weather minima are

1. Decision height (DH) or minimum decision altitude (MDA)

2. Runway visual range (RVR) or visibility

The minimum values of these two quantities are known as *weather minima* and are specified by the national licensing authorities for various airport runways and various categories of aircraft. When the traffic controller advises that RVR is above the specified minimum, the pilot may descend to the specified decision height. If by then the pilot has sighted a sufficiently large segment of the ground, i.e., the approach or runway lights, runway markings, or general features of the runway environment, to enable her or him to be confident in judgment, then the pilot may continue on to land. Otherwise the pilot must go around.

Weather Radar   An iso-echo radar, also known as a *weather radar,* is primarily used to detect thunderstorms and by inference severe weather turbulence. It is displayed on a black-and-white TV screen so that the aircraft can be navigated around the storms, if possible.

The system has a radar aerial fitted in the nose of the aircraft which can scan up to 90° left and right and tilt 15° up and down. It uses a pulse radar technique from a conical beam, where pulse returns that have been reflected back off water droplets in a cloud are interpreted and displayed by the system as black areas, and stronger returns that depict dense clouds and therefore high-intensity turbulence as *hollow centers.*

NB: *Hollow centers* are registered above a preset pulse return level. Because the equipment shows returns above a certain level differently (i.e., hollow centers) this system is known as *iso-echo circuitry.*

NB: Color weather radar uses different colors to depict differing levels of returns, and therefore different levels of storm activity and turbulence.

On a color weather radar display thunderstorms appear as *red cores* and areas of maximum turbulence as *magenta.*

NB: The quantity of the radar pulses reflected back from the storm determines the severity of the storm.

Weather radar must be used as a means to circumnavigate (avoid) storms, by turning upwind, where possible, around a storm cell.

NB: A rule of thumb would be: Below 30,000 ft avoid cells by approximately 10 to 15 nm; above 30,000 ft avoid cells by approximately 15 to 20 nm.

Do not use the weather radar to penetrate severe storms by making small changes through the middle of a cluster of storm cells. Severe wind shear can be experienced in between cells, and you may be led up a blind alley, due to one cell shielding another from your radar.

Weather Radar—Auto Gain Control (AGC)   *See* Automatic Gain Control (AGC).

WED (Water Equivalent Depth)   The water equivalent depth relates to the worst possible contamination of slush depth and density on the runway, and a given WED corresponds to a percentage increase in takeoff distance required above the dry runway distance.

The worst combination of depth and density for a WED is assumed because different slush depths and densities which, when multiplied together may produce an identical WED, do not necessarily produce the same percentage increase in takeoff distance.

A 12.5 WED with a slush depth of 25 mm produces a 46 percent TODR increase.

A 12.5 WED with a slush depth of 12 mm produces an 18 percent TODR increase.

Therefore it is important to recognize that without the knowledge of depth and density, WED information cannot be applied solely to determine the full range of takeoff parameters, and so the worst combination of depth and density is assumed.

NB: Penalties can occur on takeoff if aquaplaning does not occur at the expected speed. Instead of lifting up and aquaplaning, resulting in a reduction of drag and a lower spray, the aircraft continues to plow on and suffer increasing drag and a more damaging spray pattern, resulting in an increased takeoff run.

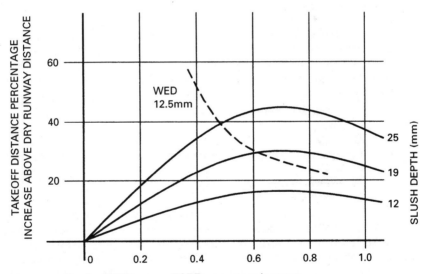

Figure 230   Graph of WED versus TODR percentage increase.

Weight of an Aircraft   If the limiting weight of an aircraft is exceeded, the following effects are experienced:

1. Performance is reduced.
    a. Takeoff and landing distances are increased.
    b. Rate of climb and ceiling height are reduced.
    c. Range and endurance will be reduced.
    d. Maximum speed is reduced.
2. Stalling speed is increased.
3. Maneuverability is reduced.

4. Wear on tires and brakes is increased.

5. Structural safety margins are reduced.

Weight—Aircraft Categories The ICAO aircraft weight categories are

Heavy (H)   136,000 kg

Medium (M)   Less than 136,000 kg and more than 7000 kg

Light (L)   7000 kg or less

NB: These weights refer to an aircraft's maximum takeoff weight. Aircraft greater than 136,000 kg, i.e., heavy, are required to be announced as *heavy* on initial contact with an ATC unit.

In other countries such as the United Kingdom, these definitions vary from the ICAO definitions, with the inclusion of a small category, as follows:

Heavy (H)   136,000 kg

Medium (M)   Less than 136,000 kg and more than 40,000 kg

Small (S)   Less than 40,000 kg and more than 7000 kg

Light (L)   7000 kg or less

NB: These weights refer to an aircraft's maximum takeoff weight. Aircraft greater than 136,000 kg, i.e., heavy, are required to be announced as *heavy* on initial contact with an ATC unit.

Weight, Altitude, and Temperature (WAT) Limits *See* WAT (Weight, Altitude, and Temperature) Limits.

Weight and Balance (Loading) The factors that determine the loading (weight and balance) of an aircraft are as follows:

1. The combined component weights of
   a. Cargo
   b. Baggage
   c. Crew and passengers, including personal effects (approximate weights can be used)
   d. Removable equipment
   e. Fuel (SG weight)

   must not exceed the aircraft's overall gross weight limitations, i.e., MTOW, ZFW, structural maximum [certificate of airworthiness (C of A)] aircraft weight, etc.

2. The distribution of the weights ensures that the cg is within its limits to longitudinally balance the aircraft. This is accomplished by adding the various component moments to obtain the total moment. The various component weights act at known positions relative to the fixed datum (arm). Each of these has a moment, which we can calculate by multiplying the individual weight by its arm from the datum. We obtain the total moment by adding all the moments of these component weights and then using the formula

$$\text{Total arm} = \frac{\text{total moment}}{\text{total weight}}$$

So we can calculate the cg position.

Carrying out the weight and balance/loading calculation for an aircraft is essential for a safe flight. In short, it comprises the following:

- Find the total aircraft weight, and ensure that *no* weight limitations are exceeded, such as MTOW, ZFW, and structural maximum (C of A) aircraft weight.
- Find the total cg moment and then divide this by the gross weight to find the length of the cg arm and thereby the position of the cg. Then verify that the cg point lies within the approved envelope.

Wet (Contaminated) Runway—Landing *See* Landing—on a Contaminated (Wet) Runway.

Wet (Contamination) Runway—Restrictions You are not permitted to take off from a wet (contaminated) runway when

1. The aircraft antiskid system is inoperative.

2. The standing water level is above a specified limit.

3. There are any other type-specific restrictions.

Wet (Contaminated) Runway—Takeoff *See* Takeoff—on a Contaminated (Wet) Runway.

Wet Start An engine *wet* start is otherwise known as a *failure to start,* after the fuel has been delivered to the engine. The causes of a wet start are normally an ignition problem. The indications of a wet start are that

- EGT does not rise.
- The rpm stabilizes at starter maximum.

The actions required for wet starts are to

- Close the fuel lever/supply as soon as a wet start is diagnosed (usually at the end of the starter cycle).
- Motor over the engine to blow out the fuel (approximately 60 s).

Wet V1 Speed A wet (minimum) $V1$ speed is the maximum speed for abandoning a takeoff on a contaminated runway. A wet $V1$ speed improves the stopping capabilities (final stop point) back to the dry-conditions level, but degrades the takeoff chances with a reduced screen height in the event of a takeoff being continued.

A recommended wet $V1$ for contaminated conditions is the dry $V1$ of 10 kn. Ergo, wet $V1$ is a lower speed than the dry $V1$.

The wet $V1$ is not a $V1$ speed as it does not imply any ability to continue the takeoff following an engine failure, and unlike a dry $V1$ this speed may be less than VMCG. Therefore a takeoff on a wet runway may result in a *risk period* between the maximum speed for abandoning the takeoff (wet $V1$) and the normal $V1$ speed during which, in the event of an engine failure, the speed is too high for a successful stop on a contaminated runway.

For a given aircraft weight on a contaminated runway, the emergency distance required is increased because of a reduced braking ability. Also a contaminated runway has a slower acceleration, and therefore the TORR is increased, which limits the stopping distance available if the takeoff is field-length-limiting. This effect is normally built into takeoff run performance graphs. If distance is limiting, the normal dry $V1$ offers the best compromise in risk associated with a continued or aborted takeoff. Namely, there is a risk of not being able to stop within the EMDA.

However, it is a remote possibility that an engine failure will occur at the worst point, i.e., after the wet $V1$ and before the dry $V1$; therefore an incident in the transition to flight in this scenario is a low risk. Furthermore, even if you lose an engine between the wet and dry $V1$ speeds, aborting the takeoff will probably only result in a runway overrun at a low speed. If there is a distance to spare, i.e., it is not field-length-limiting, then a progressive reduction in the $V1$ speed to the wet $V1$ reduces the risk associated with the aborted takeoff without unduly compromising the continued takeoff. This is so because of the increased airborne length of the effective clearway over the runway.

Whitening Effect   Whitening is a visual illusion whereby the perception of the distance to an object's surface is incorrect; i.e., a close object is perceived to be far away. A white surface or light distorting a person's visual depth acuity causes this perception.

NB: This effect may have played a part in the Air New Zealand DC10 accident in Antarctica.

Wind   Wind is the horizontal movement of air in the atmosphere that is initially driven by a difference in pressure between two places, and then it can be influenced further by a number of factors, including the earth's rotation forces, temperature, surface friction, etc. Atmospheric airflow is almost completely horizontal, with only about one-thousandth of the total airflow being vertical. However, this vertical airflow is very important to weather and to aviators, since it leads to the formation of cumuliform clouds and thunderstorms. In general, though, the term *wind* is used just in reference to the horizontal flow of air.

Wind is expressed in terms of direction and strength. *Wind direction* is the direction the wind blows *from,* and it is expressed in degrees measured clockwise from true north, with two main exceptions when the wind direction is measured from magnetic north.

- The reported wind given from the tower, either ATC or ATIS, is expressed in degrees magnetic, so that the runway direction and reported wind are relative to each other. This is extremely important when taking off or landing.

- The upper winds used for airways planning are expressed in degrees magnetic, so that the airway direction and reported wind are relative to each other.

Wind strength is measured and expressed in nautical knots.

NB: Wind direction and speed can also be expressed as *wind velocity.*

It is usually written in the form 180/25, meaning a wind blowing from 180° at a strength of 25 kn.

A wind is said to be *veering,* or said to have *veered,* when it changes its direction in a clockwise direction, say from 100/10 to 160/10.

A wind is said to be *backing,* or said to have *backed,* when it changes its direction in an anticlockwise direction, say from 160/10 to 100/10.

In general the wind in the northern hemisphere increases in speed and veers in direction with an increase of height. And in the southern hemisphere, the wind increases in speed and backs in direction with an increase of height. It normally backs in direction (veers in direction in the southern hemisphere) and decreases in speed during a descent, as you get near to the ground. At night the wind normally backs farther in direction (veers in direction in the southern hemisphere) and decreases more in speed, as you get nearer to the ground during a descent.

Wind—Anabatic   *See* Anabatic Wind.

Wind Component   To calculate headwinds, tailwinds, and crosswinds, you should use a chart (see Fig. 231) and enter the tower-reported wind; i.e., enter at the relative angle of the wind to the aircraft nose and/or runway. Follow along that line until you reach the speed for the reported wind. Then read the headwind or tailwind off the side of the chart and the crosswind off the bottom of the chart.

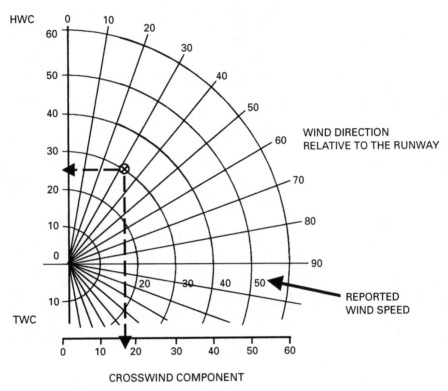

Figure 231   Wind component. Example shown: 30-kn wind at 34° to the runway, 24-kn headwind component, 18-kn crosswind component.

Wind Drift   You can calculate wind drift as follows. (1) Calculate the crosswind component. (2) Use the following formula:

$$\text{Crosswind component (point 1)} \times \frac{60}{\text{TAS}} = \text{maximum drift}$$

For example,

$$10 \text{ kn} \times \frac{60}{120} = 5\text{-kn drift}$$

Wind—Fohn   *See* Fohn Wind Effect.

Wind—Geostrophic   *See* Geostrophic Wind.

Wind—Gradient   *See* Gradient Wind.

Wind Gust—Corrections   The wind gust corrections that need to be applied to the approach speed are ½ times stable wind (minimum of 5 kn) plus full-gust value up to a predetermined, type-specific limit of approximately 20 to 25 kn. For example, given a basic approach speed of 120 kn and headwind component 20 gusting 28 kn,

$$120 + 10 \text{ (½ stable wind)} + 8 \text{ (gust value)} = 138 \text{ kn} = \text{correct approach speed}$$

Wind Gust Factors   Wind gust factors are an indication of how much wind speed change you can expect in varying wind conditions. The factor is calculated by taking the difference between the maximum and minimum wind speeds in the gusts and dividing by the mean wind speed. Thus a wind gusting from 15 to 25 kn with a mean of 20 kn will have a gust factor of 10/20, or 0.5.

Wind—Katabatic   *See* Katabatic Wind.

Windmill(ing)   A windmilling propeller is experienced when the rpm is reduced but the airspeed is maintained, which will eventually cause the blade angle of attack to become negative. When this occurs, the resultant force will act in a rearward direction, and this is known as *windmilling.*

A windmilling propeller causes a drag force, which is opposite to the direction of flight. This drag can be quite high, and in addition to the decelerating effect it can cause a large yawing moment on a multiengine aircraft with asymmetric thrust.

Windmill Engine Start   A windmill start occurs when the engine is started without the aid of the starter because the compressors are being turned by a natural airflow when airborne. This delivers the air charge to the combustion chambers where fuel and an ignition spark are introduced as usual for a stable engine relight. This effect is known as *windmilling,* and as such windmill starts are used to relight an engine when airborne.

Window (Cockpit) Heating   The purpose of heating the windows is to prevent them from breaking as a result of bird strikes. If the window freezes, a bird strike is likely to crack the window; however, if the window is heated, it is more flexible and therefore able to absorb a bird strike without cracking or breaking.

Wind Shear   Wind shear is any variation of wind speed and/or direction from place to place, including updrafts and downdrafts. The stronger the change and/or the shorter the distance within which it occurs, the greater the wind shear. However, in practical terms, only a wind change of a magnitude that causes turbulence or a loss-of-energy disturbance to an aircraft's flight path is generally considered to be wind shear. Remember that wind shear is a complex subject and is still not fully understood.

NB: Wind shear can affect the flight path and airspeed of an aircraft and can be a hazard to aviation.

Most forms of wind shear are found at low levels, i.e., below 3000 ft. Therefore the term *low-level wind shear* is used to specify the wind shear along the final-approach path prior to landing, along the runway, and along the takeoff and initial climb-out flight paths.

Low-level wind shear, i.e., near the ground, is generally present to some extent as an aircraft approaches the ground, because of the difference of speed and/or direction of the surface wind compared to the wind at altitude. This is critical in terms of aircraft safety, because the loss of energy caused by these changes can easily lead to the aircraft losing altitude and/or stalling. Low-level wind shear includes

1. CAT (clear air turbulence), which is caused by the following:
   a. Temperature inversions
   b. Surface winds being slowed down by ground surface roughness, causing a difference between the surface and gradient winds
   c. Local surface winds, i.e., land/sea breezes.
   d. Terrain-generated winds
2. Frontal passage, principally because it is characterized by a change of wind direction. (This can also be classified as a medium-level wind shear.)
3. Microburst and thunderstorm gusts. Thunderstorm gusts are associated with cumulonimbus clouds that have enormous updrafts and downdrafts, which are not necessarily strong enough to be classified as a microburst. These effects can be experienced 10 to 20 nm from the actual clouds.

   Medium/high-level wind shear can also be experienced at high or medium altitudes and include
4. CAT, in the form of jetstreams
5. Frontal passage per low-level frontal passage

Wind shear is detected by identifying a difference in wind (direction and/or speed) and/or temperature between two places, or identifying certain weather phenomena, say, CBs that give rise to possible wind shear conditions, from weather reports. This can be done in the following ways:

1. *Pilot appreciation* of differences in reported wind and/or temperature between two places and the location of certain weather and terrain phenomena. Examples include:
   - CBs in the general area (microbursts and/or thunderstorm gusts)
   - Heavy rain or even thunderstorms in the vicinity
   - Fronts (change in wind direction)
   - Virga rain (temperature inversion), e.g., rain that evaporates before it reaches the ground, absorbing latent heat out of the surrounding air, creating denser/heavier air with an associated higher pressure, causing a downdraft wind shear.
   - Land/sea breezes at coastal airports, especially at dawn and dusk periods
   - Terrain, i.e., trees, mountains, or even manmade surface obstructions, producing CAT
   These should alert the pilot instinctively to the possibility of wind shear along the flight path.

   NB: Pilots are required to report wind shear whenever they encounter it.

2. *Aerodrome equipment / reporting.* Many of the major aerodromes around the world have wind shear measuring equipment. However, a wind shear alert is issued for any airfield when aerodrome wind shear equipment (if fitted), ground observation, or pilot reports identify one or more of the following criteria:

    a. A mean surface wind of at least 20 kn

    b. A vector difference between the surface wind and the gradient wind of at least 40 kn

    c. Thunderstorms or heavy storms within 5 nm of the airfield

    d. A temperature difference between the surface and any point up to 1000 ft AGL that exceeds 10°C

3. *Aircraft warning equipment.*   Modern aircraft have wind shear warning systems, which are usually incorporated into the GPWS. This type of wind shear warning should prompt the flying pilot to initiate an immediate go-around, with full go-around thrust.

Wind shear, especially vertical up/downdrafts, will affect the lift capability of an aircraft. The two main aerodynamic principles behind this are as follows:

1. The aircraft's dynamic speed is reduced, as a result of a reduction of the wind speed.

2. Its effective aircraft weight is increased, as a result of experiencing a downdraft, which requires the pilot to recover the flight path by pitching up the aircraft. This is accomplished by deflecting the stabilizer/elevator upward (to get the aircraft to pitch up). However, deflecting the elevator/stabilizer upward creates a downward air force on the control surface, which also has the affect of increasing the aircraft's effective weight.

These two properties both subject the aircraft to turbulence and affect the aircraft's lift potential, which brings the aircraft closer to its stall.

A change in wind speed and/or direction (wind shear) is in general terms manifested as flight path overshoot and undershoot effects.

NB: The actual wind shear overshoot or undershoot effect depends on the following:

- The nature of the wind shear

- Whether the airplane is climbing or descending through that particular wind shear in which direction the aircraft is proceeding

Change in headwind and tailwinds   It is normal for wind speed and direction to change with altitude, i.e., increases in speed and veers in direction (northern hemisphere) with altitude. Therefore when climbing or descending into a decreasing headwind (or increasing tailwind), a typical wind shear situation, the aircraft will undershoot. This is true because an aircraft flying through the air will have a certain inertia depending upon its mass and its velocity *relative to the ground.* If the aircraft has a true airspeed of 140 kn and a headwind component of 30 kn, then the inertia speed of the aircraft over the ground is 140 − 30 = 110 kn. When an aircraft flies down into the calm air, the headwind component drops off reasonably fast to 10 kn, for instance. The inertia speed of the aircraft over the ground is still 110 kn, but the new headwind of only 10 kn will mean that the true airspeed has suddenly dropped back to 120 kn.

NB: Remember, lift performance is a product of TAS.

The pilot would observe a sudden drop in indicated airspeed, and the change in the performance of the airplane at 120-kn airspeed will be quite different from what it was at 140-kn airspeed. The normal pilot reaction would be to add power or to lower the nose to regain airspeed and to avoid undershooting the desired flight path. The pilot can accelerate the aircraft and return it to the desired flight path by changes in power and attitude. The greater the wind shear, the more these changes in power and attitude will be required. Any fluctuations in wind will require adjustments by the pilot, and this is why you have to work so hard sometimes, especially when approaching to land.

Conversely when descending into an increasing headwind, or decreasing tailwind, the aircraft's TAS will increase and the aircraft's lift performance will increase; therefore the aircraft will tend to overshoot.

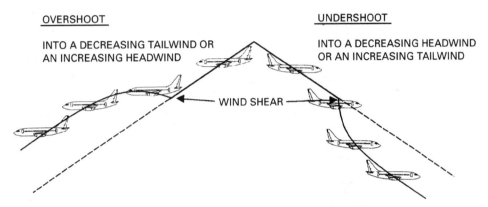

OVERSHOOT

INTO A DECREASING TAILWIND OR AN INCREASING HEADWIND

UNDERSHOOT

INTO A DECREASING HEADWIND OR AN INCREASING TAILWIND

WIND SHEAR

Figure 232   Overshoot and undershoot due to headwind and tailwind changes.

Change in updrafts and downdrafts   These changes are associated with thunderstorms and cumulonimbus clouds. The worst example is a *microburst*.

To summarize, an overshoot wind shear effect may result from flying into

- An increasing headwind
- A decreasing tailwind
- An updraft

An undershoot wind shear effect may result from flying into

- A decreasing headwind
- An increasing tailwind
- A downdraft

When flying an approach with wind shear suspected, you should increase the approach speed to compensate for the loss of energy that is common with low-level wind shear. This action will help to guard against a stall when you attempt to maintain the flight path by increasing the aircraft's attitude. You should monitor the conditions ahead in case the conditions worsened from mild to severe wind shear (microburst), especially the indicated airspeed, temperature, and lift trends. That is, an increasing headwind would increase the indicated airspeed and lift performance, for no power or attitude changes. And a decrease in the headwind would decrease the indicated airspeed and lift performance, for no power or attitude changes. The rate at which these changes occur indicates the severity of the wind shear ahead.

For mild or even moderate wind shear that causes a manageable flight path deviation (i.e., less than a single-dot deviation on the approach), it is acceptable to use normal power (for speed) and pitch (for the descent profile) corrections to regain the desired flight path.

NB: If you believe that this condition is a prewarning of a more severe disturbance ahead, then you should adopt a more aggressive attitude and initiate an immediate go-around.

For severe wind shear (which can include microbursts) a TOGA—full-power and maximum-pitch-up, go-around—should be initiated immediately.

A change in wind direction and/or speed wind shear has the following effects on an aircraft during the landing approach:

1. If the headwind component increases, then the aircraft will experience a tra sient increase in performance. This will lead to an overshoot tendency.

2. If the headwind component decreases, then the aircraft will experience a transient decrease in performance.

NB: This is a typical scenario, as you get closer to the ground, causing an increase in the rate of descent, if allowed to go unchecked. However, normally this effect is only slight and doesn't cause too much concern, although any significant wind shear drop should be treated with respect.

Loss of performance in this scenario will cause the aircraft to sink and possibly undershoot the aiming point.

Wind Shear—Warnings   Modern aircraft are provided with wind shear warnings, by utilizing air data computer-detected changes in airspeed to calculate the presence of a wind shear phenomenon, which it feeds to the GPWS that generates a "Wind Shear Wind Shear" aural and visual display on the main ADI flight instrument. Wind shear warning systems are active from the ground level to a height of 1500 ft.

A wind shear warning requires an immediate go-around at full thrust and maximum flight director pitch-up attitude, to avoid ground contact. If, however, the aircraft still does not climb, then the aircraft should be pitched up to its maximum pitch limit symbols, or "eyebrows."

The pilot's order of priority, given a wind shear, GPWS, and TCAS warning at the same time, is

1. Wind shear

2. GPWS

3. TCAS

Wind—Surface   *See* Surface Wind.

Wind—Thermal   *See* Thermal Wind.

Wind—Trade   *See* Trade Winds.

Wind—Upper   *See* Upper Winds.

Wing Design   In designing a high-speed wing, you need to consider, first, the primary requirement for economical high-speed performance in the cruise configuration. However, you also have to consider the restraints on the design of, second, the need to keep airfield performance within acceptable limits and, third, the need to give the structural people a reasonable task.

There are several interactive design areas of a wing. Some are purely for lift, others are a compromise between lift and speed, and still others are purely for speed. For the high-speed requirements of a wing, the design would focus on *sweep, thickness,* and *chamber.*

Sweep The sweep design is the pivotal feature of a high-speed aircraft wing, and it is the major divergence from the turboprop aircraft's straight wing. Because the airflow velocity vector over the upper surface on a straight wing becomes sonic at a relatively low speed, the aircraft cannot achieve the higher speeds which its jet engines can produce. Therefore a wing design was needed that delayed the airflow over the wing going sonic, and as such allowed the aircraft to maximize the high speeds from its efficient jet engines. The answer was to *sweep* back the wing. The angle of sweep is chosen so that the aircraft can cruise at an optimum Mach number.

Too much sweep produces

1. Poor oscillatory stability

2. A greater tendency for the wing tips to stall

3. Poor lift capability, especially for airfield performance

Too little sweep produces early drag rise/low attainable speeds.

Thickness of the wing A high-speed aircraft requires thin wings because a thicker wing with a larger wetted area produces greater profile drag, which is detrimental to economical high-speed cruise performance. However, a wing cannot be too thin because it is limited against the structural requirement of, say, storing gear, fuel tanks, etc., as well as the construction inability and/or economic restraints associated with manufacturing a very thin wing. Also very thin swept wings experience unacceptably high levels of aerodynamic loading.

Chamber High-speed aircraft require a minimal chambered wing because the greater the chamber, the greater the airspeed vector of the airflow over the wing and therefore the earlier the onset of sonic disturbed airflow, which is exactly what the designer is trying to avoid. Therefore in terms of achieving high speeds, a minimum degree of chamber is desirable. However, the chamber of a wing is vital to produce lift, and therefore the lift requirement of a clean wing becomes a dominant factor in this area for the designer. A compromised chamber section is selected that satisfies both the speed and the lift requirements.

The degree of sweep, thickness, and chamber used for the final high-speed wing design is dependent upon their many interactive compromises, which culminates in directly fulfilling the wing's high-speed requirement and inversely the wing's lift and structural requirements.

Winglets Winglets are aerodynamic efficient surfaces located at the wing tips. They are designed to reduce induced drag. They dispense the spanwise airflow from the upper and lower surfaces at different points, thereby preventing the intermixing of these airflows, which otherwise would create induced-drag vortices.

Yaw Dampers The purpose of yaw dampers is to (1) prevent dutch roll and (2) coordinate turns. However, the main purpose of a yaw damper system is to prevent Dutch roll. When the fin area is insufficient to provide a natural oscillatory stability, the effective fin area must be increased in some other way. This is accomplished on power-operated rudders with the yaw damper.

A yaw damper is a gyro system, sensitive to changes in yaw, that feeds a signal into the rudder, which applies opposite rudder to the yaw before the roll occurs and thereby prevents Dutch roll.

Z

ZFW Zero fuel weight is a wing loading structural maximum weight.

$$ZFW = payload + APS$$

$$= MTOW - fuel\ weight$$

Thus the maximum zero fuel weight determines the maximum permissible payload.